How do fish populations regulate themselves? Why do some fish stocks flourish and then die away? These questions have fascinated fisheries scientists for decades and in the last 20 years, answers have begun to emerge.

In this comprehensive account, David Cushing shows how the fate of fish larvae that live close to the centres of production in the sea has a crucial effect on population regulation. He shows how the timing and development of tidal fronts in particular regions have profound implications for fish and plankton production, which in turn affect fish recruitment. If recruitment of fish larvae into the pool of adult fish is insufficient, stocks may fail. It is only by understanding these processes that we can hope to recognize the implications of global climate change on marine populations. This book will be essential reading for all those interested in marine ecology and fisheries biology.

POPULATION PRODUCTION AND REGULATION IN THE SEA:
A FISHERIES PERSPECTIVE

POPULATION PRODUCTION AND REGULATION IN THE SEA:

A FISHERIES PERSPECTIVE

DAVID CUSHING

CAMBRIDGE
UNIVERSITY PRESS

Published by the Press Syndicate of the University of Cambridge
The Pitt Building, Trumpington Street, Cambridge CB2 1RP
40 West 20th Street, New York, NY 10011-4211, USA
10 Stamford Road, Oakleigh, Melbourne 3166, Australia

First published 1995

Printed in Great Britain at the University Press, Cambridge

A catalogue record for this book is available from the British Library

Library of Congress cataloguing in publication data

Cushing, D. H.
Population production and regulation in the sea : a fisheries perspective / David Cushing.
  p. cm.
Includes bibliographical references (p.   ) and index.
ISBN 0 521 38457 5 (hardback)
1. Fish populations. 2. Marine fishes – Ecology. 3. Fisheries.
4. Marine ecology. I. Title.
QL618.3.C875   1995
597'.05248'09162 – dc20   94–16146 CIP

ISBN 0 521 38457 5   hardback

WS

Contents

Preface

In 1975 I wrote *Marine ecology and fisheries* (Cambridge University Press). The text presented here is quite different. The earlier book tried to link disciplines that appeared to be diverging. In the thirties they were subsumed in the study of marine biology. In the last decade or so, fisheries scientists have turned their attention to communities (described as multispecies), to larval life in the plankton and to juvenile life on the beaches and in the midwater. They have become interested in the broader population problems and the biological oceanographers now need estimates of fish stocks to complete their models.

The present text arose from long interest in the regulation of fish populations, which remains an intractable problem. In 1955, I became responsible for the study of the stock of herring which supported the East Anglian herring fishery which had been recorded since the seventh century. In 1955, the fishery started to decline and it did not recover. Recruitment overfishing was diagnosed in 1963, but I did not succeed in promoting a recovery probably because at that time people did not believe that excessive fishing could reduce the recruitment to the adult stock. Since then the phenomenon has become commonplace. There is a delightful children's book called *Old Winkle and the seagulls* by Elizabeth and Gerald Rose (Faber and Faber, 1960) which describes these events and shows the scientist slinking down to the station.

I stayed and continued to study the regulation of fish populations from a number of viewpoints, generally in an examination of the biology of fishes, where that was possible. Many people have helped me. Dave Mills, Kevin Flynn, Colin Purdom, Bob Dickson, Keith Brander and Peter Bromley read some chapters and I am very grateful to them. Irene Gooch drew the pictures. Brian Rothschild, Joe Horwood, Bob Dickson

and John Shepherd have instructed me and I hope that they have
enjoyed this as much as I have.

David Cushing
Lowestoft

1
Some new discoveries in marine biology

Introduction

There have been many advances in marine biology in the last decade or so, but fisheries biologists are interested in only some of them. For example, the tidal fronts were discovered in the seventies and their physical description is of direct interest because they affect marine fishes. As shown below, our view of production in the temperate summer seas has been radically changed. Further, the development of the fronts reflects that of production itself in certain regions and, much more important, its timing.

The picoplankton and the microbial food loop are of great interest because they represent fractions of production that are not diverted towards the higher trophic levels; but most copepods and fish larvae need food organisms more than 5 μm across, larger than many of the organisms in the microbial food loop.

Tidal fronts

Fronts are boundaries between water masses, for example in estuaries, upwelling areas or along the equatorial complex. Some can be seen, such as the edge of the Gulf Stream, and Uda (1959) described a front or 'siome' as a noisy place with *accumulations* of material. They are readily detected by infra-red radiometry from an aeroplane or satellite (Legeckis, 1978). Tidal fronts are the boundaries between stratified and tidally mixed waters and in temperate waters they are usually present only in summertime. Le Fèvre and Grall (1970) discovered the Ushant front, which extends round the coast of Brittany and which was the subject of so much later work by Pingree and his colleagues. Simpson (1971) noticed that an area of weak tidal streams in the western Irish Sea was strongly stratified in summer, and Pingree *et al.* (1974) described a persistent and convergent front between Guernsey and Jersey. Simpson and Hunter (1974) surveyed the western Irish Sea with ships and an airborne radiation thermometer flown at a height of 150 m. They proposed that a tidal front was formed where the energy loss from the seabed due to turbulent tidal motion was balanced by the rate of heat input at the surface (which generates stratification); it is expressed by the fraction h/U^3 (or the Simpson/Hunter parameter), where h is the depth of water and U is the observed tidal velocity at the surface (assumed proportional to the amplitude of tidal velocity at the seabed). The cube of the tidal velocity represents the mixing energy of the tidal flow. The tidal front is found at a critical value of the Simpson/Hunter parameter, about 80. Pingree (1978*b*) developed a turbulent energy dissipation rate, E; when $E > -1.0$ the waters are mixed and when $E < -2.0$ the waters are stratified. Hughes (1975) developed a stratification parameter, the potential energy needed to mix the water column, and charted the positions of fronts in the English Channel and to the west of the British Isles. In a second paper (Hughes, 1976), he used his stratification parameter to predict the positions of nearly all the fronts around the British Isles. With a numerical model of $\frac{1}{4}°$ mesh, Fearnhead (1975) charted the positions of the same fronts. Pingree (1975) described the position of the Ushant front with continuously recording instruments; he also described the advance of the thermocline from deeper to shallower water in spring as the rate of heat input at the surface overcame the dissipation of energy from the seabed. It is important to realize that the tidal front is essentially convergent.

Figure 1.1 shows a vertical section across the Ushant front in chlorophyll (*a*) and in temperature (*b*) (Pingree *et al.*, 1975). There is a concentration of chlorophyll at the thermocline and also one in the top 10 m on the stratified side of the front. At first sight the latter is difficult to explain if the front were merely convergent. Pingree and his colleagues noted that, whereas the lower layer below the thermocline was stirred by the tide, the upper was mixed by wind stress and by nightly convection. The thermocline in such a region occurs where the wind can no longer stir excess heat downwards against the tide. Pingree *et al.* (1976) described the outburst in chlorophyll in the Celtic Sea in spring and the chlorophyll in the Ushant front in summer (Figure 1.2). They also noticed that there is a high stock of chlorophyll in the thermocline all through the summer. Such observations have been confirmed by Savidge (1976), Simpson *et al.* (1979), Beardall *et al.* (1982)

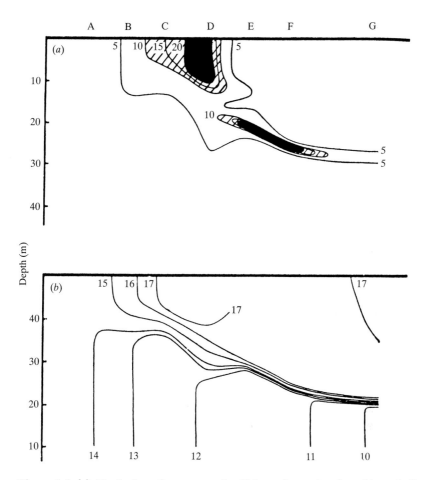

Figure 1.1 (*a*) Vertical section across the Ushant front showing chlorophyll, mg m^{-3}. (*b*) Temperature. (After Pingree *et al.*, 1975.)

and Richardson *et al.* (1985). Pingree *et al.* (1977) described the mean vertical stability throughout the year with the Brunt–Väisälä period (Figure 1.3), which expresses the buoyancy in the layer above the thermocline; the shaded area indicates the stock of chlorophyll in the thermocline during the summer; that is, production continues throughout the summer there. Holligan and Harbour (1977) make the same point, but note that there were diatoms in the spring outburst, with dinoflagellates in the summer thermocline (Figure 1.3*b*). This was a most important discovery: after the spring outburst in temperate waters, production continued in summer within the thermocline.

(a)

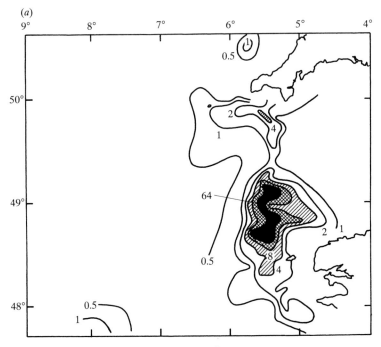

Figure 1.2 Charts of chlorophyll (mg m^{-3}) (*a*) in the Ushant front in summer and (*b*) in the Celtic Sea front in spring (Pingree *et al.*, 1976).

Pingree *et al.* (1977) described the Ushant front continuously at 2 m and at 20 m. They found that as the spring tides approached, the mixed area off the French coast became enlarged and the front was displaced to the north as the thermocline was eroded from below by increased mixing at spring tide. As the tidal cycle shifts back and forth, the production cycle is sustained from the region of tidally mixed waters. Simpson *et al.* (1977) charted the western Irish Sea front with a batfish survey and confirmed the results with infra-red satellite images. They developed a model of both tidal and wind mixing and showed that the geographical positions of the fronts are primarily determined by the tidal streams and that therefore their positions are predictable.

Pingree and Griffiths (1978) described a fine mesh (5 nm by 5 nm) numerical model of the M2 tide for the area 42° N to 62° N by 12° W to 13° E. Figure 1.4(*a*) shows the distribution of their stratification parameter, $S = \log_{10}[h/C_dU^3]$, where C_d is the drag coefficient on the sea bed and here U^3 is the vertically averaged amplitude of the tidal velocity. This is the fullest description of the tidal fronts in the waters around the British

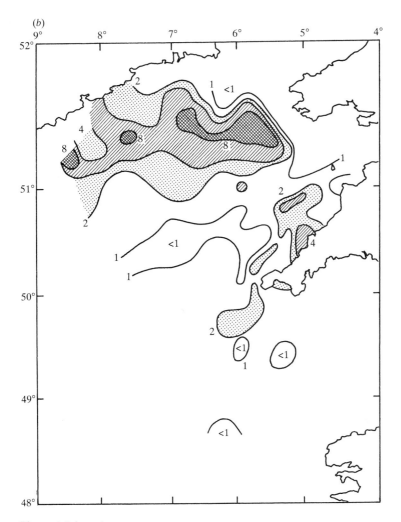

Figure 1.2 (*cont.*)

Isles, including the Ushant front. Figure 1.4(*b*) shows the distribution of fronts in the Gulf of Maine, with a model of tidal and summertime wind mixing (Loder and Greenberg, 1986; note that $h/D \approx h/U^3$). Such are permanent structures in summertime. In the waters around the British Isles, the tidal fronts occur in depths of about 100 m. In regions where the tides are not so strong, fronts would be found in shallower water.

James (1977) assumed that the vertical eddy viscosity varied with the Richardson number of the flow and that the horizontal eddy viscosity

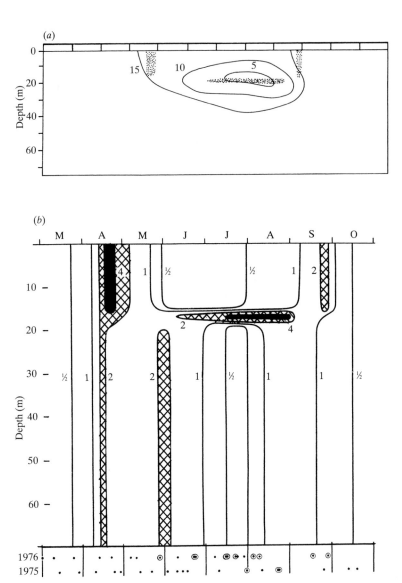

Figure 1.3 (*a*) The mean vertical stability throughout the year in the Ushant front shown by the Brunt–Väisälä frequency (after Pingree *et al.*, 1977). The shaded area shows the chlorophyll in mg m^{-3} in the thermocline in summer; (*b*) (after Holligan and Harbor, 1977) is the same in more detail. Sampling dates are shown at the foot of the figure.

was linked to the bottom friction. His model showed that a flow occurred along the front with the well-mixed water to the right and that there was a slighter flow normal to the front towards the stratified region. Further, there was upwelling on the well-mixed side. This would generate production and it could be translocated into the stratified water.

Pingree (1978a) summarized the development of the spring outburst in a diagram (Figure 1.5). The ordinates are, in depth and in critical depth, both normalized to the Secchi disc depth; the abscissa shows the stratification parameter, S. If $S = 4$, stabilization starts in March and if $S = 1.5$, it starts much later and the surface mixed layer reaches the surface at the front; this accounts for the advance of the thermocline from deeper water in spring. In Figure 1.5 are summarized the effects of depth, irradiance and the stratification parameter. It is a static or geographical statement and so does not describe events in time and it excludes the effect of winds of more than Force 6.

Pingree *et al.* (1979) described baroclinic eddies in the Ushant front, which are 20 to 40 km across and which persist for a few days. They were detected by an infra-red satellite image and by a ship survey with continuous measurements of temperature, salinity and chlorophyll. At a tidal front, warmer water at the surface overrides the cooler water below, which sinks in the convergence. The cyclonic eddies are formed as the sinking cold water is stretched and anticyclonic eddies have been observed, if less frequently (Pingree, 1978b, 1979). Indeed, Loder and Greenberg (1986) expect both upwelling and downwelling. James (1983) made a three-dimensional model of the tidal front in shallow seas and suggested that eddies resulted from instabilities in the form of vortex pairs, because any cyclonic eddy has an anticyclonic pair. The model shows that both upwelling and downwelling can occur. Hence there may well be devices that generate production in the instabilities of the convergent fronts. Loder and Platt (1985) investigated four processes in nitrate flux: (a) that owing to cross-frontal flow, which was very low indeed; (b) that owing to the baroclinic eddies, which constituted a fifth overall; (c) that owing to the springs/neaps tidal cycle, which constituted a third; and (d) that owing to the vertical mixing through the thermocline, which constituted half. Such are the processes that might continue production at a tidal front.

The study on George's Bank was continued by Horne *et al.* (1990). When the front becomes established, there is an eastward flow along the northern edge. Under tidal influence, the front moves back and forth

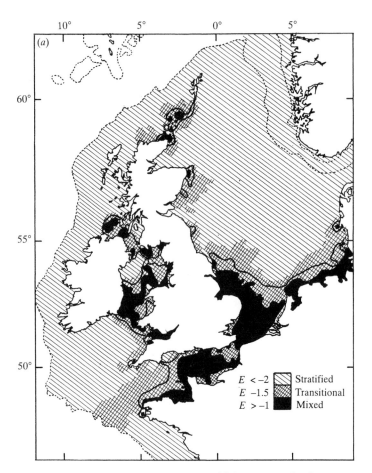

Figure 1.4 Distributions of tidal fronts (*a*) in summer in the waters around the British Isles (after Pingree and Griffiths, 1978); (*b*) in the Gulf of Maine (after Loder and Greenberg, 1986).

over a distance of about 10 km and so there is a cross-frontal flow that moves nitrate southward. The radiocarbon measurements show a common productivity in the mixed water, at the edge of the Bank and off the Bank. But the nitrate demand in the mixed water is only one-quarter of that at the front. There is a northerly flux of nitrate at 13 m of 0.01 mmol N m^{-2}, but a southerly one at 33 m on the Bank of 0.28 mmol N m^{-2}. The total supply of nitrate to the front is about four times the demand. In general, nitrate is consumed in the frontal zone twice as quickly as in the mixed area. As estimates of POC (particulate organic carbon), PON (particulate organic nitrogen), chlorophyll and radiocarbon are all about

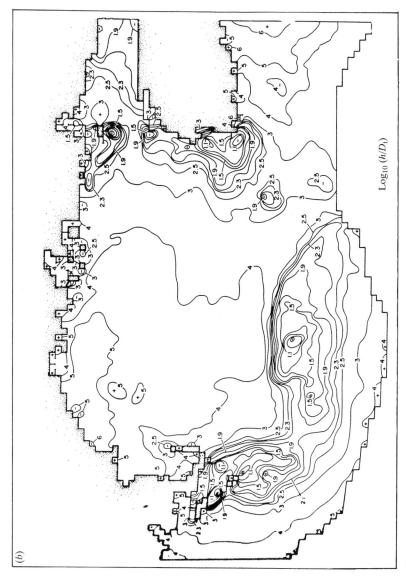

$\text{Log}_{10} (h/D_1)$

Figure 1.4 (cont.)

$$S = \log_{10}\left[\frac{h}{C_D|u|^3}\right]$$

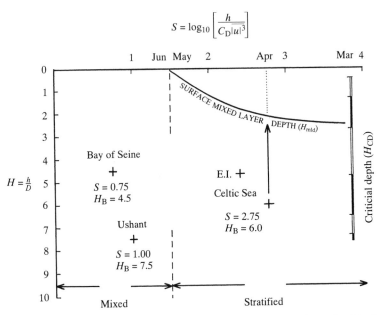

Figure 1.5 Development of the spring outburst as function of critical depth, transparency and the stratification parameter, S (Pingree, 1977). h is the depth of water and D is the depth at which the Secchi disc is no longer visible, an index of transparency. H_{CD} is the critical depth normalized by the depth at which the Secchi disc is no longer visible.

the same, there must be a greater transfer of energy to higher trophic levels at the front.

Because the front is convergent, Okubo (1978) suggested that chlorophyll had accumulated there. The same point was made by Le Fèvre (1986) and it must be partly true. However any accumulation must lead to increased production; with constant reproductive rates, the stock would be increased and hence production also. Fogg *et al.* (1985) studied the Isle of Man front with traditional methods and found that bacteria in the stratified waters were five to ten fold more active, with increased uptake in the stratified waters. Similarly, there was high uptake of glucose and urea, all of which indicates high production in the community of the stratified water. Holligan *et al.* (1984a,b) came to similar conclusions with continuous sampling and radiocarbon measurements.

The result shown in Figure 1.3 is of some importance. It had been believed that in temperate waters there was little, but sporadic production between the spring and autumn outbursts (for example, see works of Cushing, 1974, 1975a). From Figure 1.3, it can be seen that production continues within the thermocline until the autumn outburst. This

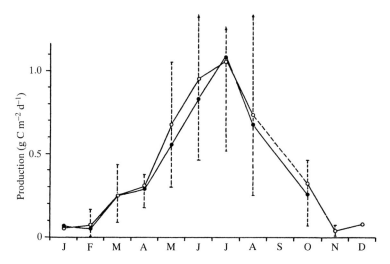

Figure 1.6 Production in g C m^{-2} d^{-1} in the Baie de Roscoff (○ 1964–67, ● 1964–69) (after Grall, 1972).

discovery was made with a pump taking continuous samples as it was lowered. Production at the thermocline close to the front may be low because of shading but, away from it, it is adequately lit (Holligan *et al.*, 1984a).

Production also continues in some well-mixed waters to a summer peak, after which it declines. Figure 1.6 (Grall, 1972) shows the annual production in g C m^2 d^{-1} (g C is grams carbon) in the coastal waters off Roscoff, with a peak in July; Kremer and Nixon (1978) show a similar production cycle in Narragansett Bay, as does Diwan (quoted by Raymont, 1980) in Southampton Water. Fransz and Gieskes (1984) found a single production cycle in the Dutch coastal waters in the Southern Bight of the North Sea. Such events suggest that the production cycle in protected and mixed waters is driven by irradiance, but Fransz and Gieskes showed that a distinct spring and autumn outburst occurred offshore. This area is open to heavy winds and sea from the north (Dickson and Reid, 1983). Under the strong winds, attenuation is increased (Lee and Folkard, 1969). Then the reproductive rate of the algae is reduced and so is (D_c/Z), where D_c is the compensation depth and Z is the depth of water. Then μ is less than G, where μ is the algal reproductive rate and G is the grazing mortality of the algae; thus, the stock of algae becomes reduced. The single production cycle is then broken into distinct spring and autumn outbursts.

Production in the shelf seas depends profoundly on the tidal turbulence. Pingree (1978*a*) wrote that the wind stress did not match the effect of tide in waters <100 m (in British seas) until it reached Force 6 (1.1 to 1.4 m s^{-1}), but spring is a season of strong winds. Sverdrup's original formulation (1953) showed that production at Weather Ship M in the Norwegian Sea proceeds in an intermittent manner. A similar result emerged from the work of Williams and Robinson (1973) at Weather Ship I in the North East Atlantic. Further, Colebrook (1982), using material from the continuous plankton recorder network, showed that the spring outburst in the North East Atlantic is nearly over before the seasonal thermocline becomes fully established. Dickson *et al.* (1988*a*) showed that the time of onset of the spring outburst in a region of summer stratification in the western North Sea from 1948 to the late seventies was delayed by about a month under the increased stress of northerly winds (and particularly gales), presumably because the depth of mixing exceeded the critical depth for longer (see Chapter 5 for a fuller exposition). Hence, production in shelf seas is delayed or advanced by the incidence or not of strong winds in spring.

The tidal fronts are complex and dramatic structures. They are convergent and hence accumulate stocks of algae. This in itself increases production, but other processes enhance it: upwelling, cross-frontal flow, the springs/neaps tidal cycle and vertical mixing through the thermocline. So the tidal fronts are highly productive, with high transfer to higher trophic levels. Therefore, fish stocks would be expected to exploit them. Iles and Sinclair (1982) showed that the early autumn spawning herring released their larvae in the region of some fronts (for example, Georges Bank and south west Nova Scotia and perhaps in the northern North Sea, where fronts sometimes survive for a time before the equinoctial gales). Similarly the western mackerel stock spawns off Ireland at the shelf break front, and in most upwelling areas there are fronts, or even arrays of fronts, that must play a part in the biology of the upwelling fishes.

Another consequence of the work on tidal fronts was a revision of the nature of production in temperate waters. It continues throughout the summer within the thermocline, which must contribute to the survival of summer spawning fishes. In the tidally mixed waters, production continues throughout the summer. Many spring spawning herring release their larvae in tidal waters and so into a production cycle that continues steadily for six months or so.

The start of production in temperate shelf seas

Inshore of 100 m in the waters around the British Isles and with winds less than Force 6, the water column is mixed by the tides (Pingree, 1978*a*); where the tides are weaker, the position of the front is in shallower water. Here, the depth of water is also the depth of mixing, and production starts when the critical depth exceeds the depth of water and the subsequent progress of production is governed by the ratio of compensation depth to depth of water (Cushing, 1972). The variation in time of onset is governed by irradiance and attenuation. Differences in irradiance (light intensity at the surface of the sea) are due to the clouds and to the average elevation of the sun (to which the effective daylength in the euphotic layer is linked); differences in attenuation are due to silt stirred up from the seabed as noted above. With wind strength greater than Force 6 and with clear skies, the time of onset of production is an inverse function of depth of water and the sun's elevation. It is delayed at all depths by cloudiness. As the wind strength increases above Force 6, production is delayed also by an increment in attenuation. Both factors may, but need not, work in conjunction.

Between 100 m in depth, where the tidal fronts are likely to be found in summer (in the waters around the British Isles), and the shelf break, production starts in a critical depth that exceeds the depth of mixing (as originally suggested by Sverdrup, 1953). The continuation of production depends on the progress of stabilization, which may be intermittent, as suggested earlier. Pingree (1975) has shown that the development of the thermocline starts near the shelf break and proceeds inshore to the 100 m isobath as the influence of tidal mixing from the seabed increases. Variation in the time of onset of production is due to differences in wind strength, convection and irradiance as illustrated in Figure 1.5.

Probably the most important factor governing the onset and progress of production in temperate waters in spring is the incidence of gales. They increase attenuation in the tidally mixed waters and increase the depth of mixing where the thermocline has started to develop (and where they may destroy the developing and transient thermoclines). The variability in time of onset of production is probably greater in the deeper water because of the variation in wind strength and irradiance, to which is added that owing to strong winds and gales. A further source of variability is the deepening of the summer thermocline in weakly stratified water under the gales. In the shallower water the main

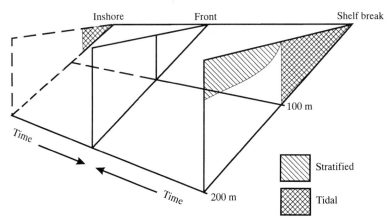

Figure 1.7 Development of production in time, from the shore and from the shelf break towards the position of the front.

variation is due to irradiance and then attenuation when the wind blows more strongly.

Thus, variability in time of onset of production is probably less in the tidally mixed waters than in the potentially stratified waters. Production starts in the shallowest waters first and later as the water deepens towards 100 m, where the tidal fronts occur. It also starts earlier at the shelf break and later towards the depth of the potential fronts. So this region is the last at which production starts on the temperate shelf. If one compares the two regions in shelf seas, inshore and offshore of the 100 m isobath, production must start first close inshore and at the shelf break. It starts last at the position of the potential front (Figure 1.7). Hickling (1939) charted the surface phosphorus in the western English Channel in April and found the high values in the region of the Ushant front, with low values up Channel and also towards the shelf break, where production was already under way.

The variability in the time of onset of the spring outburst in the shelf seas might depend inversely on depth in tidally mixed waters and the reverse is true in stratified waters. Then in coastal waters with a single production cycle, as for example in the southern North Sea, the variability may be quite low. Because most shelf seas are open, the general variability is greater.

Differences in the time of onset of production with depth or with the progress of stratification may play considerable parts in the generation of recruitment to fish stocks. If there is basis to the match/mismatch hypothesis (Cushing, 1974, 1990*b*), some sources of variation could perhaps be investigated in the changes in time of onset of the production cycle.

The discovery of the picoplankton

The picoplankton comprises cells in the range 0.2 to 2.0 µm in diameter and include bacteria, cyanobacteria such as *Synechococcus*, and the smallest eukaryotes. They were discovered with the use of the epi-fluorescence microscope. With acridine orange as a stain, free-living bacteria were described (Francisco *et al.*, 1973; Daley and Hobbie, 1975); until this discovery, it had been thought that in the sea bacteria lived only attached to particles. The very small cyanobacteria were discovered by the fluorescence of phycoerythrin (Johnson and Sieburth, 1979; Waterbury *et al.*, 1979).

The small autotrophs do not sink (Takahashi and Bienfang, 1983). The pigments (phycoerythrin and phycobilin) are possibly protective because the cells can be swirled by turbulence from strong light to weak light in the surface-mixed layer. Waterbury *et al.* (1987) showed that in the sea the cyanobacteria increase at a remarkably steady rate for periods of up to six weeks until grazing supervenes at 10^5 cells ml^{-1}; in the turbulent sea, cells change their depths continuously and the irradiance they receive is averaged. Because of the sizes of the genome and of the photosynthetic apparatus, Raven (1987) found that the smallest autotrophic cell must be at least 0.3 µm in diameter.

At such a size with maximal surface to volume ratio, there is the highest rate of uptake of nutrients and the highest absorption of radiant energy. The photosynthetic efficiency of cells less than 1 µm in diameter is twice as high as that of cells greater than 1 µm (Platt *et al.*, 1983) and the growth rate is maximal at relatively low irradiances (El Hag, 1986), but the smallest cells would be expected to leak more into the radioactive filtrate of the assay (Raven, 1987).

The picoplankton live in all waters, fresh and salt, oceanic and inshore. They are abundant, 10^3 ml^{-1} to 10^5 ml^{-1} in the sea and up to 10^6 ml^{-1} in fresh water. The eukaryotes are less abundant by about an order of magnitude and some live below the 1% light level (Wood, 1985). The density of bacteria is about 1 to 5×10^6 ml^{-1} in offshore waters (Azam *et al.*, 1983). Phytoplankton release 15% to 40% of organic material as exudates (Fogg, 1983), on which the bacteria subsist. Indeed, Lochte and Turley (1985) showed that the production of bacteria in the Isle of Man front was of the same order as that of the primary production, at least for a short period.

Although picoplankton are present in all oceans they predominate in the oligotrophic ocean, if only because the larger cells are less abundant there. The cells of the picoplankton produce a fair proportion of the total production in the sea, which was not appreciated until the last decade. The picoplankton are too small to be of direct use to the food chains that lead to the fish stocks, but they are of indirect interest in that the small cells often comprise a fair to large proportion of chlorophyll. Therefore the quantities of chlorophyll must be separated by sizes of cell.

The microbial food loop

Azam *et al.* (1983) proposed that there was a 'microbial food loop', a new food chain in the sea and in fresh water. The protozoa and heteroflagellates feed on the bacteria, and the cyanobacteria on the smallest eukaryotes. Although cladocerans and appendicularians can take bacteria, they do not control the populations of free-living ones (Riemann and Bosselmann, 1984). The small cells of the picoplankton are eaten by choanoflagellates, cryptophytes, dinoflagellates and chryso-monads (Fenchel, 1982*a–d*; Sherr and Sherr, 1984). The protozoa and heterotrophic flagellates are themselves eaten by ciliates, which are in turn eaten by copepods. Ciliates do not eat anything smaller than cells of 1 to 2 μm across (Anderson and Fenchel, 1985). The protozoa and heterotrophic flagellates are about 5 to 7 μm across and Fenchel (1986), using Lighthill's (1976) description of flagellar movement, showed that they could filter up to 5.7×10^{-5} ml h^{-1}. He also found that they grow at 0.15 to 0.25 h^{-1}. Andersen and Fenchel (1985) made a predation experiment in large aquaria and found that the protozoa took 47 to 73 cells animal^{-1}, enough to control the bacterial numbers in the sea; Goldman and Caron (1985) came to similar conclusions. Because the respiratory demand of the small cells is high (see below), the algae and animals of the microbial food loop predominate in stratified waters.

The nature of production in the spring outburst and in the upwelling areas

It is a truism that production in upwelling areas in subtropical seas and in the spring outburst in temperate waters have common processes. Margalef and his colleagues provided a rationale for distinguishing three

Table 1.1. *Margalef's rationale for the succession of the three groups of algae in the spring outburst and in upwelling areas*

Group	Divisions d^{-1}	Surface/Volume	mg dry wt/PPU	%Chl/dry wt
I	1 to 2	0.7 to 0.8	0.005 to 0.013	3
II	0.6 to 1.0	0.4 to 1.0	0.012 to 0.016	1.5
III	0.2 to 0.4	0.2 to 0.3	0.039	<1.5

Note:
PPU: plant pigment units, an early and primitive method of estimating chlorophyll.

Table 1.2. *Ratios of respiration to the maximal rate of photosynthesis*

Group	r^*
Diatoms	0.04 to 0.08
Green flagellates	0.08 to 0.25
Dinoflagellates	0.25 to 0.80

groups of algae in the spring outburst and in upwelling areas (Duran *et al.*, 1956; Margalef *et al.*, 1955; Margalef, 1958; Margalef, 1967). There is a common succession of algae in both systems, first the small diatoms (up to 10 μm in diameter), Group I, bigger diatoms (Group II) and then dinoflagellates (Group III). Table 1.1 lists some characteristics of the three groups.

The small diatoms have relatively high division rates with high chlorophyll and high surface to volume ratios. They can exploit irradiance more fully than their successors in the productive season. Since the fifties, when Margalef was working on succession, improved methods of sampling have revealed that the small and delicate green flagellates are important in the sequence of succession. Harris (1978) established the ratio, r^*, of respiration to the maximal rate of photosynthesis for each of the three algal groups as shown in Table 1.2.

Harris also found a link between r^* and the production ratio (ratio of the depth of the euphotic layer to the depth of mixing) (see Table 1.3). Thus the small diatoms survive when the production ratio is low and with their high division rates can start the production cycle in the turbulent upwelling areas or in the transient thermoclines of the spring

Table 1.3. *Relationship between r* and the production ratio for three algal groups*

Production ratio	r*	Group
0.1	0.07	Small diatoms
0.25	0.102	Diatoms
0.50	0.155	Green flagellates
0.75	0.208	Dinoflagellates
1.00	0.26+	Dinoflagellates

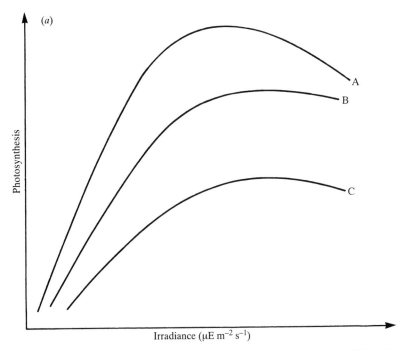

Figure 1.8 Photosynthesis/irradiance curves for (*a*) small diatoms, (*b*) medium-sized diatoms and green flagellates, (*c*) dinoflagellates and (*d*) cyanobacteria. In (*a*) and (*b*), curves A, B and C are at increasing temperatures (after Cushing, 1989).

outburst. In a fully stratified system they are succeeded by the green flagellates and then by the dinoflagellates. Of course, the succession of algal groups is often more complex, particularly in lakes (Munawar and Talling, 1986). Figure 1.8 shows the relationships of photosynthesis and irradiance for (*a*) small diatoms, (*b*) medium-sized diatoms and green

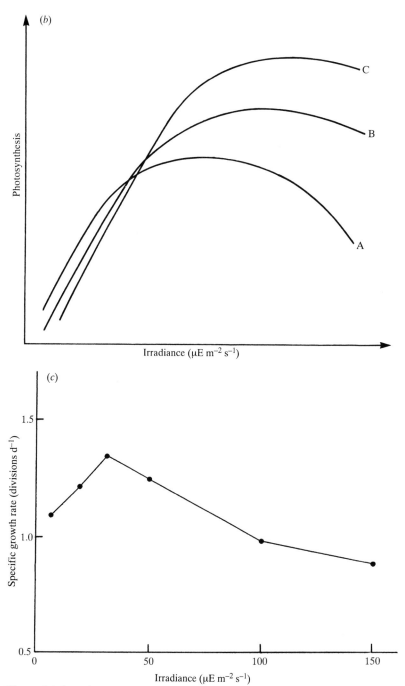

Figure 1.8 (*cont.*)

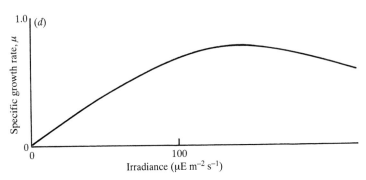

Figure 1.8 (*cont.*)

flagellates, (*c*) dinoflagellates and (*d*) cyanobacteria. Figure 1.8(*a*) and (*b*) also show the photosynthesis/irradiance curves at different temperatures. As the temperature rises, photosynthesis per unit irradiance for the small diatoms declines, whereas for the green flagellates and the medium-sized diatoms it increases. So, as the production cycle advances the small diatoms must be replaced by larger diatoms and by green flagellates.

The spring outburst and the subtropical upwelling system starts with the small diatoms. Production in both systems is intermittent and the small diatoms are adapted to the high rates of mixture. But as stratification becomes less intermittent, the small diatoms are succeeded by green flagellates that photosynthesize more efficiently in the slowly warming waters. This account is taken from the article by Cushing (1989).

The herbivorous copepods of the traditional food chain feed on cells greater than 5 μm in diameter (Bartram, 1981; Harris, 1982*b*; Frost *et al.*, 1983). Cushing (1989) suggested that there was a difference in structure between the microbial food loop and the traditional food chain. The small diatoms, characteristic of spring outburst and upwelling area, are in general larger than 5 μm in diameter and support the traditional food chain from copepod to fish. The microbial food loop is based on bacteria and cyanobacteria, which are eaten by protozoa and heterotrophic flagellates, themselves eaten by ciliates, which are in their turn exploited by copepods. Although the loop is found in all waters, it predominates in fully stratified waters and particularly in the oligotrophic ocean. Figure 1.9 shows the difference between the microbial food loop and the traditional food chain. The difference in structure is that between

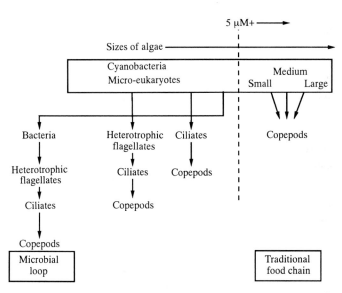

Figure 1.9 The difference in structure between the microbial food loop and the traditional food chain (after Cushing, 1989).

production in the regions and periods of intermittent thermoclines or turbulent upwelling areas and that of fully stratified waters.

The microbial food loop is long, of five steps to fish, in contrast to the two steps of the traditional food chain. The transfer of energy to fish is much less in the microbial food loop, which might explain why fisheries are in general based on spring and autumn outbursts in temperate seas and in upwelling areas. Here the phrase 'upwelling area' includes not only the strictly coastal upwelling system, but all the complex arrays of squirts, filaments and hammerheads that lie offshore, up to 300 to 500 km. Of course, the transfer efficiency is probably greater in the stratified waters, where nothing is spilled or wasted as in the upwelling systems or the spring outburst.

Conclusion

Marine biologists have revealed many remarkable phenomena in the last decade or so, but only those of direct relevance to fisheries have been treated here; some, such as the remarkable work of Strickler and his colleagues on the feeding mechanisms of copepods, will be noted below. Physicists can now work at the scales in which the plants and animals live and in which their populations develop. For this reason, the study of tidal fronts has become important to the study of fisheries.

The study of the fronts has led to a closer evaluation of production in the shelf seas, in particular, on the variation with depth of the time of onset of production as a function of irradiance, attenuation and wind strength. Because fish often spawn just before the spring outburst in temperate seas or in the plumes from capes in the upwelling areas, the closer study of the nature of production is of greater importance to fisheries biologists.

2

The role of nutrients in the sea

Introduction

The part played by nutrients in the sea may appear a little remote from the practice of fisheries science. Since the time of Atkins (1923, 1925*a,b*) and Hentschel and Wattenberg (1931), fisheries have been associated with the presence of nutrients in spring and in the upwelling areas. In the last decade the discovery of fronts of various types has reinforced interest in nutrients (for example, see Le Fèvre, 1986). Further, the study of nutrient uptake had made great progress since Dugdale's (1967) use of Michaelis–Menten kinetics. Again, the analysis of events in the oligotrophic ocean where production continues in the surface layers in the absence of nitrate has raised questions on how that production is sustained. In the last decade, free-living bacteria and cyanobacteria have been discovered in waters both fresh and marine and the microbial food loop has been well described, as noted in the last chapter. Dissolved nutrients are taken up by bacteria and algae and both are eaten by microflagellates, protozoa and ciliates, which provide food for the copepods and other grazers such as salps. All heterotrophs regenerate ammonia. Before these processes were described, an experiment in a bottle of natural sea water from which zooplankton were filtered could be interpreted in a simple manner, revealing an uptake of nutrient or an increment of radiocarbon. But it is now clear that the bottle, at least in the oligotrophic ocean, contains at least a three-stage ecosystem and any simple interpretation must be difficult to make. Where the traditional food chain (diatoms, copepods and fish) exists, as in spring outbursts in the North Atlantic or in upwelling areas, such bottle experiments may remain of use.

From the spatial correlation of production and nutrient lack portrayed by Hentschel and Wattenberg and the temporal one shown by

Atkins, it has been assumed that production is limited by nutrient lack. Atkins' three papers set out the quantitative link between the decrement of phosphorus and the increment of algal stock, the temporal correlation between the two and the effect of stratification on the development of the algal stock.

Cullen (1991) and Cullen *et al.* (1992) distinguished two forms of limitation which were named after Liebig and Blackman. The first, Liebig, is the limit to the quantity produced, which Atkins had used in his agricultural model. The second, Blackman, is the limit to the rate of production, specific growth rate or nutrient uptake and forms the basis to any model or dynamic study. (Flynn, K.J. (personal communication) tells me that stress can also play a part, affecting growth (or uptake) until it is rate limited. But we need to know at which point stress is yield limiting.) The first definition was expressed in the production of algae in a bottle which stops after nutrients run out (for example, see articles by Ketchum, 1939*a,b*). In its simplest form this means that production in the sea is a simple and direct function of nutrient uptake, because the algae are produced until the nutrient is depleted and then they are eaten. But grazing starts before the nutrients are exhausted and, indeed, in the oligotrophic ocean they are probably eaten almost as they are produced. As nitrate is taken up by the algae, it is transferred to animal flesh by grazing and part is regenerated as ammonia. Indeed, ammonia is excreted by all parts of the ecosystem (from bacteria to the blue whales), but the larger animals may excrete urea. Another form of the Liebig definition is the geochemical one where 'plant activity (and in the case of silicate, animal activity as well) is efficient enough to extract these six constituents almost completely from the surface waters' (above the oceanic thermocline) (Broecker and Peng, 1982); the six constituents are silicate, nitrate, phosphate, zinc, cadmium and germanium. There is also a group of intermediate elements that become partially depleted in the surface layers. Broecker and Peng describe a model with two reservoirs, deep and surface separated by the oceanic thermocline; rivers provide the input and sedimentation to the sea floor, the output. Upwelling is balanced by downwelling (in a broad and long-term sense) and the circulation is in a steady state. This form of nutrient limitation sets a maximum to production in the ocean, but within a year that maximum will not be reached because fish and other animals live for many years and organic sedimented material may take hundreds or even thousands of years to return to the surface.

The study of nutrient kinetics

Mackereth (1953) showed that *Asterionella formosa* in Windermere in north-west England stored enough phosphorus before lake overturn to drive all the production of *Asterionella* in spring and early summer. This meant that the temporal and inverse correlation observed by Atkins (1925*b*) between production and the apparent nutrient uptake need not hold. The early models, however, needed a restraint on production such as nutrient lack, probably because the death of algae due to grazing was poorly formulated and the animals did not get enough to eat (Cushing, 1958). However, one needs to distinguish between the rate of uptake, which may slow the dynamic processes, and the quantity available, which should set an absolute limit, if it were ever fully used. The rate of uptake must be described if any model is to be constructed because the algal division rate must be reduced from time to time in the patchy ocean. Dugdale (1967) and Eppley and Coatsworth (1967) introduced the Michaelis–Menten equation to describe the uptake of nutrients by cells:

$$V = V_{max}S/(K_s + S),$$

where V is the rate of nutrient uptake, V_{max} is the maximal rate of uptake, S is the external nutrient concentration, and K_s is the half-saturation coefficient, the quantity of external nutrient at which V_{max} is halved (it is somewhat difficult to establish experimentally). It has been linearized by plotting (S/V) against S. It depends on the external nutrient, so that, if there is none, there is no uptake. The production observed in the oligotrophic ocean cannot be supported by uptake, as described by this equation, if the nitrate level is zero, but there may be continuous uptake of ammonia. Droop (1973) noted that at steady state:

$$\mu = \mu_{max}/(K'_s + S),$$

where μ is the specific growth rate, μ_{max} is the maximal specific growth rate and K'_s is the half-saturation coefficient (but $K'_s \neq K_s$).

Droop (1968) introduced the idea of a cell quota, an internal cell store. He wrote:

$$\mu = \mu'(1 - K_q/Q),$$

where Q is the cell quota, K_q is the subsistence quota at which $V = 0$, and μ' is the specific growth rate at infinite quota. The equation describes the rate processes of phosphorus, iron and vitamin B_{12}, but not those of silicon or nitrate (McCarthy, 1981); indeed, the cell quota for nitrate is that for the whole cell, only a small part of which can be accessible. Goldman and McCarthy (1978) introduced the ratio (K_q/Q_m), where Q_m is the maximal cell quota, and the quota ratio expresses the proportion of subsistence quota to maximal cell quota. Then:

$$\mu_{max} = (1 - K_q/Q_m).$$

The quota ratio has the following values:

Si, N, P,   0.20
Fe, vitamin B_{12} 0.01–0.02.

Goldman (1980) noted that K_q depends on cell size, but that the quota ratio is independent of cell size and that K_q, Q_m and the specific growth rate all depend on temperature. Further, the cell quota is maximal when the specific growth rate is maximal. So if cells divide quickly, they have a store that ensures further divisions, which explains Mackereth's observations on *Asterionella*. If external nutrients fail the cells divide more slowly until more nutrient becomes available. This is the true function of nutrient limitation especially in the oligotrophic ocean. Much of the above account is taken from McCarthy (1981).

Li (1983) noted that some methods of linearization introduced bias and recommended non-linear methods of fitting. Shuter (1978) tabulated estimates of K_q (for nitrate and phosphate) for prokaryotes and eukaryotes, ranging in volume from $0.39\,\mu m^3$ to $17\,000\,000\,\mu m^3$. He derived equations as follows:

$$K_q(N) = 0.214\,v^{0.71} \text{ and } K_q(N) = 1.627\,C^{0.87},$$

where v is volume in μm^3 and C is carbon in $\mu mol/l$. As (K_q/Q_m) (N) $= 0.20$, analogous equations can be derived for Q_m. The kinetics of nutrient uptake has been the focus of an enormous amount of work and the brief summary given here does no more than provide a base for the understanding of the models

The Redfield ratio

Redfield (1934) noticed that the proportions of carbon, nitrogen and phosphorus in sea water and in the plankton existed in the same proportions, that is by mass:

C	N	P	
100	16.7	1.85	in the sea
100	15.4	1.88	in the plankton

Samples were taken from surface, intermediate and deep water in the North and South Atlantic and the Barents Sea; 'it is as though the seas had been created and populated with animals and plants and all of the nitrate and phosphate which the water contains had been derived from the decomposition of this original population'. The Redfield ratios, C/N or N/P are readily available from observations in the sea and in the plankton; with recent methods, carbon can be measured as readily as nitrogen or phosphorus. Dr Flynn (personal communication) tells me that the C/N ratio differs from group to group; for example, for bacteria, C/N (by mass) is 4, for cyanobacteria it is 4 to 5, for dinoflagellates it is 5 and for many other algae it is 6.5 to 7. Hence the field measurements have become a little difficult to interpret. Ryther and Dunstan (1971) showed that in coastal waters, nitrogen is exhausted first (perhaps because phosphorus is regenerated quickly). Further, addition of ammonia* generated greater algal growth in bottles than did phosphorus. This was the initial evidence that nitrogen in the sea should be considered as the main limiting nutrient. In lakes, phosphorus is considered to be the limiting nutrient.

Goldman *et al.* (1979) noticed that nutrients were always low where there was little seasonal variation in the depth of the mixed layer, that is in the oligotrophic ocean. They referred to the constant Redfield ratio in sea water, in the plankton and in the rate at which the constituents are taken up. They grew the chrysophyte *Monochrysis lutheri* under phosphorus limitation, the chlorophyte *Dunaliella tertiolecta* under nitrogen and phosphorus limitation and the diatom *Thalassiosira pseudonana* under ammonia limitation. In Figure 2.1, the N/P and C/P ratios are shown as functions of specific growth rates. In each figure the Redfield ratio is shown as a broken line. Under phosphorus limitation (Figure 2.1(*a*)), the specific growth rate decreased with increased Redfield ratio; Figure 2.1(*b*) shows that under nitrogen and phosphorus limitation the same increase in the Redfield ratio takes place. In Figure 2.1(*c*), under

---

* Ammonia/ammonium: ammonia is used throughout, in lieu of more detailed information.

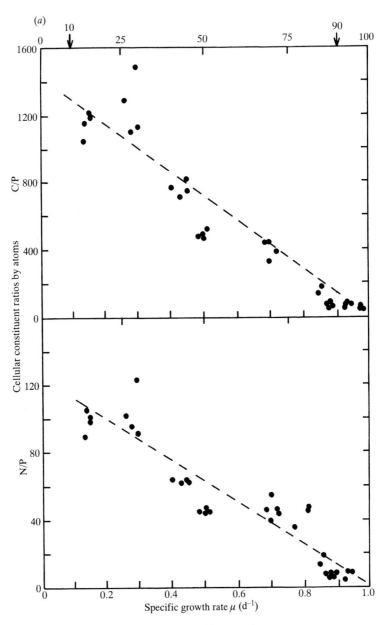

Figure 2.1 The Redfield ratio and specific growth rates:
(*a*) the decrease in specific growth rate of *Monochrysis lutheri* under phosphorus
limitation, increasing the C/P and N/P ratios; (*b*) the decrease in specific growth
rate of *Dunaliella tertiolecta* under nitrogen and phosphorus limitation as the
ratios C/P and N/P decrease; (*c*) the increase in specific growth rate of
Thalassiosira pseudonana under ammonia limitation with increasing C/P and
N/P.

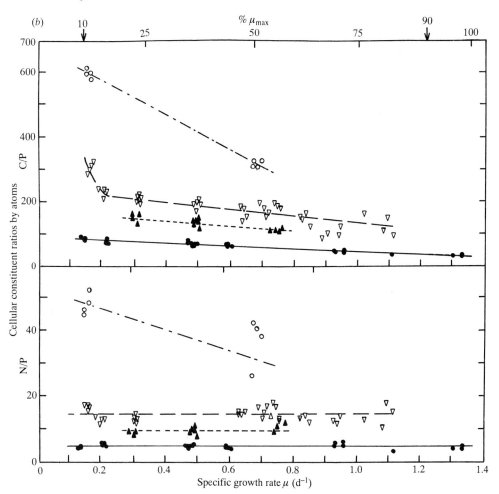

Figure 2.1 (*cont.*)

ammonium limitation, the N/P and C/P ratios are low and the specific growth rates are reduced. If the Redfield ratio obtains in the oligotrophic ocean, the specific growth rate should be maximal and if the ratio is not at the level expected, the specific growth rate should be reduced. Bienfang and Takahishi (1983) examined cells less than $3\,\mu m$ in diameter from oligotrophic waters and showed that from the increments of chlorophyll in the absence of grazers, the division rates amounted to 1.3 to 2.5 doublings d^{-1} (but the estimate from chlorophyll may overestimate the rate of doubling) and that enrichment had no effect, presumably because the cells were dividing at the maximal rate. Furnas

(c)

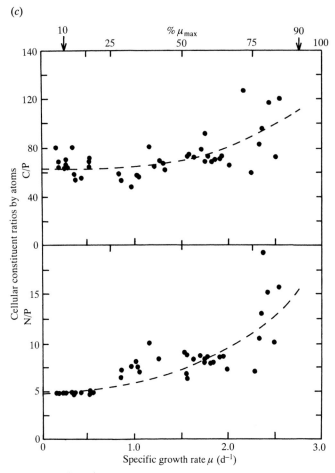

Figure 2.1 (*cont.*)

(1982) devised a method of measuring the algal division rate *in situ* using a diffusion chamber with nutrients free to pass through the membranes. He reviewed the measurements of reproductive rates in the sea (Furnas, 1990) and in 1991 reported the results from an array of diffusion chambers in depth in oligotrophic shelf waters north of Oahu; he found that maximal division rates in diatoms were high and the nitrate nitrogen in the water amounted to less than $0.2 \, \mu\text{mol} \, l^{-1}$.

Goldman *et al.* (1979) drew attention to the analogy with continuous culture, in which nutrients are added at a fixed dilution rate and their experiments were made in such a system. A steady state is reached in which biomass produced is balanced by its cropping, that is the algal division rate is equal to the mortality due to grazing. If the nutrients are undetectable (save by the more recent electrochemical methods), the algae must take them up rapidly. The authors also pointed out that the transport of nutrient by gradient transfer across the thermocline is about 1% to 10% of the demand of the phytoplankton. There are many physical mechanisms by which nitrate could be augmented, but, across the oligotrophic ocean as a whole, the deficit can perhaps be satisfied only from dissolved organic material such as amino acids and ammonia from the whole animal ecosystem.

Goldman *et al.* noted that the algae could obtain their nutrient demand in a small fraction of their doubling time. So the ammonia or organic nutrient excreted can be absorbed quickly. The important point is that the volume sampled by a cell is very much smaller than that demanded by the autoanalyser, which is about 10 ml. If the cells are dividing at a maximal rate, the cell quota is also maximal. Then the function of uptake from the excretion of the animals (from bacteria upward) is to augment the cell quota, so transient limitation is minimized.

Dortch and Postel (1989) have compared external and internal nutrients (cell quotas expressed as $\mu mol/l$) in a transect at 1 m from shore to deep water off Oregon:

Depth	Station	External N (μm)	Internal N (μm)	Chl a ($\mu g/l$)
9	13	27.08	0.298	0.95
27	15	5.00	0.359	1.20
41	17	0.50	0.557	5.12
54	19	0.18	0.734	4.90
93	21	0.021	1.408	12.53
147	23	0.020	0.241	0.39
880	25	0.260	0.071	0.17

Note (a) External N comprises nitrate, nitrite and ammonia; internal N comprises nitrite, ammonia and amino acids, presumably in vacuoles in large cells (K. Flynn, personal communication). (b) These samples were not from the oligotrophic ocean, but it is interesting to note that, where nitrate was very low (Stations 21 to 25), the internal nutrients were five to six fold greater than the quantities of ammonia and nitrite observed externally.

New and regenerated production

Dugdale and Goering (1967) distinguished 'new' production from regenerated production; the first is derived from nitrate and the second from ammonia and organic sources. With Broecker's (1974) box model, Eppley and Peterson (1979) suggested that regenerated production would continue forever in the euphotic zone of the stratified ocean, if there were no loss. That loss as fish catch or as sediment to the seabed must balance the new production. It can only be replaced as nitrate is transferred to the euphotic layer (by, for example, upwelling – in the broad sense – or input from the rivers). But replacement can only take place over a long time, longer than some periods of climatic change. Eppley and Peterson introduced the f-ratio, production from nitrate as a proportion of total production from nitrate and ammonia, all estimated by ^{15}N measurements (but cells using nitrate may take up $^{15}NH_4^+$ more rapidly than cells growing on ammonia (Dr K. Flynn, personal communication). Figure 2.2(a), (b) shows the f-ratio as function of total production in g C m^{-2} yr^{-1} estimated from radiocarbon measurements in the central North Pacific, eastern Mediterranean, Southern Californian Bight, eastern tropical Pacific, Costa Rica Dome and the Peru upwelling. As the eutrophic system progresses to an oligotrophic one, f declines to low levels. Eppley and Peterson used this quantity to estimate total production. Platt and Harrison (1985) found it difficult to believe that f was never greater than 0.5. They used ^{15}N measurements in the Bedford Basin (Nova Scotia) and plotted them on the ambient measurements of nitrate (where ammonia is less than 0.1 mol/l) (Figure 2.2(c)). This result differs from that of Eppley and Peterson in that f = 0.8 to 0.1 as the system moves from eutrophic towards oligotrophic conditions. Because the system is highly variable and intermittent, the whole distribution is used and if production is linked to nitrate (or carbon) the value of f is higher than that from the mean.

New production is expressed by f, and r (= 1 − f) expresses the effect of grazing; indeed, Longhurst and Harrison (1989) used r as a relative index in depth in the water column. In the spring outburst of the temperate North Atlantic or in an upwelling system, new production is followed by regenerated production as the bloom is reduced by grazing. Sources of new production may also be found under different physical conditions at fronts, warm core rings, with internal waves, Langmuir circulations and so on.

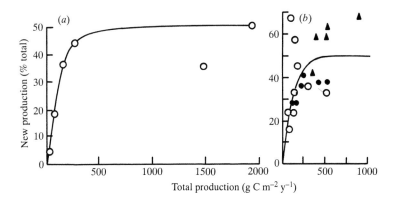

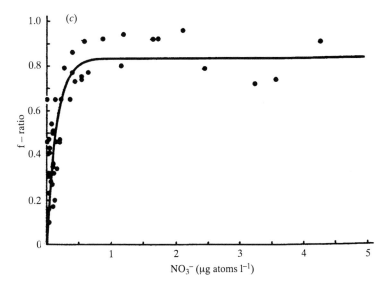

Figure 2.2 The f-ratio.
(*a*), (*b*) The dependence of the f-ratio ($NH_4/(NH_4 + NO_3)$) on the rate of production in $g\,C\,m^{-2}\,yr^{-1}$: (*a*) is from the central North Pacific and eastern Mediterranean; (*b*) from the upwelling areas between California and Peru (after Eppley and Peterson, 1979).
(*c*) The dependence of the f-ratio on nitrate in the Bedford Basin (after Platt and Harrison, 1985).

The agricultural model and the predator–prey model

Cushing (1975*a*) described an agricultural model of nutrient limitation, where production depended only on the quantity of available nutrient, so the algae were produced to the nutrient limit and were then eaten.

An alternative model proposed that production was limited by grazing to some degree and the loss of nutrients during the spring outburst represented a transfer to animal flesh and faecal pellets (which sank quickly to the seabed). Flynn (1989*b*) called the alternative the predator–prey model and, since the ecosystem is now known to be much more complex, he noted that, if production becomes nutrient limited, then the limiting step is in the regeneration. Indeed the question to be asked is not why production is limited by nutrient but why it is not limited by grazing.

Butler *et al.* (1979) had noticed that the total dissolved organic nitrogen remained high throughout the summer in the western English Channel and equal to the quantity of dissolved inorganic nitrate in the spring. Flynn and Butler (1986) showed that the dissolved organic nitrogen comprised urea and dissolved free amino acids amongst other compounds. Algae can take up lysine and arginine, although only half as quickly as nitrate (Flynn and Syrett, 1986; Flynn and Wright, 1986; Wheeler and Kirchman, 1986); perhaps it is the dinoflagellates that take them because they divide rather slowly. Flynn and Fielder (1989) and Flynn (1990) have found that the glutamine/glutamate ratio (Gln/Glu) indicates nitrogen deficiency; a high ratio (>0.3 to 0.5) may indicate the formation of proteins.

Flynn (1989*a*) reviewed the evidence for nutrient limitation. He noted that the transfer from dissolved inorganic nutrients in winter to dissolved organic nutrients in summer is really a change from an immature to a mature ecosystem, in Margalef's (1968) terms, or from new production to regenerated production. The mature system is a very complex one of three to five levels, with continuous regeneration in each. Flynn also pointed out that an upwelling zone cannot be fertilized from below, because the upwelled water displaces that at the surface, and the presence of nitrate merely indicates new water and immaturity in the ecosystem, or the first stage in the sequence of succession in the spring outburst in temperate waters of the North Atlantic. Flynn (1989*b*) suggested that the two models represent extreme positions and that production is really limited by too much grazing or by too little (which is too little regeneration). Further, nutrient limitation arises when predators cannot grow fast enough or where predator and prey become mismatched in time or in space.

Figure 2.3 illustrates such a mismatch in an experimental ecosystem. An autotroph, *Dunaliella tertiolecta*, was grown using ammonia and a predator, *Oxyrrhis marina*, regenerated ammonia. But the predator

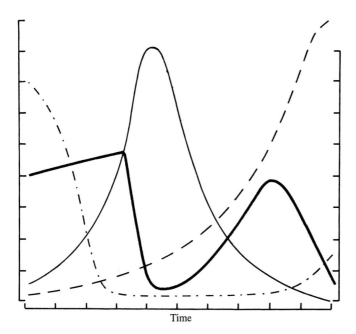

Time

Figure 2.3 A predator–prey interaction. *Dunaliella tertiolecta* (——) was grown on an ammonia source (–·–·–) and was eaten by *Oxyrrhis marina* (– – –). The glutamine/glutamate ratio (▬▬) declined sharply as ammonia became limiting, but recovered as ammonia was perhaps regenerated by the predator and taken up by the algae to be incorporated into protein. The evidence for the last step resides in the Gln/Glu ratio (after Flynn, 1989*b*).

responded to the production of the autotroph too slowly and the ratio of Gln/Glu fell rapidly, indicating nutrient limitation. Subsequently the Gln/Glu ratio increased as ammonia released from the predators was taken up by the algae and converted to protein as noted above. Such relationships may occur at all levels: protozoa on cyanobacteria, bacteria or small eukaryotes; ciliates on protozoa; copepodites on somewhat larger algal cells. Because the ocean is patchy, that is with high variance of samples, there are many opportunities for such matches and mismatches, and nutrient limitation may be transient. By the same token, there are opportunities for matching predator and prey at any level in the ecosystem.

This view of nutrient rate limitation arises from the realization that any experiment involving the addition of a nutrient in a bottle in a time course is a perturbation of an ecosystem. Because winter inorganic nitrate in temperate waters becomes converted to dissolved organic nitrate in summer in the stratified euphotic layer, this traces the development of an immature ecosystem to a developed one, in the sense of Margalef (1968). So one's view of nutrient limitation always reflects ecosystem processes. It is no longer the prime determinant of production, but a minor brake as a result of the patchiness of the ocean or a fail-safe mechanism if the system breaks down.

In his study of Canadian lakes, Vollenweider (1975) established differences between oligotrophic and eutrophic lakes. The annual quantities of nitrate and phosphate are larger in eutrophic waters and they vary more slowly with time and so depend on external sources, the rivers. In the lakes, Vollenweider was able to measure the turnover rates: they are high in the oligotrophic waters, whereas the annual quantities are low. The analogy with processes in the sea is obvious. In eutrophic lakes, dissolved inorganic phosphorus and total phosphorus are about equal, but it is only one-tenth of the total phosphorus in the oligotrophic ones. During the summer, phosphorus is used 20 times over in oligotrophic lakes. Vollenweider and Harris (cited by Harris, 1986) plot the logarithmic ratios (C/P) and (N/P) from particulate material. Figure 2.4 shows marine observations from the ocean and those of Goldman *et al.* (1979) plotted about the so-called depletion line, $(C/P) = 5.2 \ (N/P)^{1.18}$. Observations from eutrophic waters lie towards the bottom left of the diagonal and those from the oligotrophic ones lie along the top right. The Redfield ratio in weight is indicated as R. Phosphorus depletion as departure from that ratio requires observations to be distributed towards the top right of the depletion line. Nitrogen depletion would be shown by observations lying parallel to and above the depletion line. If one contrasts the observations of Goldman *et al.* (1979), which were based on deliberate experimental depletion as a departure from the Redfield ratio, with observations at sea, two out of 16 observations show possible nitrate depletion (but it must be recalled that the C/N ratio varies from group to group).

The structure of the oceanic ecosystems

From the work of Goldman *et al.* (1979) and Furnas (1982, 1990, 1991) it appears that the ecosystem in the oligotrophic ocean may be running quickly at or near maximal algal division rates. It is in a quasi-steady state and the daily increment of production must be eaten because the

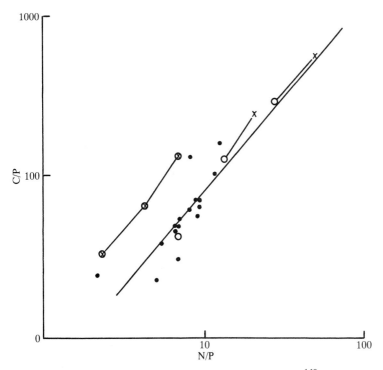

Figure 2.4 Vollenweider's depletion line, $(C/P) = 5.2(N/P)^{1.18}$. Phosphorus depletion appears at the top right with the observations of Goldman *et al.* (1979) and nitrogen depletion appears on the left and parallel to the depletion line of phosphorus, as shown by the observations of Goldman *et al.* (1979). The marine observations (•) are those of Harris (1986). With phosphorus limitation in culture, N/P and C/P increase and observations appear on the line at the top right. If N/P deceases, C/N and C/P increases, so observations on nitrogen limitation appear along a line parallel and above the phosphorus line of depletion.

numbers remain the same in time. The quantities of nitrate in the surface or near-surface waters are so low as to be negligible and account for only a very small part of the daily production. It is a regenerative system and presumably runs on ammonia excreted by all animals from bacteria to the vertically migrating fish and larger crustacea. Further, the cell quotas are probably high enough to accommodate the transient lacunae in the system. It will be of some interest to examine some of the other ecosystems in the world ocean.

Ocean Weather Station P lies about 1200 miles (1930 km) west of Vancouver Is. in 50° N and is considered to represent the Alaska gyral

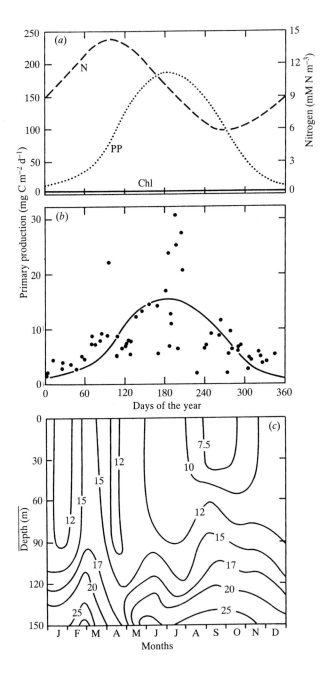

Figure 2.5 Events at Ocean Weather Station P in the Alaska gyral: (*a*) modelled annual cycle of primary production in mg C m^{-2} d^{-1} and of nitrogen in the mixed layer with chlorophyll held at 0.3 mg m^{-3}; (*b*) primary production at Ocean

and the deeper parts of the Bering Sea (Parsons and Anderson, 1970; Larrance, 1971; Anderson and Munsen, 1972; Saino *et al.*, 1979). Extensive biological and physical sampling has been made there since 1961. There is a permanent halocline at 100 to 130 m that prevents deep winter mixing and there is a seasonal thermocline at about 30 m and between May and September, the mixed layer depth being less than the critical depth. There is no seasonal variation in chlorophyll, but there is a marked change in primary production as modelled by Frost (1987) from January to June by a factor of about 15 (Figure 2.5(*b*)); Figure 2.5(*c*) shows the average distribution of nitrate in mmol N m^{-3} from 1965 to 1971 (Anderson *et al.*, 1977). Because the increment in production is continuously removed, it must be eaten (Parsons and Lebrasseur, 1968).

Minas *et al.* (1986) distinguished two forms of production in different regions, high nitrate/low chlorophyll (HNLC) and high chlorophyll/low nitrate (HCLN). There are three HNLC regions, the subarctic North Pacific, the equatorial Pacific and the Antarctic (outside the Antarctic Peninsula). The SUPER program in the Alaska gyral during the eighties (Miller *et al.*, 1991) examined the major biological processes. They reiterate that the obvious explanation of the HNLC seas is that the algal cells were eaten as quickly as they grow. The system produces 0.24 to 1.30 g C m^{-2} d^{-1} and Booth *et al.* (1982) showed that the cells were growing as fast as permitted by irradiance and temperature. Dagg and Walser (1987) found that *Neocalanus*, the predominant large grazer, did not eat enough phytoplankton to satisfy their respiration and so must eat the micrograzers. Micrograzing by the protozoa was estimated by the Welschmeyer and Lorenzen (1984) method of measuring phaeopigments derived from chlorophyll in the grazer's gut less phaeopigment derived by photodegradation outside the gut; the protozoa took 5 to 10 times as much as the macrograzers. Barlow *et al.* (1988) questioned the degradation of phaeopigments so the grazing had been underestimated. An important point is that these small animals can reproduce very quickly (Banse, 1982; Fenchel, 1982*a–d*), much faster than the phytoplankton; so in their turn they must also be eaten quickly, perhaps by *Neocalanus*. The dilution method (Landry and Hassett, 1987) decreases the density of grazers but not that of the phytoplankton, so the algal

Figure 2.5 (*cont.*)
Weather Station P averaged from 1961 to 1967 together with the curve fitted from Frost's (1987) model; (*c*) distribution of nitrate at Ocean Weather Station P in mmol N m^{-3} by depth and season (after Frost, 1987).

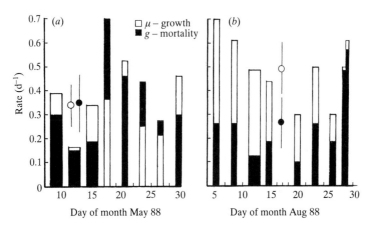

Figure 2.6 The balance of specific growth rate and grazing rate in the Alaska gyral (after Miller *et al.*, 1991).

division rate, μ, is maximal at the greatest dilution; both μ and g, the algal mortality due to grazing, are estimated in the same system. Figure 2.6 shows that μ and g do match to some degree, which is sufficient to explain the HNLC system in the Alaska gyral. It is likely that the faecal pellets of the micrograzers are broken down to nitrate in the euphotic layer, but 10 μmol of nitrate are used during the summer; because ammonia inhibits the uptake of nitrate (Dortch and Conway, 1984; Dortch (1990) pointed out that preference for ammonia is perhaps of more importance than inhibition), there must be considerable variation within the quasi-steady state of the Alaska gyral.

Frost (1987) has examined the production processes at Ocean Weather Station P with a two-layer model (Figure 2.5(*a*)). The Jassby–Platt dependence of photosynthesis on irradiance was used and the carbon/chlorophyll ratio was varied seasonally by a factor of 3. Respiration was taken to be 10% of P_{max}, the maximal rate of photosynthesis (Steeman Nielsen and Hansen, 1959). The temperature dependence of photosynthesis was taken from Eppley (1972). The half-saturation coefficient of the Michaelis–Menten relationship was 1 μmol N/l; from Figure 2.5(*b*), this means that there is some nutrient limitation, but not much. Nitrate was mixed through the thermocline at a very low rate. Grazing was estimated as the production less the mixing loss, for nearly all the algae were eaten; excretion of ammonia was taken as 0.3 of the quantity of food eaten. The curve in Figure 2.5(*b*) shows the fit of the model to the data.

The major copepods are *Neocalanus plumchrus* and *N. cristatus*. The filtration rate needed to satisfy the quantity grazed ranged from 6.5 to 72.6 mg C d^{-1}, but the greatest rate from experimental results and the quantities of *Neocalanus* amounted to 0.4 to 3.61 mg C d^{-1}. Frost then examined the grazing of the microzooplankton, 0.02 to 0.2 mm in length, and found that the filtration rate did satisfy the quantity grazed; it depended on the winter production of the small herbivores and the estimated mortality rates in winter, but the original supposition of Parsons and LeBrasseur was demonstrated.

The algal cells comprised *Synechococcus*, flagellates, cryptomonads, dinoflagellates and diatoms (Booth *et al.*, 1982), which implies that the spring outburst of the traditional form does not take place where the stock of chlorophyll increases sharply during the period of transient thermoclines, when the depth of mixed layer shallows in spring. It is not the quasi-steady state production system of the subtropical anti-cyclones because there is a marked seasonal cycle in primary production, if not in chlorophyll. Frost examined the role of *Neocalanus* and suggested that the animals were partly herbivorous and partly carnivorous in that they grazed on the smaller herbivores; however, many of these may be too small for *Neocalanus*. Frost's paper is of considerable importance because it is the first that shows the prime importance of grazing in the productive process.

Frost (1991) extended his model. Average advection at the single station was estimated as 10% d^{-1}, diffusion as 2% and with an average algal division rate of 0.57 d^{-1},

$$\delta P/\delta t = (0.57 - 0.10 - 0.02 - g)P = (0.45 - g)P.$$

His grazing hypothesis states: 'in the nutrient rich areas of the open sea, phytoplankton net specific growth rate and specific mortality tend towards approximate balance that might be perturbed, but not fully disturbed by forcing events external or internal to the pelagic food web'. Figure 2.7 shows the fit of the model to observations of (*a*) chlorophyll m^{-2} for 200 days, (*b*) primary production in mg C m^3 d^{-1}, and (*c*) chlorophyll mg m^{-2} throughout the year. It follows that the grazers must be insulated from *destabilizing* predation and because the blooms are absent, the predators are food limited. Frost notes that the large diatoms tend to be absent from the surface layers at Ocean Weather Station P, but perhaps the ocean is not turbulent enough there.

Smetacek *et al.* (1990) has revised our ideas of how the Antarctic ecosystem works. The earlier view was based on the work of the

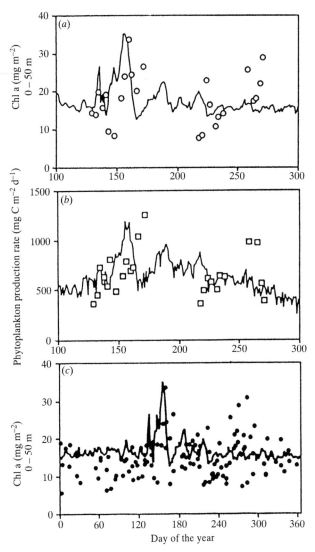

Figure 2.7 The fit of Frost's (1991) model to annual observations at Ocean Weather Station P: (*a*) chlorophyll (m^{-2}); (*b*) rate of primary production ($mg\,C\,m^{-2}\,d^{-1}$); (*c*) chlorophyll ($mg\,m^{-2}$ for the whole year).

Discovery in the region around South Georgia where the blue whales used to live. Here there was a simple food chain of diatoms, krill and whales. Smetacek and his colleagues showed that the blooms are found only in the restricted areas from the Antarctic peninsula to the South Orkneys and South Georgia. Elsewhere the Antarctic region is not very productive, dominated by flagellates. There is a deep mixed layer and the Secchi disc can be seen as far down as 60 m, which represents very clear water indeed. All the macronutrients are in excess at depth and at the surface. Indeed the Antarctic has been described as an oligotrophic ocean in a sea of nitrate.

The communities comprise nanoflagellates, small pennate diatoms, heterotrophic nanoflagellates and ciliates, all of which are characteristic of the oligotrophic ocean or the summer production in the temperate seas. The larger diatoms are of course found in the Bransfield Strait, where production is locally high. Sedimentation is low in the pelagial, 1 to 8 mg m^{-2} d^{-1}, characteristic of Peinert's retention system (see Chapter 3), where production is fuelled by ammonia. In contrast, the sedimentation in the Bransfield Strait can reach nearly 2000 mg m^{-2} d^{-1}. The interesting question is: how does this system work with a rather deep mixed layer? The water is very clear and, as the stock of phytoplankton is low, there can be no decoupling of phytoplankton from the zooplankton.

The zooplankton comprises copepods and some large ones that overwinter in diapause with one generation per year. Smetacek *et al.* gives a new view of the krill biology. The swarms have long been known to forage in open water, but in winter the krill feed on ice algae as the frazil ice is forming. They congregate under the ridges where the production is higher. Then when the ice melts and production increases in the stratified water they switch from ice scraping to the filtration of algal cells. They take the production as it develops and no blooms appear except in the special regions off the Antarctic Peninsula.

The Antarctic ecosystem is one that, in general, is driven partly by the persistent deep mixed layer even at the height of the austral summer and partly by the efficient transfer of small cells to the euphausiids. Like the North Pacific, it is probably a system dominated by the grazers and the transfer of material to animal flesh. Nutrient limitation does not play a part, except transiently.

Parsons and Lalli (1988) compared production in the Alaska gyral at Ocean Weather Station P with that in the North Atlantic at Ocean Weather Station I (with some information from Stations B and J). The difference between the two oceans is illustrated in Figure 2.8; between January and April, the water is isothermal (and the mixed layer depth may be greater than 200 m) at I whereas the water at P is stratified as noted above. Figure 2.8 shows the seasonal changes in chlorophyll by depth at P and I. Primary production continues at a low rate in the

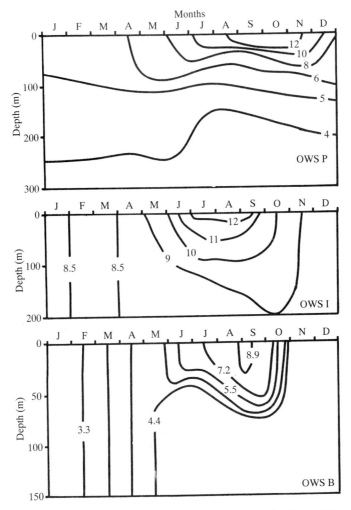

Figure 2.8 Contrast between the North Pacific and the North Atlantic.
(*a*) The temperature distributions by depth and season at Ocean Weather
Stations P, I and B (after Parsons and Lalli, 1988).

North Pacific winter, whereas in the North Atlantic there is none. In the
North Atlantic, there are two peaks of chlorophyll, spring and autumn,
whereas at P there is a low level of chlorophyll throughout the year.

At P the species composition comprises mainly haptophytes, (prymnesiophytes), chrysophytes and prasinophytes and many coccolithophorids; diatoms play only a minor part. In the North Atlantic, the
diatoms predominate in the spring bloom and dinoflagellates in that of

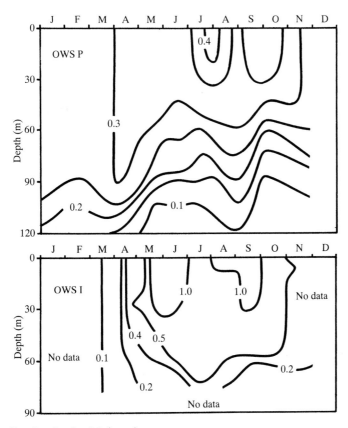

Caption for fig. 2.8 (*cont.*)
(*b*) The seasonal changes in chlorophyll by depth at Ocean Weather Stations P and I (after Parsons and Lalli, 1988).

the autumn; coccolithophorids occur during the spring bloom alongside the diatoms. The North Pacific herbivores comprise the microzoo-plankton and the neocalanids; the latter spawn only once a year. In the North Atlantic, calanids such as *Calanus* are much smaller and need food in the spring bloom; grazing is therefore mobilized with the second generation. The quantity of copepods in the two oceans is about the same and the difference between the two oceans is that the bloom in the Atlantic results from the decoupling of the production of phytoplankton and the zooplankton. The surface levels of nitrate fall in the Atlantic by midsummer to less than $1\,\mu$mol/l, as compared with $>5\,\mu$mol/l in the Pacific.

In the North Atlantic, the new production of the spring bloom driven by nitrate is transferred to the regenerated one driven by ammonia. There is no nutrient limitation in the Northern Pacific and that in the North Atlantic takes place in the brief hiatus between new and regenerated production. In the oligotrophic ocean, production is in a quasi-steady state, where the daily increment of production is eaten and where the nitrate in the surface layers is very low, but production continues and the algal division rates may be high. Models of production must include functions of nutrient limitation because the latter must occur within the structure of patches in time and space. Further, the lack of nutrient control in the oligotrophic ocean has not been shown; indeed the models of Fasham *et al.* (1990) demand it, as will be shown in Chapter 3.

It has always been a little difficult to demonstrate the existence of nutrient limitation in the sea, but Platt *et al.* (1992) have shown that the shape of the photosynthesis/irradiance curve changes as nitrate decreases. They worked in the north Sargasso Sea between April and September between 1983 and 1990, that is under post spring bloom conditions and during the summer oligotrophic phase. Between 5 and 20 April 1990 the vessel worked in the same water mass, as shown by a free-floating buoy. The slope at the origin of the photosynthesis/irradiance curve (normalized to biomass) declined by a factor of about 3 and was correlated with the decline in nitrate after the spring bloom. So the photosynthetic system ran more slowly. Head and Horne (1993) showed that fucoxanthin, indicating the presence of diatoms, declined during this period but 19-butanoyl-oxy-fucoxanthin, indicating 'prymnesiophytes such as *Phaeocystis*, increased a little. As nitrate declined, diatoms were possibly replaced by *Phaeocystis* and the photosynthetic system became less efficient. A succession occurred and nitrate declined and the question arises is the succession a consequence of the decline in nitrate or is the decline a consequence of the succession?

In recent years the low chlorophyll level in the HNLC regions has been attributed to iron deficiency (Martin and Gordon, 1988; Martin *et al.*, 1989, 1990a,b; de Baar *et al.*, 1991). These authors used enrichment experiments to show that iron levels apparently limit the rate of production in the Alaska gyral, the equatorial Pacific and in the Antarctic. There has been some discussion on the interpretation of these experiments (Banse, 1991a,b,c; Cullen, 1991). If phytoplankton grows in a bottle for some days, the nutrients decline. Because the algal division rate, μ, depends on the nutrient concentration (as in the Michaelis–Menten relationship), it must decline as the nutrients are diminished. If

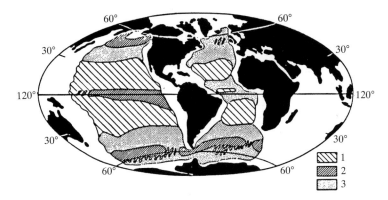

Figure 2.9 Banse's (1992) three domains of productive systems in the world ocean. (1) Subtropical gyres where nutrients are very low indeed and there is little seasonality in the production cycle. (2) The HNLC areas, high nutrient low chlorophyll. (3) The regions where the spring bloom dominates with low nutrient in summer.

the bottle is enriched with a nutrient, μ is sustained for longer and more material is produced, but only for a limited period. This must happen even if the initial sample of sea water is nutrient replete. Hence, an enrichment experiment tells one little of the nutrient condition of the sea. It describes the dependence of μ (or the quantity produced) on nutrient concentration, but not very well, probably because of variations in cell quotas as species succeed each other in the bottle.

Banse (1992) reviewed the role of grazing in the production of algae in the sea. He extended the concept of HNLC regions to the world ocean and distinguished three domains (Figure 2.9). Domain 1 is in the subtropical gyres where the nutrients are very low in the euphotic zone and there is little seasonality in the production cycle. Domain 2 is characterized by lack of seasonality and high nutrients: the subarctic North Pacific, the equatorial Pacific and the Antarctic (excluding the area around the Antarctic Peninsula). In domain 3, the North Atlantic, the equatorial Atlantic and the region between 30° and 40° latitude in all oceans, there are pronounced seasonal cycles and nutrients are reduced after the spring bloom.

In domains 1 and 2 the algal stocks appear to remain in a quasi-steady state and Banse formulated the productive processes:

$$\mu = a + m + s + g,$$

where μ is the algal division rate, a is the rate of divergence from below,

Table 2.1. *The vital parameters in the production system in Domains 1 and 2*

	μ	a	m	s	g
Subantarctic (summer)	0.25 (0.50)	0.04	0.05	≥ 0.02	0.15
Subarctic (Alaska gyral)	0.50 (0.80)	<0.01	−0.03	0.01	~0.45
Subtropical gyre	1.20 (1.70)	<0.01	<0.01	≥ 0.01	~1.20
Equatorial upwelling	1.0 (1.4)	0.04	0.02	0.01	~0.90

Note:
The figures in brackets are the maximal values of μ for the temperatures in each region (from Eppley, 1972).

Table 2.2. *Rates of uptake in a warm core ring in the Atlantic of carbon, nitrogen and phosphorus (pC, pN and pP) for twelve days (McCarthy and Nevins, 1986)*

	21 April	23 April	26 April	1 May	4 May
pC	200	41	34	81	110
pN	21.0	5.2	8.2	16.0	24.0
pP	1.6	0.52	0.62	0.91	1.2
f-ratio	0.74	0.62	0.66	0.53	0.64

Note:
Means pC: pN, 6.8; pC: pP, 104.3; pN: pP, 15.3. f = 0.64.

m is the mixing rate across the thermocline, s is the sinking rate, and g is the grazing rate. Table 2.1 gives the values cited.

Banse gives good reasons for the estimates of a, m and s. It follows that the reason for the HNLC systems is that the algal division rates are matched by the grazing rate and that nutrients play little part in such processes.

Warm-core rings are generated poleward of the western boundary currents. Off the Gulf Stream, they are large circular structures up to 100 km across and they persist for several months, as they move

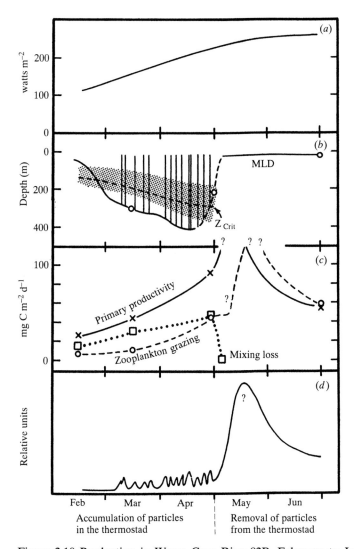

Figure 2.10 Production in Warm Core Ring 82B, February to June 1986: (*a*) absorbed short wave radiation; (*b*) mixed layer depth and critical depth: the vertical lines indicate times when the thermostad was gaining heat, i.e. when transient thermoclines might have appeared; (*c*) euphotic zone, rates of primary production, grazing mortality and mixing loss; (*d*) euphotic zone, phytoplankton biomass.

southwestward seaward of the 200 m depth contour and inshore of the Gulf Stream (Brown *et al.*, 1986). They have also been described off the Kuroshio and the East Australian Current. Those from the Gulf Stream derive from the nutrient-poor Sargasso Sea. Yet Bishop *et al.* (1986)

showed that during a period of 12 days there were sharp increases in carbon, nitrogen (as nitrate, ammonia and urea) and phosphorus.

The uptake rates were measured with isotopes, ^{14}C, ^{15}N and ^{33}P and were expressed as mmol $m^{-2} d^{-1}$. Table 2.2 gives the results, together with the total amount of nitrogen produced, the average elemental ratios and the average f-ratio. The elemental ratios are close enough to the Redfield ratios to preclude nutrient rate limitation, and the f-ratio suggests that new production predominated. McCarthy and Nevins wrote that this was due to deep convective mixing following a storm (Evans *et al.*, 1986).

Figure 2.10 (Bishop *et al.*, 1986) shows a more general picture of the processes. The observations of McCarthy and Nevins were made at the end of April and beginning of May, when the critical depth exceeded the mixed layer depth. Subsequently the bloom may well have been eaten. With apparently simple methods, rates of production were measured, the nutrient status established and the degree of ecosystem maturity established. From the methods of Goldman *et al.* (1979), estimates of algal division rates should emerge from the measures of elemental ratios and thence loss rates, if a study of population dynamics were needed.

Conclusion

Much work has been done on the part played by nutrients in the study of production in the sea. This is because the simplest view of limitation is the Liebig form in the quantity produced. The Blackman form, limitation in specific growth rate, μ, or in the rate of uptake, is perhaps more difficult to assess, because it is hard to measure μ in the open sea. But the most important point is that the algal cells are nearly always being eaten and converted in part to ammonia. Then the Blackman limitation is delayed and somewhat more is produced than is expected from the simple Liebig view of the decrement of nitrate. Indeed, in the oligotrophic ocean of the subtropical gyres, the Liebig view is impossible because the algal stock remains at the same level all the time, with no nitrate in the surface layers and $\mu = g$. These are the reasons why there are not many decisive demonstrations of nutrient limitation in the sea.

3

A view of production in the sea

Introduction

Understanding of productive processes in the sea has increased very much in the last decade or so. There has been much discussion on radiocarbon measurements and on the nature of productive processes in the sea. Considerable advances have been made in the construction of models that describe such processes; they have been stimulated by the prospect of satellites that will fly in the mid-nineties and by the need to estimate the loss of carbon in the biological pump. This is the mechanism by which carbon is transferred below the mixed layer to animals and sediments. It is at the centre of studies of climate in which the heat store of the ocean plays such an important part.

A major lacuna in the study of oceanic ecosystems is our understanding of grazing. The first difficulty is that the smaller copepods have not been well sampled and the second is that the feeding mechanisms of copepods were wrongly described as a filtration rather than a direct capture. Many of the mechanisms have been elucidated by Strickler and his colleagues (see below), but much remains to be done, particularly on the smaller animals. A more general point is that the growth and mortality of zooplankton in the sea has not been well described.

The loss of diatoms from the euphotic layer has been described in some detail. The reason is that the sinking rate can be estimated directly from measurements at sea. Then, given the loss rates and algal division rates, the quantity grazed can be derived, which improves our understanding of grazing mechanisms. A general point is that it might be desirable to estimate the vital rates at sea rather than from laboratory measurements.

Oceanic production

Methodology

The traditional ways of estimating production are simple, by measuring uptake of ^{14}C, ^{15}N or the evolution of oxygen, all in short periods of time. Peterson (1980) described the technical changes up to the end of the seventies. Until the free bacteria and the picoplankton were discovered, the water in sample bottles was drawn through 0.45 μm filters (rather than 0.20 μm filters, which would have retained these microorganisms) and so the older estimates are probably too low. The experimental bottles may include an ecosystem of three or more levels of organisms and the increments measured are, of course, generated by the autotrophs; but the protozoa (and/or the heterotrophic flagellates) eat the picoplankton and perhaps some of the nanoplankton, so a proportion of production is underestimated (unless the time of exposure is very short indeed).

Riley *et al.* (1949) measured oxygen production in the Sargasso Sea, the results being much higher (by a factor of 2 to 3) than those with the radiocarbon measurements (Nielsen and Jensen, 1957). The question was raised again by the Dutch, amongst others. Gieskes *et al.* (1979), with a series of bottles of increasing size (30 to 3800 ml) found that the carbon assimilated increased by a factor of 11 with size, minimizing potential pollution. Postma and Rommets (1979), from the diurnal variation in particulate organic carbon, found an increment from 0.45 to 1.00 g C m^{-2} d^{-1}. On the same cruise, Tijssen (1979) found a diurnal increment of oxygen evolution of from 0.55 to 1.10 g C m^{-2} d^{-1}. Fitzwater *et al.* (1982) showed that the radiocarbon measurements, at least in oligotrophic waters, were underestimated by a factor of 2.6 because of trace metal contaminants. As a consequence, a programme was established (Eppley, 1982) to investigate the planktonic rate processes in the oligotrophic ocean. The first result came from Marra and Heinemann (1984), who found no difference between clean methods (that is, with elimination of trace metal contaminants) and conventional ones, but the water was sampled with a Teflon bottle, which implied that the difficulty lay with the method of collection. Laws *et al.* (1984), with clean equipment, showed that the growth rates of phytoplankton in oligotrophic waters (estimated with five distinct methods) ranged from 0.66 to 2.1 d^{-1}, much higher than had been thought in the previous decade. Williams *et al.* (1983), with clean methods, found no difference between radiocarbon measurements and those from light and dark oxygen bottles (light bottles estimating photosynthesis less respiration, dark ones respiration only). Marra and Heinemann (1987), again with clean methods, and blue filters (rather than neutral density filters) found a production of 0.456 g C m^{-2} d^{-1}, compared with 0.315 g C m^{-2} d^{-1} in the same region previously. Grande *et al.* (1989) compared light and dark oxygen methods, the radiocarbon method and the $^{18}O_2$ method (the rate at which labelled oxygen is evolved from labelled water) in Teflon bottles; they found that the light and dark oxygen bottles yielded results comparable with those of $^{18}O_2$, but that estimates from deck incubators were up to 60% less. Martin *et al.* (1987) suggested that estimates of primary production resulting from clean techniques be revised upward by a factor of 2.6. Laws *et al.* (1990) replaced the neutral density filters of deck incubators with blue-violet ones and found that production rates in the deep blue sea were doubled. In fact, they recorded values of 0.777 ($\pm$0.219) g C m^{-2} d^{-1} in the oligotrophic waters north of Hawaii. So the production in the oligotrophic ocean has been revised upwards. Riley *et al.* (1949) were probably right, but the important conclusion is that the oligotrophic ocean is fairly productive, perhaps approaching 1 g C m^{-2} d^{-1}. In the deep, clear, euphotic zone, production is fairly rapid but dispersed, and all the higher trophic levels are also dispersed to the very limits of efficiency.

A quite different approach was made by Jenkins (1982); the downward carbon flux was estimated from the oxygen utilization rate (OUR) in depths below the euphotic zone, dated by $^3H/^3He$. This represents an estimate of P_c, the sedimented material, which in the long term must balance the new production, P_{new}; the estimate was 55 g C m^{-2} yr^{-1} and it was sustained by an 18-year time series (Jenkins and Goldman, 1985). The salient point is not its relation with new production but the fact that the estimate was high. Martin *et al.* (1987) measured the downward organic carbon fluxes (free from zooplankton) with free-floating traps and they concluded that the oxygen consumption agreed with clean radiocarbon measurements. Platt *et al.* (1989) distinguished *in vitro* estimates (^{14}C, ^{15}N, O_2) of one hour to one day from those in bulk (OUR, P_c that sedimented into traps, the accumulation of oxygen below the euphotic zone and the flux of nitrate into the euphotic zone, lasting from days to years). They note that the upper limit is the optimal conversion of photons absorbed by the photopigments; the lower is the depletion of winter nitrate above the seasonal thermocline. They wrote:

$$P_{new} + P_r = P_T,$$

where P_{new} is new production, P_r is regenerated production and P_T is total production.

We should recall that the photosynthetic quotient for new production, $PQ(P_n) = 1.8$ and that for regenerated production $PQ(P_r) = 1.2$.

Platt *et al.* suggested two hypotheses:

(a) $P_{new} > P_T(in\ vitro)$;
(b) $P_{new}\ (bulk) > P_T(in\ vitro)$.

Today, the first hypothesis depends on Jenkins' (1982) estimate, but no local radiocarbon measurements were cited. The second hypothesis depends upon the estimate of f; $P_T = P_{new}/f$, where f is the f ratio described in the last chapter. But the estimate of f should be properly averaged in depth and in time, as noted in Chapter 2 and Figure 2.2(*c*). This brief discussion illuminates a major difficulty. The samples are very small in number as compared with the size of the ocean and, much more important, the changes in space and in time are very difficult to sample properly. The only way in which this can be overcome is by estimating production on an oceanic scale, by satellite. This point will be returned to later, when Platt's work on the information to be derived from satellite data is described.

The part played by grazing

In earlier models of production, for example those of Steele (1974) and Walsh (1975), the mortality due to grazing was expressed as a simple and correct function of the biomass of the zooplankton. But today, there is an unstated conflict between the model of Evans and Parslow (1985), in which the control of the spring bloom is attributed to grazing, and those of Walsh and his colleagues (see Walsh, 1983), in which the algal cells sink to the seabed. The resolution of this conflict is considered below.

The discovery of the picoplankton and the microbial food loop has changed our perception of the processes of grazing. Cushing (1989) proposed that there were two distinct ecosystems in the sea, that of diatoms, copepods and fish, the traditional food chain, found in weakly stratified water, and that of the microbial food loop, after the thermocline has become fully established. The herbivores of the traditional chain are the copepods that depend on particles larger than 5 μm in diameter (Bartram, 1981; Harris, 1982*b*; Frost *et al.*, 1983). In the microbial food loop there are three levels of grazers: copepods, ciliates and heterotrophic flagellates, of which the latter feed on particles $<5\,\mu$m in diameter. Much energy is retained within the photic layer, but the downward transfer below the mixed layer depth is carried out in the main by radiolarians, coccolithophores and the larger faecal pellets. In the upwelling areas, equatorward of 40° lat. and in the spring and autumn outbursts of temperate and high latitude waters, new production is generated in the traditional food chain and much of the material ends below the euphotic layer.

The smaller copepods are the most numerous and probably exert the greatest grazing impact, but recently Nicholls and Thompson (1991) have shown that they have been poorly sampled. Wiborg (1948) had found that most stages of copepods in the cool waters of the Norwegian Sea were retained by mesh sizes of 145 μm. The spread of selection of the meshes of plankton nets is rather broad (Saville, 1959; Heron, 1968). Nicholls and Thompson measured the selectivity of six meshes on a Lowestoft multi-purpose sampler (Beverton and Tungate, 1967): 275 μm, 198 μm, 124 μm, 90 μm, 60 μm and 35 μm; the larger mesh sizes were 90% efficient, but that at 60 μm was 87% efficient and that at 35 μm was only 55% efficient in filtration. At 275 μm, of all the naupliar stages of *Calanus*, >50% escape. For *Paracalanus*, >50% of all stages of nauplii and copepodites escape from meshes between 60 μm and 275 μm in size (Figure 3.1). The small copepods examined by Wiborg were much larger than those observed by Nicholls and Thompson. Indeed Saville (1959) had noticed that *Oithona* was poorly sampled by the mesh sizes of a normal plankton net. Omori and Ikeda (1984) noted that nets with meshes $<200\,\mu$m were liable to clog in coastal or shelf waters (and those $<100\,\mu$m in the open sea), therefore the smaller animals, such as *Paracalanus*, should be sampled by pump or water bottle and not by net. Most samples of zooplankton have been taken with meshes of about 100 μm to 200 μm and the smaller and much more numerous animals have not been sampled properly. If the impact of grazing is to be estimated as a proportion of primary production, either all stages of grazer should be counted and included or the primary production should be partitioned by sizes of animals sampled.

The expression of grazing used in most models is based on the work of Butler *et al.* (1969), Frost (1972) and Mullin and Brooks (1970), and can be briefly written: $I = 0.314\ W^{0.7}$ where I is ingestion in μmol N and W is the weight of the grazer in μmol N. Huntley and Boyd (1984) showed that the assimilation, $A = aC^{b}W^{n}$, where C is food concentration in μg C l^{-1}, n and b are powers of temperature and W is weight in μg C.

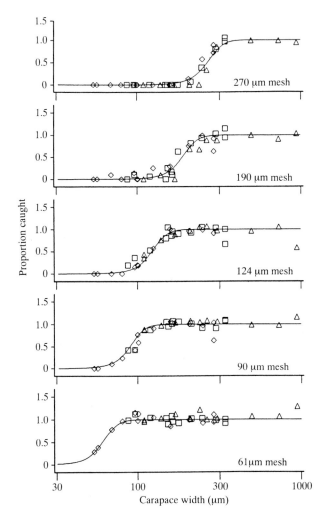

Figure 3.1 Mesh selection of plankton nets. The animals sampled were *Calanus* spp. (△), *Pseudocalanus minutus* (□), *Paracalanus parvus* (◇). The curve fitted to the data is:

$$\frac{\text{Number caught }(i)}{\text{Number available }(i)} = \frac{1}{1 + \exp\{-8.9 \times (R - 1)\}}$$

where R is the body width of the ith stage divided by the mesh size.

Respiration (in μg C animal h^{-1}), $R = kW^m$ where k and m are also power functions of temperature. Then the maximal growth rate.

$$G_{max} = aC_c{}^b W_n - kW^m,$$

where C_c is the food density at which the animals are fed *ad libitum*. We suspect that growth is often food limited and the only way to detect this is to estimate the growth rate and to compare it with the maximal for a given temperature (see Chapter 6). An interesting point is that the growth of zooplankton in the sea is often estimated with a simple exponential, where a von Bertalannfy curve might be more appropriate because the oldest copepodites and adults do not grow much. Dagg and Turner (1982) estimated ingestion, or grazing impact, by applying the Mullin and Brooks formula by sizes of animals taken in a $102\,\mu$m mesh and by sizes of algal cell (5 to $124\,\mu$m in diameter) in 16 channels of a Coulter counter. It is often expressed as a proportion of primary production, but the animals eat only the stock of algae (i.e. production minus quantity grazed). Grazing experiments have often been executed in dim light in which the algal cells are assumed not to divide. Radiocarbon experiments were assumed to estimate productivity directly, with grazers excluded. Today neither assumption can really hold. In such an experiment, with the algal quantities at time t_1 and t_0, $P_1 > P_0$, the increment in radiocarbon, $P_1 - P_0 = P_0\{\exp(\mu - G)t - 1\}$, where μ is the algal division rate and G the mortality due to grazing. The filtration rate, $F = VG/n$, where V is the volume of the experimental vessel and n the number of animals. Then the ingestion, $I = (V/n)[G/(\mu - G)](P_1 - P_0)$. Eliminating V/n, $I/(P_1 - P_0) = G/(\mu - G)$. On the original assumptions of no algal division rate in the grazing experiment or a radiocarbon experiment with no grazers, $I/(P_1 - P_0) = G$, which is the justification for the expression of ingestion as a proportion of primary production. It might be desirable to write $G = I(\mu - G)/(P_1 - P_0)$.

There are problems in the estimation of grazing. The mortality, or volume swept clear, of cells (counted and identified) is estimated in the traditional grazing experiment in dim light. There is nothing wrong with the single observation, but it would be desirable to spread the samples of such observations much more broadly, by depth, time of day, and sizes of cells eaten. A resistivity counter can be used to make an analogous estimate, but living cells cannot be distinguished from inert particles. The fluorescence of phaeopigments in the gut estimates ingestion, but the method requires full conversion of chlorophyll, which may not occur. Cells can be labelled with radiocarbon for short periods, which may not represent the grazing throughout the day. (This account is taken from Paffenhöfer, 1988). Thus, it appears that the methodology of experimental grazing studies is at the moment incomplete.

In the analysis of productive systems there are three important points: (1) the smaller copepods should be sampled with water bottles because they escape from most mesh sizes in current use and the nets with smaller meshes may well become clogged; (2) the microzooplankton is sampled with water bottles and its grazing capacity must be estimated properly because of its importance particularly in the oligotrophic ocean; (3) if ingestion must be expressed as a proportion of primary production then it must be estimated by the right sizes of cells, for example >5 μm if the grazers are copepods. Also, the algal division rate should be estimated. Further, it might be desirable to calculate the processes in time as Welschmeyer and Lorenzen (1984, 1985; see below) did rather than take single shots at a very dynamic system.

The modern analysis of copepod feeding started with Friedman and Strickler (1975), who described in calanids the chemo-receptors for which the setae are the sensillae. Then Purcell (1977) gave an account of life at low Reynolds' number (Re)

$$Re = \rho \, v \, L / \mu,$$

where ρ is the density of water, μ is the dynamic viscosity, L is the linear dimension of the object and v is the relative velocity of the fluid across the solid object. When $Re = 0.01 - 0.1$, flow is laminar across the object and the fluid is viscous within a relatively large boundary layer.

Figure 3.2 shows a copepod with the motion core, M, through which fluid is drawn, the viscous core, V, at the edge of which lies the boundary layer and the sensory core, S. CA is the capture area where feeding actually takes place and FS the sensory fields on the first antennae which delineate the sensory core. There are chemoreceptors on the first antennae and on the mouthparts. This description is taken from Strickler (1985): 'When an eddy decreases to the size of a copepod, it will dissipate quickly; there is not enough energy to maintain spin but plenty of friction due to velocity'. In other words the cascade of turbulence ends in such dimensions. Within the viscous fluid, algae can perhaps be detected by the flow of extracellular molecules and, in principle, the animal can detect the cell at a distance. The major consequence of life at low Reynolds numbers is that filtration is

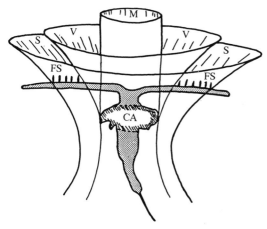

Figure 3.2 The feeding mechanism of the copepods; M is the motion core, V is the viscous core, S is the sensory core, FS is the sensory field on the first antenna and CA is the capture area (after Strickler, 1985).

inconceivable for it is 'like pushing a fork through honey' (Strickler, 1985).

Alcaraz *et al.* (1980) described a method of observation. They glued *Eucalanus crassus* to a dog hair, itself linked to a micro-manipulator. With a camera working at 500 frames s^{-1} and an exposure of 111 μs, they obtained a resolution of 3 μm; the system was lit by a xenon laser. Koehl and Strickler (1981) used this system with dyes in the water and showed that the copepod flaps its appendages to carry the water past, so they are paddles and not filters. The second maxilla is used actively to capture parcels of water with food in them. Strickler (1982) showed that the animal captures a volume of water with the 'fling and clap' mechanism. It uses mechanoreceptors as it sinks, in the 'hop and sink' mode, which saves energy (Haury and Weihs, 1976). A neutrally buoyant animal dies because it cannot search for food (Strickler, 1985).

Paffenhöfer *et al.* (1982) described the capture of spherical and elongated cells. Cowles and Strickler (1983) found that feeding was intermittent. Price *et al.* (1983) showed that the larger and less numerous cells are selected, which implies relatively distant detection. An important point is that there is active rejection. Strickler (1985) noted that search takes tens of seconds, encounter takes seconds and capture occurs in milliseconds. The 'fling and clap' mechanism draws the algal cells into the capture area at 1 cm s^{-1}. For a *Eucalanus crassus* female, 345 ml^{-1} can flow through the capture area; a cell can be detected 0.8

body lengths ahead, 430 ms before capture. For this animal there is an upper limit of volume captured of 410 ml d^{-1}. Higher values can be found from 'volume swept clear' experiments, which estimate mortality. Paffenhöfer and Lewis (1992), with *Eucalanus pileatus* feeding on *Thalassiosira weissflogi* at low cell densities, established that the perceptive range is as much as $450 \mu\text{m}$ and that the average daily volume swept clear amounted to 995 ml per animal (and a maximum of 1565 ml per animal); they used the Strickler method to make their estimates which are considerably higher than those obtained by earlier methods. But what is needed is a theoretical model that combines the descriptions of behaviour with the results of experimental studies.

Gerritsen and Strickler (1977) formulated the encounter rate of predator and prey, Z_p, in terms of their respective velocities, v and u. The predator is at the centre of a sphere of encounter radius, R, which is really the perceptive range of the predator; note that u is the mean velocity of the prey population

$$Z_p = \{(\pi R^2 N_H)/3\} \{(u^2 + 3v^2)/v\} \quad v \geq u$$
$$= \{(\pi R^2 N_H)/3\} \{(v^2 + 3u^2)/u\} \quad u \geq v$$

where N_H is the uniform distribution of prey per unit volume. The encounter rate is the number of prey entering the encounter sphere per unit time. Then the proportion eaten, taking into account handling time, degree of hunger and attack efficiency, is an estimate of mortality; but the essential information needed is the velocity of the animals.

Rothschild and Osborn (1988) proposed that the encounter rate increased with turbulence (Figure 3.3) and confirmation was given by Sundby and Fossum (1990). Rothschild and Osborn developed the Gerritsen and Strickler theory with turbulent motion. Figure 3.3 shows predator velocity and prey velocity at three levels of uncorrelated root mean square turbulent velocity, $w = 0$, 0.1 and 0.3 cm s^{-1}. Figure 3.3(*a*) shows the contact rates with no turbulence and Figure 3.3(*b*) and (*c*) shows increments in contact rates with two rates of turbulence. Costello *et al.* (1990) examined the thesis experimentally with high speed infrared cinematography and created artificial turbulence. The four behaviour patterns, slow swim, fast swim, break (or sink) and groom, were observed under periods of pre-turbulence, turbulence and post-turbulence. They found that all three active behaviours increased with turbulence, but the most interesting point was that the proportion of slow swimming remained high and even increased in the post-turbulent

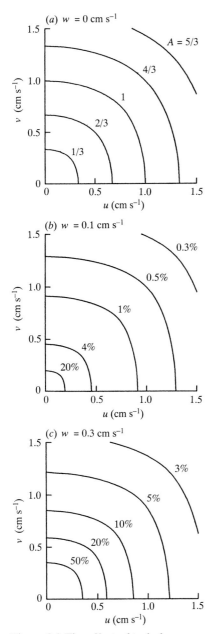

Figure 3.3 The effect of turbulence on encounter rates (Rothschild and Osborn, 1988): predator velocity and prey velocity are shown at three levels of uncorrelated root mean square turbulent velocity, $w = 0$, 0.1 and 0.3 cm s^{-1}; (*a*) shows the contact rates with no turbulence and (*b*) and (*c*) show the increments in contact rates with two rates of turbulence, $w = 0.1$ and 0.3 cm s^{-1}.

period. Marrasé *et al.* (1990) estimated the encounter rate (rate at which algal cells enter the capture area), the foraging efficacy index (feeding encounter rate/control encounter rate, in water with no cells) and the effective encounter rate (encounters when slow swimming). They showed that the turbulent diffusion was of the order observed at sea. They found that encounter rate increased with turbulence in high and low food densities and that the foraging efficacy index decreased with turbulence in high and low foods. But the effective encounter rate increased with turbulence at low food, and at high food effective encounter rates were the same with turbulence or without. In both papers, Strickler and his colleagues have shown that there is a real response to turbulence, but that it is more complex than expected. If encounter rate increases with turbulence, most of the experimental measurements are probably underestimates. Haury *et al.* (1992) investigated the effects of shear and turbulence on the distributions of animals in the sea; some variables were related to the degree of turbulent dissipation.

The major problems facing zooplanktologists today is to estimate growth and mortality at sea. Growth is well described in physiological or laboratory terms, but not many attempts have been made to model the growth of copepods at sea, perhaps because they grow by discrete stages and do not grow as adults. Mortality is hard to estimate, because the predatory pressure may change during the period of production of a life stage. Wood and Nisbet (1991) gave methods of overcoming this difficulty with the use of a cubic spline together with a cross-validation spline.

Strickler and his colleagues have made major discoveries of the way in which copepods feed in their sticky sea. It is now known how the food organisms are selected and captured and evidence emerges of a longer encounter radius than would have been thought possible in earlier decades. The Gerritsen and Strickler model is valuable because it can be used as a basis to describe the death rates of the algae and the rates of capture, ingestion and growth of the zooplankton. The simple model is not enough and other effects such as turbulence can be readily attached.

The same model can be used to describe the mortality of the zooplankton animals as indicated above but would need estimates of velocity, which can perhaps be obtained acoustically.

Models

An important model of production in the sea in the last decade or so is that of Evans and Parslow (1985). The mixed layer depth (MLD) separates an upper and lower layer in the ocean and it varies in depth seasonally. The algal cells live in the upper turbulent layer and their growth is averaged throughout the MLD because the time spent at a given depth is short compared with the time for a cell to divide because the mixed layer is turbulent. Their growth is expressed by the Jassby and Platt (1976) equation relating photosynthesis to irradiance. Nutrients are actively mixed across the boundary in winter and they diffuse across it in summer. Algal growth is restrained by nutrient lack in a Michaelis–Menten manner ($K_s = 1$ mmol N m^{-3}) and the mortality due to grazing is expressed in another Michaelis–Menten formulation ($K_g = 1$ mmol N m^{-3}). The same expression with a 50% grazing efficiency in ingestion and a carnivore mortality is used to estimate the production of herbivores. Because the MLD changes in time, the densities of organisms at each trophic level become concentrated or diluted. The onset of grazing is arranged by the concentration of herbivores as the MLD shallows; it should be remembered that many herbivores migrate seasonally and nightly through the thermocline. An important characteristic of the model is that at the end of the seasonal cycle the initial conditions are recreated.

Figure 3.4(*a*) shows the annual cycle of the phytoplankton and of the herbivores (as at Flemish Cap); the ordinate is transformed by the square root, in mmol N m^{-3}. In Figure 3.4(*b*) the annual cycle of the MLD is illustrated together with that of the photosynthetic rate in depth. Comparing the two figures, the MLD shallows from 80 m to 25 m in about three weeks and during this period the spring outburst increases by a factor of 20. This is a succinct expression of a difficult period, that of the transient thermoclines, as the seasonal thermocline struggles to establish itself against the spring storms. A second comparison shows that the peak of the spring outburst is controlled by self-shading and subsequently the decline expresses the transfer of material to the herbivores. The control of the spring outburst is density dependent and the transfer of energy to herbivores follows as the MLD shallows. This assumes that the grazers live within the mixed layer. The spring outburst, at least in the North Atlantic, comprises mainly diatoms and they are eaten by copepods and euphausids, which migrate from below the MLD each night. Indeed *Calanus* overwinters at 600 m in the

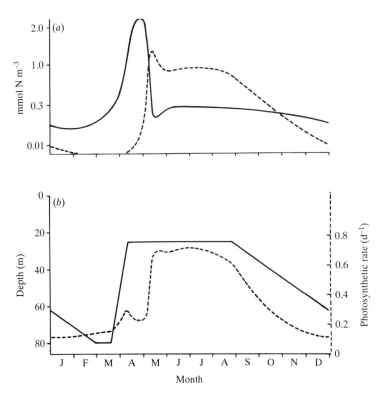

Figure 3.4 The Evans and Parslow (1985) model: (*a*) annual cycle of stocks of algae (——) and of herbivores (– – –) at Flemish Cap in mmol N m⁻³; (*b*) annual cycle of mixed layer depth (——) off Bermuda and of photosynthetic rate (– – –).

Norwegian Sea (Ostvedt, 1955) and the adults rise to the surface at about the time of the spring outburst. The Evans and Parslow description provides a useful metaphor of the processes involved. The spring outburst is controlled by the grazers if the herbivores can develop quickly enough.

Evans and Parslow also showed that, if the MLD remained at 80 m all through the year, the basic pattern of the spring outburst remained essentially the same, but it occurred earlier. This means that the predominant effects are the algal growth rates and the grazing by the herbivores. The model was then reduced by leaving out the nutrient limitation and the self-shading and the MLD was kept at 80 m. The spring outburst remained but it was reduced by grazing more quickly with greater production of herbivores. To achieve this there was greater winter production of herbivores. Analysis of these results reveals that the spring outburst is not the result of stratification or indeed of any parameter change, but merely of *rapid* change. This is effectively the increase in algal division rate in spring in contrast to the lagged production of the herbivores. If the onset of the spring outburst occurs rapidly in seas that calm quickly, it must be limited by self-shading and the herbivores start to graze late. If the onset proceeds slowly, the herbivores may well control the outburst before the density-dependent process of self-shading occurs.

There are three consequences of the model of Evans and Parslow. The first is that the predominant parameters are merely the division rates of the algae and the grazing rates of the herbivores, which is the real conclusion from the last chapter. The second consequence is that the most important variable is the time at which the MLD shallows, which is a happy metaphor for a very complex series of processes. Much of the spring outburst is well under way before the seasonal thermocline is established in the model and in the ocean (see Colebrook, 1982), so Atkins' (1925*b*) thesis that stratification is necessary is dismissed (which implies that the result of the FLEX experiment, where stratification dominated, was fortuitous). The third consequence is that nutrient limitation does not play much part in the development of the spring outburst because the peak is determined by the self-shading mechanism; its function is that of a trimmer or fail-safe procedure.

Fasham *et al.* (1990) have introduced a model that is derived partly from that of Evans and Parslow in that the physical environment is described by changes in the mixed layer depth. But they diverge from Evans and Parslow in driving the biological system by the algal division rate based on Eppley's (1972) relationship between maximal algal division rate and temperature, to which all the modifiers of that rate are attached, for example diurnal, seasonal and latitudinal differences in surface irradiance, in self-shading and in nutrient limitation as in other models. The use of the algal division rate is an important device because the measure of radiocarbon per unit of chlorophyll is perhaps not very reliable. It is also a return to population dynamics.

Fasham *et al.* (1990) have devised a novel model. The object of the model is to estimate the losses from the mixed layer during the year because of the part played by such losses in the biological pump. This is the mechanism by which material is sequestered below the mixed layer depth. It is complex, but the loss of carbon dioxide to the deep ocean is at the centre of the problem of global warming. The model has seven compartments: dissolved organic nitrogen, bacteria, phytoplankton, nitrate, ammonia, zooplankton and detritus and the stocks are defined by the rates that flow between the compartments (the flow model was devised by Fasham, 1985); in other words, the stocks in the boxes are created by the rates, which is very different from the earlier ecosystem models. The photosynthesis/irradiance relationship is described by the Smith (1936) equation and nutrient limitation by the Michaelis–Menten equation is modified to allow cells to take up ammonia in preference to nitrate as new production switches to regenerated production (Dortch, 1990). Consequently, during the spring bloom the quantity grazed delays nutrient limitation. Self-shading plays a more important part, a significant form of density-dependent regulation.

The grazing animals eat phytoplankton, bacteria and detritus and their preferences are taken into account in so far as they are known. Figure 3.5 shows the fit of the model to Menzel and Ryther's (1960) data off Bermuda in 1958, 1959 and 1960 with sinking estimates of $10 \mathrm{~m~d}^{-1}$ and $1 \mathrm{~m~d}^{-1}$ (*a*) in $\mathrm{mmol~N~m}^{-2} \mathrm{~d}^{-1}$ and (*b*) in $\mathrm{mmol~N~m}^{-3} \mathrm{~d}^{-1}$. The difference between the two is the consequence of the changes in the mixed layer depth; the spring outburst is shown in the former but the summer production continues in the latter. Figure 3.6(*a*) shows the changes in phytoplankton, zooplankton and bacteria in nitrogen for the two sinking rates and in quantities m^{-2} and m^{-3}. Figure 3.6(*b*) shows annual changes in herbivory, bacterivory and detritivory, absolutely and proportionally, the losses and the loss rates. The main value of this model is that the main processes of new and regenerated production are described and that the losses from the photic layer in the biological pump can be assessed. But the model could be adapted for more particular purposes, for example the analysis of the spring bloom as function of the physical and biological parameters.

Recently, Fasham *et al.* (1993*a*) have compared model production cycles off Bermuda and at Ocean Weather Station India. The most remarkable result was that, although the oligotrophic ocean off Bermuda appeared to be nutrient limited, at Ocean Weather Station India it was not. Figure 3.7 shows the time series of the limiting factors in the two areas: off Bermuda the nitrate and ammonia decline during the summer months when irradiance is high; at Ocean Weather Station India the oligotrophic ocean follows the course of the irradiance with no nutrient limitation at all except for a transient period when ammonia replaces nitrate.

The General Circulation Models, GCMs, of the atmosphere have now

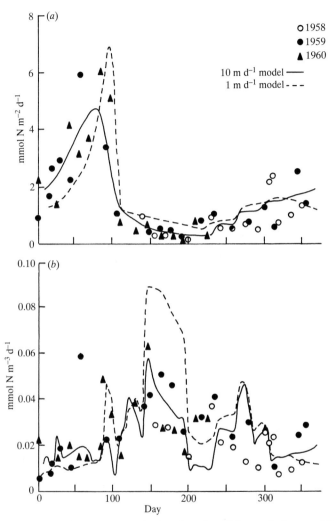

Figure 3.5 (*a*) The modelled annual cycle off Bermuda in mmol N m^{-2} d^{-1}, with the data of Menzel and Ryther (1960); (*b*) annual cycle off Bermuda in mmol m^{-3} d^{-1} (after Fasham *et al.*, 1990). In both, sinking rates of 10 m d^{-1} and 1 m d^{-1} are modelled.

developed into coupled GCMs that are linked to oceanic processes down to the MLD. Then the model of Fasham *et al.* is well adapted to describe biological processes in the ocean under the influence of weather from year to year and of climatic change on a long time scale. In an entirely different direction, Platt and Sathyendranath (1988) gave a useful algorithm that converts chlorophyll measurements in the upper

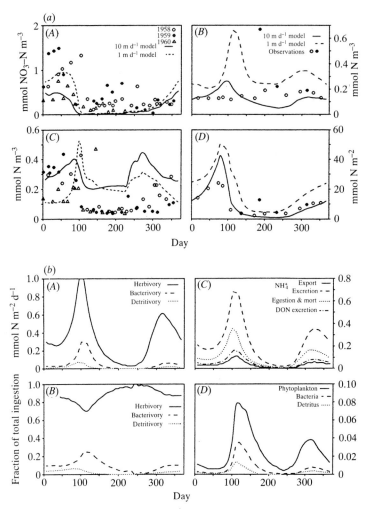

Figure 3.6 (*a*) Modelled output in nitrate (A); in phytoplankton nitrogen (B); in bacterial nitrogen m^{-3} (C) and m^{-2} (D); all in mmol N and with sinking rates 1 and 10 m d^{-1}. (*b*) Zooplankton dynamics: (A) total ingestion, (B) proportional ingestion, (C) total losses, (D) clearance of stocks (after Fasham *et al.*, 1990). DON is dissolved organic nitrogen.

part of the photic layer as observed by satellite to production in the whole of that layer (the satellite sees only part of the photic layer in the deep ocean). Platt *et al.* (1990) have derived an analytical model of the dependence of photosynthesis on irradiance that is suitable for incorporation into the GCMs and for use with satellite material. Then all the

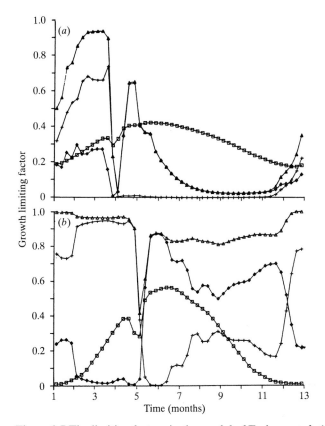

Figure 3.7 The limiting factors in the model of Fasham *et al.*, 1990 (after Fasham *et al.*, 1993) off Bermuda (*a*) and at Ocean Weather Station India (*b*) in the North Atlantic. Off Bermuda production is limited between June and December by lack of nitrate and ammonia. In the North Atlantic nutrient limitation occurs only transiently as nitrate is replaced by ammonia. □, light intensity; +, nitrate; ◇, ammonia; △, total nitrogen.

mass of data that will become available when the new satellites fly can be converted directly to production.

Platt *et al.* (1990) presented an exact solution for the daily integral of photosynthesis by phytoplankton in a vertically homogeneous water column. Their equation is based on the formulation of Platt *et al.* (1980):

$$P = BP_m^B(1 - \exp[-a^B I/P_m^B])$$

where P is the instantaneous rate of primary production, B is the photosynthetically active biomass, I is irradiance, a^B is the slope at the origin normalized to biomass, and PB_m is the assimilation number.

This method can be used directly in the GCMs and the satellite material could be harnessed with the earlier algorithm and so the two very large-scale procedures could be linked.

Extending their observations, Platt *et al.* (1992) have revised the view of the critical depth put forward by Sverdrup (1953). Following Smetacek and Passow (1990) they reiterated that losses must be taken into account. They noted that the Sverdrup criterion shows where and when net phytoplankton can possibly grow but that it cannot describe the rapid accumulation of biomass during a bloom. They showed that higher accumulation rates occur at the lower values of KZ_m, the optical depth, where K is the vertical extinction coefficient of the photosynthetically active radiation and Z_m is the depth of the mixed layer. They tabulated critical depths in metres by days and latitude for different estimates of loss.

Brock *et al.* (1993) have applied this thesis to the condition in the Arabian Sea, which becomes quite eutrophic during the period of the South West Monsoon. According to the model used, the base of the euphotic zone lies below the mixed layer throughout the Arabian Sea at the ends of the intermonsoons. Then during the monsoons the material produced is brought to the surface by Ekman pumping across the Arabian Sea north of the Findlater jet. This explanation provides a solution to one of the difficult problems of biological oceanography.

Elliott and Clarke (1991) have developed a model of seasonal stratification on the north-west European Shelf. It is a two-layered model of surface heating, wind mixing and tidal stirring. It was validated with a large number of XBt (expendable bathythermograph) samples and it describes the positions of fronts and the regions of tidal mixing as well as the seasonal stratification. A development by Woods (1991) has run the model at separated grid points on the shelf, assuming no advection, and she has been able to describe the onset of the spring bloom in the North Sea and in the waters to the west of the British Isles in some spatial detail. Such work has obvious value for the study of the link between fish stocks and the changes in primary production from year to year.

Woods and Onken (1982) followed the trajectories of individual plankton organisms at hourly steps as they rise and fall through the upper ocean entering or leaving the mixed layer, a Lagrangian method. The diurnal variation in the depth of the mixed layer was described throughout the year and as a consequence, algal cells were detrained and lost (see The loss of diatoms in the spring bloom, below). Wolf and Woods (1988) extended this form of model to describe the transition in latitude from the region of the spring bloom in the North Atlantic to the oligotrophic ocean, which was confirmed by Straas and Woods (1988); both demonstrated the chlorophyll layer rising with latitude using the Kiel Sea Rover, a continuous sampler of the upper ocean at all depths.

Recently, Woods and Barkmann (1993*a,b*) have developed their Lagrangian models. The first is the 'Plankton Multiplier', which may explain the positive feedback between carbon dioxide and temperature observed since the end of the last Ice Age. The idea is that the nitrate available for production becomes reduced a little as the mixed layer depth becomes shallower. Their second paper described the onset of the spring bloom in the North Atlantic. The algal cells in the surface mixed layer behave differently from those in the thermocline, because the irradiance they receive has different effects. In spring, the compensation depth tends to track the thermocline. In winter, the algae continue to grow, if slowly, because they receive radiant energy in the turbulent mixed layer. A surprising consequence is that the critical depth, as formulated by Sverdrup, tends to occur later than the start of production. The conclusions are the consequence of the Lagrangian methods employed.

Earlier models of the productive processes in the sea were to some extent experimental, in that the purpose was to show that reasonable results could be obtained. Today, the models in use will probably survive as basic descriptions of the mechanisms at work. Of course, they will become modified as more is learnt in the future.

The loss of diatoms in the spring bloom

Introduction

From the presence of diatomaceous ooze on the seabed in the upwelling areas and off the Antarctic Peninsula, it has long been known that diatoms sink. The actual quantities sedimented have been measured in numerous and detailed trap observations in coastal waters and in the ocean. The physiology of sinking has long been investigated. There are two reasons for this examination: (a) to understand the population processes and to establish whether the quantity grazed can match the

estimates from grazing experiments, and (b) to illuminate part of the
processes that constitute the biological pump.

The sedimentation of diatoms

Hart and Currie (1960) found diatomaceous ooze on the seabed in the
Benguela upwelling. Deuser *et al.* (1980, 1981) found a sedimentation
rate of $3.5 \, \mathrm{g \, C \, m^{-2}}$ in 60 days at 1000 m, of which about one-fifth
comprised organic carbon. This is about one-hundredth of the primary
production. However, Betzer *et al.* (1984) at the equator found that the
flux amounted to about 10% of the primary production. Billet *et al.*
(1983), Lampitt and Burnham (1983) and Lampitt (1985), with time
lapse cameras, showed a seasonal cycle of phytodetritus on the seabed
between 1000 m and 4000 m at stations in the North East Atlantic.
Barnett *et al.* (1984) confirmed the observations with a multiple corer;
the aggregates were up to 12 mm in diameter. Rice *et al.* (1986) found
phaeophorbides in the phytodetritus. The existence of 12% of polyun-
saturated acids implies that the cells did not pass through the guts of
herbivores, so they must have sunk without being eaten. Takahashi
(1986) found that diatom aggregates in the Alaska gyral sank at
$175 \, \mathrm{m \, d^{-1}}$.

Trent *et al.* (1978) and Silver *et al.* (1978) made the first quantitative
estimates using divers with hand-held rings. Up to 489 aggregates per
litre were recorded in near-surface waters and up to 7.5 per litre in
deeper seas. They included chain-forming diatoms, appendicularian
houses, faecal pellets, mucus, intact diatoms, dinoflagellates, coccolithophores and protozoa. McCave (1984) described the physics of aggregation of suspended particles. Fine material is aggregated by Brownian
motion and larger pieces pick up particles by shear-controlled coagula-
tion; the shear depends on the turbulent energy in the water column.
Organic material may provide a sticky substrate, but cannot increase the
frequency of collisions. Such aggregates sink at rates of up to $368 \, \mathrm{m \, d^{-1}}$.
Riebesell (1991*a*) showed that the number of aggregates followed the
stock in the spring bloom and that the sizes increased towards the end of
the bloom. Riebesell (1991*b*) found that the proportion of aggregates by
diatom species increased at the end of the bloom. Alldredge and
Gotschalk (1989) found that intact diatoms in the aggregates were
actively synthesizing. Further, nitrate and ammonia within the aggregates were high. Such is one of the major mechanisms by which organic
material sinks to the seabed. It is of particular importance to the heavier

cells such as diatoms and the calciferous organisms, such as radiolaria, foraminifera and coccolithophores.

Simulation models were made off Peru, in the Bering Sea, in the Gulf of Mexico and in the Mid-Atlantic Bight (Walsh, 1983). It was concluded that most of the primary production was not eaten, but sank as phytodetritus. Malone *et al.* (1983) showed that the flux of phaeophytin to the benthos was low (about 1%) and concluded that neither grazing nor the flux to the benthos could match the primary production in spring and so phytodetritus must be exported off the shelf to the slope. The effect of the microzooplankton was not then known and the primary production was not partitioned by size of grazer. From all the simulations, loss by grazing in all areas amounted to about 27% of the primary production.

Export to the shelf was disputed by Rowe *et al.* (1986) because the pelagic microbial consumption was not estimated and because there was a lag between production and consumption. Export to the slope was low, as shown by traps on the slope bed. Again, throughout the year, consumption matched production and so there would not be much available for export.

Walsh *et al.* (1988*b*) simulated the spring bloom in the Mid-Atlantic Bight to describe the export. The effect of wind stress was modelled in three dimensions between Cape Hatteras and Nantucket Is. to describe the circulation. It was run daily between 28 February and 27 April 1979 at three depths and it was started by the chlorophyll distribution observed by satellite in that year. The maximal algal division rate ranged from 0.234 to 0.390 d^{-1} and the nutrient half-saturation coefficient $K_s = 0.1\,\mu g$ $NO_3 - N\,l^{-1}$. An exponentially increasing grazing stress (doubling every ten days or so) was used. Fluorometers were moored south of Long Is. which showed that the algae sank suddenly at day 78 to 82 into the middle layer and in the bottom layer they increased steadily from day 78 to 90. This observation was simulated with a sinking rate of 20 m d^{-1} at all three levels for the whole period, which expresses the fact that diatoms sometimes sink more quickly than would be expected from experimental observations. Of primary production during the whole period, 66% was eaten at all three depths (and 21% was exported). From this considerable study, it appears that not only do the diatoms sink suddenly, but they may also do so continuously.

Studies in the Kieler Bucht and open Baltic

Smetacek *et al.* (1978), with multi-depth sediment traps (Zeitschel *et al.*, 1978), described the spring outburst in the traps in 1975 of the order of 0.1 to 0.3 g C m^{-2} d^{-1}; there were very few faecal pellets in the traps. A very detailed study each day at 11 depths for ten days in April was made by von Bodungen *et al.* (1981). The stock increased in the euphotic layer between 20 April and 23 April after the wind had dropped, but a sharp increase in stock occurred below the 1% light level (at 25 m), at which the cells cannot grow, and above the halocline (between 40 and 60 m), so this increment must have sunk. The proportion sunk was perhaps a third of the stock on those three days (Figure 3.8). Between February and June, for a depth of 25 m, Peinert *et al.* (1982) linked primary production and sedimentation. The peak of algal stock occurred on 21 March (7.3 g C m^{-2}). On the 20 and 21 of March there was a very sharp increment in sedimentation after a storm. During the period of the spring bloom, 10 g C m^{-2} sedimented. The spring bloom in the Kieler Bucht occurred before the numbers of herbivores grew very much.

To re-examine the work of Peinert *et al.* (1982) in a little more detail,

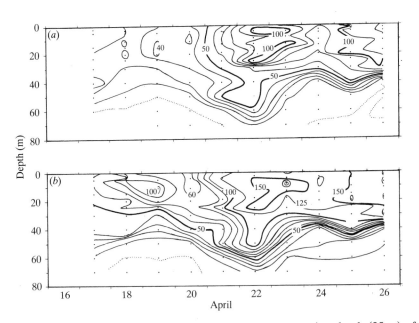

Figure 3.8 The downward movement below the compensation depth (25 m) of the population of the diatom *Skeletonema costatum*, after the wind dropped in the western Baltic (after von Bodungen *et al.*, 1981).

Table 3.1. *Production and loss in the Kieler Bucht in the first half of 1980* (after Peinert *et al.*, 1982)

	Production ($g\,C\,m^{-2}\,d^{-1}$)	Loss ($g\,C\,m^{-2}\,d^{-1}$)
Winter	1.17	1.17
Spring	21.44	20.93
Early summer	16.84	16.52

assume that $\delta P/P = \mu$, the algal reproductive rate, where δP is the increment of carbon estimated by the radiocarbon method and P is the carbon content of the phytoplankton stock; the cells were counted with the Utermöhl method and sized. The loss rate,

$$L = \mu - (1/t)\ln(P_1/P_0),$$

where t is the time in days between observations, and P_0 and P_1 are the estimates of phytoplankton carbon at the beginning and end of the period between observations. Production, P_r, was estimated as follows.

The rate of change of stock, P, is given by:

$$dP(t)/dt = (\mu - L)P,$$

$$P(t) = P_0\exp(\mu - L)t.$$

Then production:

$$P_r = P_0\int_0^t \mu\exp(\mu - L)t\,dt$$

$$= P_0\mu\int_0^t \exp(\mu - L)t\,dt$$

$$= P_0\mu[\exp(\mu - L)t/(\mu - L)]_0^t$$

$$= \{\mu/(\mu - L)\}P_0\{\exp(\mu - L)t - 1\}$$

$$= \{\mu/(\mu - L)\}(P_t - P_0).$$

Similarly, it can be shown that the quantity lost:

$$L_t = \{L/(\mu - L)(P_t - P_0)\}.$$

Stock, production and loss in $g\,C\,m^{-2}\,d^{-1}$ for the whole period between February and June are shown in Figure 3.9. The loss is high just after the peak of the spring outburst because the estimate of μ may have been a little low. For three periods, winter, spring and early summer, the total production and loss is summarized in Table 3.1.

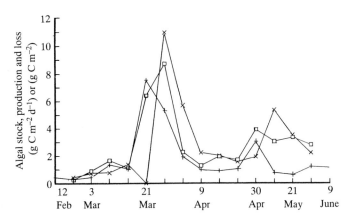

Figure 3.9 The algal stock ($+$), production ($\square$) and loss ($\times$) observed in winter, spring and early summer in the Kieler Bucht (Cushing, 1992; after Peinert *et al.*, 1982).

Perinert *et al.* (1982) suggested that the integral of radiocarbon measurements during the spring bloom amounted to $17.6\,\mathrm{g\,C\,m^{-2}\,d^{-1}}$, close to $21.4\,\mathrm{g\,C\,m^{-2}\,d^{-1}}$ in Table 3.1. They estimated the quantity sedimented at about $10\,\mathrm{g\,C\,m^{-2}\,d^{-1}}$ during the spring bloom; the loss as estimated in Table 3.1 was about twice as much. The protozooplankton reached a peak of $0.6\,\mathrm{g\,C\,m^{-2}}$ on the 3 April and a proportion of the algal cells may have been eaten.

Other studies on sedimentation

The sedimentation in depth during the spring bloom as part of the FLEX experiment on the Fladen ground in the North Sea in 1976 was reported by Davies and Payne (1984). The sedimentation rate was $50\,\mathrm{mg\,C\,m^{-2}\,d^{-1}}$ before the spring bloom and during the bloom itself it amounted to $185\,\mathrm{mg\,C\,m^{-2}\,d^{-1}}$. The primary production was $1.9\,\mathrm{g\,C\,m^{-2}\,d^{-1}}$ (Gieskes and Kraay, 1980; Weichart, 1980), so about one-tenth of the primary production sank to the seabed. The sedimented material was separated into faecal pellets, phytodetritus, diatoms, crustacean remains and sand grains; the diatoms were whole cells and the number of faecal pellets was low. Forsskahl *et al.* (1982) found that in the entrance to the Gulf of Finland about one-quarter of the phytoplankton production reached the traps.

The siliceous cells of diatoms are relatively heavy (1020 to $1250\,\mathrm{kg\,m^{-3}}$) and senescent cells sink three times as fast as healthy ones

Table 3.2. *Loss rates in Dabob Bay (near Seattle) and in the Central Pacific gyre*

	Phaeopigment loss (%)	Chlorophyll loss (%)	Cell sinking rate $(\mathrm{m\,d}^{-1})$
Dabob Bay	22.84	2.24	0.51
Pacific gyre	0.866	0.097	0.146

(Reynolds, 1984). The heavily silicified resting spores can survive for a year or so in deep water and can sink at up to 16 m d^{-1} (Hargraves and French, 1983). Davis *et al.* (1980) found that resting spores of *Leptocylindrus danicus* appeared when the division stopped. Welschmeyer and Lorenzen (1984, 1985) used sediment traps, at depth and just below the euphotic layer, in Dabob Bay and the Central Pacific gyre. Losses of phaeophorbide by sedimentation were estimated, as was the loss of phaeophorbide by photodegradation. The daily chlorophyll loss rate was raised by the depth of the euphotic zone to give the cell sinking rate in m d^{-1}. In the deep traps the phaeopigment represents the quantity grazed by the macrozooplankton and that in the upper traps (less the quantity degraded) represents the quantity grazed by the microzooplankton. Then grazing rates could be calculated and from the differences in time of the quantities of chlorophyll, the division rates of the algae were estimated to be 0.05 to 0.92 d^{-1}. The major results of this work are shown in Table 3.2.

The remainder of the loss is due to grazing. This model was applied to a study of the spring bloom in Auke Bay, Alaska, at about 80 m by Laws *et al.* (1988), who found that about 40% of the daily production in Auke Bay sank, and that 26% was taken by microbial grazing and 32% by macrozooplankton grazing.

Increments in lipids, polysaccharides or in gas vacuoles probably do not account for the sinking of diatoms (Reynolds 1976*a,b*; Walsby and Reynolds, 1980). *Fragilaria crotonensis* sank when growth stopped; uptake and the population start to decline when the growth rate becomes less than the sinking rate.

A population of *F. crotonensis* in a Lund tube in Blelham Tarn in north-west England was studied by Reynolds (1983). Stability was broken every two weeks and nutrients were maintained at a high level. In June and July, the algal division rate was about 0.7 d^{-1} and the rate of sinking about 0.3 d^{-1}, so the production of the diatoms was greater than the quantity sunk by nearly two orders of magnitude over the period. In August and September, a quiescent period, algal division rates fell to $0.1 \, d^{-1}$ and the sedimentation rate rose to about $0.7 \, d^{-1}$, so the greater proportion of the stock sank. Reynolds suggested that under calm conditions and in bright light (300 to $600 \, \mu E \, m^{-2} \, d^{-1}$) the cells may suffer from photoinhibition.

Sedimentation is much greater in spring blooms and in upwelling areas than in systems based on regenerated production (Smetacek *et al.*, 1990). The two systems were named 'loss' and 'retention' (Peinert *et al.*, 1982). They are the two forms of ecosystem described by Cushing (1989) on quite different criteria. The main transfer to the seabed, or to below the mixed layer, occurs in the 'loss' systems, in new production, spring blooms and in upwelling areas. In regenerated production, material is retained because the faecal pellets are eaten in the upper layers.

The continuous loss of diatoms in Loch Striven

The spring bloom in Loch Striven, on the west coast of Scotland, was studied by Marshall and Orr (1929/30). They counted the cells from centrifuged water samples at six intervals of depth and showed quite clearly that the diatoms (nearly all *Skeletonema costatum*) sank below the compensation depth towards the end of the spring bloom. From 14 March to 29 April the water column was mixed and after stratification it was isothermal on the 12 April, after which it stratified again. Table 3.3 shows the original material in numbers of cells per 20 ml. On the 16 and 26 of April, the population sank, peremptorily.

Consider samples of diatoms at intervals of time, t, and of depth, z. P_0 and P_1 are the initial and final numbers in that interval of time at the same depth.

Let

$$P_1 = P_0 \exp(\mu - g - s_1)t,$$

where P_0 is the initial number of cells ml^{-1} of *Skeletonema costatum* at z_1, P_1 is the final number at the same depth, μ is the instantaneous diatom division rate, g is the instantaneous rate of mortality due to grazing, and s_1 is the instantaneous rate of loss due to sinking.

$$P'_1 = P_0 \exp(\mu - g + (s_1 - s_2))t,$$

where P'_1 is the final number of cells at the depth z_2 ($>z_1$) at the end of the interval of time. It is assumed that the effects of turbulence are

Table 3.3. *Numbers of* Skeletonema costatum $20\,ml^{-1}$ *by depth and in time in Lock Striven in 1928*

Depth (m)	14.3.28	19.3.28	22.3.28	26.3.28	29.3.28	2.4.28	
0	2182	8100	7600	9000	64000	73000	
2	2246	7300	8300	11300	68000	61000	
5	2191	7300	7800	13500	57000	133000	
10	2607	7600	6800	2900	41000	8000	
20	579	1900	3400	1800	2000	1200	
30				1500	1050	115	
Mean	1961	6440	6780	6667	38842	46053	

Depth (m)	4.4.28	6.4.28	9.4.28	12.4.28	16.4.28	19.4.28	26.4.28
0	180000	128000	9000	225000	14200	1120	101
2	248000	199000	106000	220000	99000	2015	138
5	60000	510000	176000	169000	24000	1400	3072
10	2100	56000	152000	83000	13800	306	18400
20	1250	3100	77000	10000	10900	400	17800
30	890	1280	3500	2400	4700	1008	19600
Mean	82040	149563	100750	118233	27767	1042	9852

Note:
The data for 19 April appear to be low.

common to these intervals of depth and time and that $(\mu - g)$ is constant in the same intervals.

$$(1/t)\ln(P_1/P_0) = \mu - g - s_1$$

and

$$(1/t)\ln(P'_1/P_0) = \mu - g + s_1 - s_2$$

and

$$\mu - g - s_1 - \mu + g - s_1 + s_2 = 2s_1 - s_2.$$

Then

$$(1/t)\ln(P'_1/P_0) - (1/t)\ln(P_1/P_0) = 2s_1 - s_2,$$

an index of sinking, which is an instantaneous rate. The sinking indices were averaged for all depths. Then the average index is:

$$(2s_1 + s_2 + s_3 + s_4 + s_5 - s_6)/6.$$

Woods and Onken (1982) used a Lagrangian model to describe the spring bloom as function of irradiance and the physics of the mixed layer. Because the depth of the mixed layer changes from day to night, cells can become irrevocably lost below it. If the maximum depth of the mixed layer in the daytime does not increase faster than the sinking rate, many cells may be lost. Indeed, all could be eventually lost if the day depth of the mixed layer decreased quickly enough; it decreases in spring from week to week, but the cells are also eaten. The Woods and Onken mechanism could be enough to describe the continuous loss. The peremptory loss at the end of the spring bloom may be due to spore formation, senescence or lack of silica (Walsby and Reynolds, 1980).

Production, the quantity grazed and the quantity sunk may be estimated. Following Fasham *et al.* (1990), the maximal algal division rate μ_{max}, was taken from Eppley (1972) as function of temperature T, in degrees Celsius: $\log \mu_{max} = 0.0275\ T - 0.07$. The maximal division rate was assumed to occur at $300\,\mu E\ m^{-2}\ d^{-1}$ (Platt and Jassby, 1976, for coastal marine phytoplankton; Reynolds, 1984, spring phytoplankton in Crose Mere). Marshall and Orr measured the compensation depth, D_c, frequently during the period of observation. At D_c, $I_0 = \exp - k\ z = 0.01$, where k is the attenuation coefficient and z is the depth of water in metres and I_0 the irradiance at the surface. Let $I_0 = 300\,\mu E\ m^{-2}\ d^{-1}$; (much less than the irradiance under a clear sky at midday) but effects of photoinhibition are unknown and the low surface irradiance may express them. The irradiance at depth z, $I_z = 300 \exp - k\ z$ in $\mu E\ m^{-2}\ d^{-1}$. Again, following Fasham *et al.* (1990), the division rate at depth, μ_z, was estimated with the Smith (1936) equation:

$$\mu_z = (\mu_{max} a I_z)/(\mu_{max}^2 + a^2 I_z^2)^{1/2}$$

where a is the slope of the photosynthesis/irradiance curve at the origin ($=0.025\ d^{-1}$). With this method, μ_z was established at intervals of depth of 1 m from the surface to D_c and were averaged for the euphotic layer. Then the daily average, $\mu' = 0.7\,\mu_z$ ($0.63 - 0.77$) (Vollenweider, 1965). The depth of mixing, D_m, was taken as the depth of water or the depth of the thermocline, z_t. If $D_c > z_t$, μ' was averaged to z_t; if $D_c < z_t$, μ' was averaged to D_c. Then μ was reduced by the appropriate fraction at the time of observation.

Phosphorus was estimated as the pentoxide and the quantities were raised by 1.35 to correct for salt error (Cooper, 1938) and converted to mmol m^{-3}. Eppley *et al.* (1969) found a half-saturation coefficient of 0.4 mmol N/l (0.025 mmol P/l) for *S. costatum*. The simple Michaelis–

Menten relationship was used. The maximum algal division rate at a given temperature (for each depth and date of observation) was reduced by phosphorus limitation and by self-shading, expressed in the dependence of attenuation coefficient on stock (see below). The grazing rate is given by $\ln(P_1/P_0) - (\mu - s)t$, where s is the sinking rate as the averaged sinking index. Then, for each interval between observations for t_0 and t_1, with the methods described above, production, P_r, quantity grazed, Q_g, and the quantity sunk, Q_s, were estimated, i.e.:

$$P_r = [\mu/(\mu - g - s)](P_1 - P_0),$$

$$Q_g = [g/(\mu - g - s)](P_1 - P_0),$$

$$Q_s = [s/(\mu - g - s)](P_1 - P_0).$$

Figure 3.10 shows the quantity sunk for a period of about six weeks during the spring bloom; it is initially very low and rises during the spring bloom, but in the later stages, as may be seen from Table 3.3, the loss rate at the lowest level predominates. Figure 3.10 also shows the history of the stock in numbers of cells ml^{-1}. The quantity sunk was highest during the peak period of the spring bloom, a result also noted in the work of Peinert *et al.* (1982). There is a positive relationship between the quantity sunk and the stock of algae ($r^2 = 0.56$; $p = 0.01$).

Figure 3.11 shows (*a*) standing stock (*b*) production and (*c*) quantity grazed. A peak is reached after 24 days on 6 April. The surface layer was

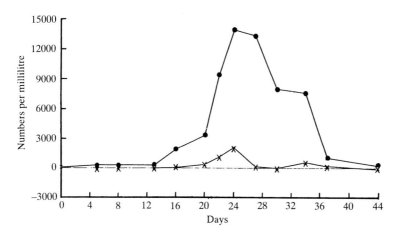

Figure 3.10 The loss of diatoms, *Skeletonema costatum*, during the spring outburst in 1928; both the algal stock (●) and the quantity sunk (×) are shown, in cells ml^{-1} (after Cushing, 1992; after Marshall and Orr, 1929/30).

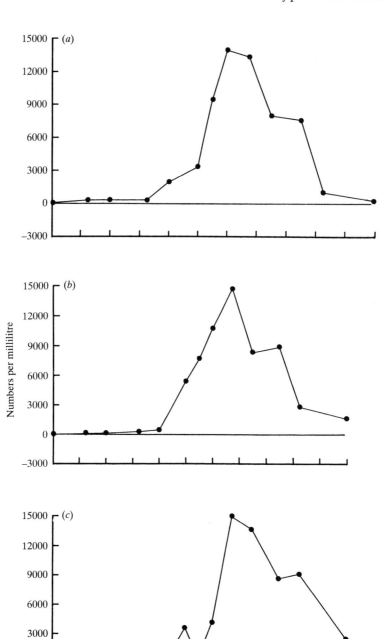

slightly warmer for the first eight days, but the compensation depth was at 13 m and the water column was mixed on days 13 and 16. The thermocline on day 20 was probably responsible for the sudden subsequent increment in stock, but until day 20 the increase must have been due to irradiance.

The peak of production occurs later than peak stock. The quantity grazed remained low until day 20, reached a peak by day 27, after which it declined as the stock was eaten. The quantity sunk was less by about an order of magnitude. During the period, 64 000 cells ml^{-1} were produced, 59 000 cells ml^{-1} were eaten and 5000 cells ml^{-1} were sunk in a continuous manner, but at the end of the bloom, 2790 cells ml^{-1} sank during the peremptory loss.

There was an inverse relationship ($r^2 = 0.46$; $p = 0.05$–0.01) when the attenuation coefficients were plotted against the stock; with a quadratic fit, $r^2 = 0.51$ ($p = 0.05$–0.01). The effective division rate was reduced by self-shading from about 0.34 to about 0.27, a reduction to 79%.

Diatoms sink rapidly to the deep seabed in masses soon after the spring outburst in temperate waters. Walsh and his colleagues (1988a) asked how much of the diatom production sinks and how much is eaten. Smetacek and his colleagues (1978) showed that much of the spring bloom ended in the sediment traps in the Kieler Bucht. It now appears that the quantity grazed is often but not always greater than the quantity lost by sinking.

The Woods and Onken mechanism is probably responsible for the continuous loss of stock during the spring outburst. The quantity grazed is greater than that sunk by about an order of magnitude. The ratio might be less with larger cells, which sink more quickly, and different physical processes may alter the detrainment below the mixed layer.

The self-shading mechanism is density dependent. Self-shading, grazing and nutrient lack may control the spring outburst, but the most important conclusion is that independent estimates of the quantity grazed can now become available. Then the study of grazing may advance.

Figure 3.11 Stock, production and losses in Lock Striven in 1928 in cells ml^{-1}: (a) the stock of diatoms, predominantly *Skeletonema costatum*; (b) the production of diatoms; (c) the quantity grazed (after Cushing, 1992).

Conclusion

The underlying theme of this chapter is the estimation of loss rates during the spring outburst (and, by extension, to the upwelling areas). The models of Evans and Parslow and of Fasham *et al.* comprise distinct advances on their predecessors. The first is the use of changes in the mixed layer depth as a metaphor of the physical factors during the development of the spring outburst. The second is the recognition that self-shading of the algal populations is a density-dependent control of some importance. Further, Fasham and his colleagues used a direct measure of the algal reproductive rate rather than the complex proxy, increment of carbon per unit of chlorophyll. Also, they developed a model in which stock is estimated from the vital rates between entities, rather than the other way round. The Lagrangian models developed by Woods and his colleagues have changed some perceptions of algal sinking, of the position of the compensation depth and of the nature of the critical depth.

One of the unsolved problems in the study of the planktonic algae, particularly during the spring bloom, is the separation of loss rates into those due to sinking and those due to grazing. Three studies have illuminated the problem: (a) the discovery of an echo of the spring bloom in the fall of phytodetritus to traps above the bed of the deep ocean, (b) the measurement of sedimentation of diatoms and other organic material into traps in the Kieler Bucht, (c) the estimation of the losses on many regions of the continental shelves by Walsh and his colleagues showing that in the early eighties less than a third of the loss in the spring bloom was due to grazing (later the proportion was revised to two thirds). This is a more reasonable estimate because the pelagic stocks of fish comprise a greater biomass than do the demersal ones.

There are two important requirements. First, in the study of the biological pump it is necessary to estimate properly the losses through the mixed layer depth, of diatoms from the spring bloom and calciferous organisms at other periods. The second requirement is to estimate the losses due to grazing, particularly during the spring and autumn blooms (and upwelling systems), which provide the food for growing fishes. For both purposes, understanding of why diatoms sink and how herbivores graze is needed.

Diatoms appear to sink in two ways: first by detrainment in the Woods/Onken mechanism, which continues throughout the spring out-burst, and second, by an unspecified physiological mechanism (perhaps photoinhibition). Given the potential sinking rates of cells of given sizes, it should be possible for physicists to estimate detrainment from the daily rates of change of the mixed layer depth. The physiological changes that occur at the end of the spring bloom will be examined more fully. The development of behavioural studies by Strickler and his colleagues will lay the foundations of a theory of grazing activity. Recent work by Paffenhöfer and Lewis (1992) has revealed rates of grazing much higher than was imagined in earlier work. It should be possible to elaborate a theoretical structure of detection, capture and ingestion. It would be desirable to estimate sinking and grazing independently and to match them to the observed total loss rate.

4

Hydrographic containment and stock structure

Introduction

The idea of a unit stock originated in the need to manage identifiable groups of fish in the sea. Sometimes the identity of the stock is self-evident. Usually, additional evidence is needed, for example, the distribution of fish recovered from taggings on the spawning ground. In temperate seas fish usually return to the same spawning ground each year at the same season, much as do the Pacific salmon, and this may well be true in the upwelling areas and in the subtropical ocean. The identity of the stock is sustained, even if there is some immigration and emigration. That of the spawning group of plaice in the Southern Bight of the North Sea will be examined in relation to the tidal streams and to other spawning groups in the waters around the British Isles. The containment of this spawning group in the tidal waters forms the basis of its circuit of migration.

The development of genetic methods raised the hope that the unit stocks established by the fisheries biologists might be shown to be genetically based. A short history of this development will be given and the use of genetics assessed.

Hydrographic containment

Cushing (1968) proposed that a fish stock uses tides or currents in different ways at different seasons in such a way that it is retained within a region. The concept was based on Harden Jones' (1968) triangle of migration (Figure 4.1). Fish migrate (with selective tidal transport or a current/counter current system) apparently against a stream from the feeding ground to their spawning ground, and the spent fish are carried back in that current to the feeding ground, which is often in deeper

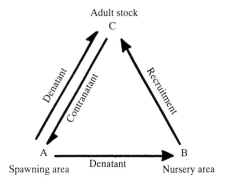

Figure 4.1 The triangle of migration (Harden Jones, 1968)

water. At the same time the larvae are drifted in the same current and by behavioural mechanisms they find their way through shallower water to their nursery ground which often lies inshore. The spent fish migrate to deeper water and the larvae are carried inshore to shallower grounds. After a period of two or more years, the immature fish leave the nursery ground and join the adults either on the spawning ground or on the feeding ground. The two distinct migrations, that of the adults from feeding ground to spawning ground and back, and that of the larvae and juveniles to the nursery ground and thence to the adult stock, comprise the circuit of migration by which the stock maintains its identity from generation to generation and its isolation from its neighbours in so far as wind, tide or current permit.

As an example, let us examine the plaice within the Southern Bight of the North Sea, which has been studied for a long time. Plaice spawn in the North East Atlantic from Iceland to the Barents Sea in the north and to the western Mediterranean in the south, but principally in the waters around the British Isles. In the southern North Sea, the fish spawn in the German Bight of the North Sea, on the Flamborough Off Ground off the Yorkshire coast of England, in the Southern Bight and in the eastern English Channel; together these groups comprise a unit of management (there is a separate group in the western English Channel; see Figure 4.2). The hydrographic containment of this group of spawners will be described.

Containment is based on the spawning ground and on the time of spawning. Figure 4.2 (Harding *et al.*, 1978) shows the distribution of stage 1 eggs in the southern North Sea and English Channel; the spawning groups are distinct although the whole distribution is con-

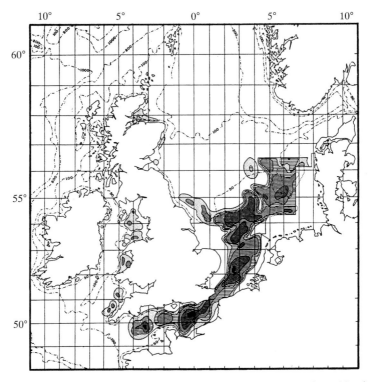

Figure 4.2 Distribution of stage 1 eggs of plaice in the southern North Sea and eastern English Channel (after Harding *et al.*, 1978). Contours at 0.1, 3.0 and 27.0 eggs m^{-2}.

tinuous. Figure 4.3(*c*) (Simpson, 1959) shows the charts of stage 1 eggs made between 16 and 28 January for seven years between 1921 and 1950; the distributions have remained the same on subsequent surveys (Harding *et al.*, 1978; Jaworski and Rijnsdorp, 1989). The spawning ground in the Southern Bight has remained in the same position since 1921 when it was first fully sampled. Cushing (1969) found that the mean *peak date of spawning* of the Southern Bight group sampled between 1921 and 1967 was 19 January with a standard deviation of less than a week (and a standard error of about two days). Rothschild (1986) noted that the low standard error implies a restricted time of spawning, but the range of spawning time over two or three months implies that some eggs will be spawned at the most favourable time for them. Not only did the plaice spawn on the same ground but they did so at the same time each year.

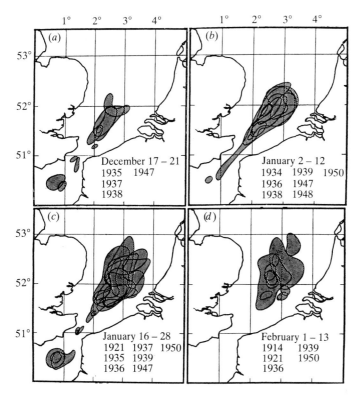

Figure 4.3 Charts of the distribution of stage 1 plaice eggs from December to February between 1921 and 1950 (after Simpson, 1959).

Figure 4.4 shows the temporal distributions of stage 1 eggs of seven species including the plaice, together with the development of the spring outburst in the Southern Bight, as indicated by the 'greenness' from the continuous plankton recorder (1948–83); Gieskes and Kraay (1977) showed that this was a reasonable estimate of the net phytoplankton on which the copepods probably depend. Recall that fish larvae, like copepods, take particles >5 μm in diameter (see Chapter 3). Each fish species spawns for up to two or three months and the spring outburst develops rather slowly, as might be expected in a tidal region which is rather turbid. Plaice and sandeel larvae feed on *Oikopleura* but most of the other fishes feed on the nauplii and copepodites of small copepods such as *Pseudocalanus*, which arise from the spring outburst. The plaice larvae hatch about three weeks or a month after spawning, depending on the sea temperature, and food is then available for them.

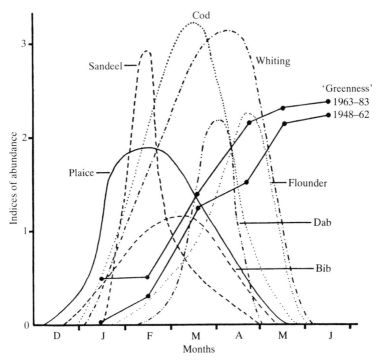

Figure 4.4 The distribution in time of the stage 1 eggs of seven species: plaice, sandeel, cod, whiting, flounder, dab and bib, and the development of the spring outburst in the Southern Bight (from the continuous plankton recorder network) (after Cushing, 1989).

The plaice larvae drift north-easterly from the spawning ground (Figure 4.5; Cushing, 1972) in the clearer water towards the shallows where production starts early. The larvae live in the lower part of the water column (Harding and Talbot, 1973). Most may enter the Dutch coastal water by an inshore flow, at certain states of the tide (Dietrich, 1953), that was named the Texel Gate (Cushing, 1972). Then some settle off the Dutch coast (Harding *et al.*, 1978) and migrate through the Texel Gate to the nursery ground (Figure 4.6; Zijlstra, 1972). Most larvae do not settle offshore but move into the Wadden Sea where they metamorphose and settle on their nursery grounds. Those from the Flamborough ground find a nursery in Filey Bay on the coast of Yorkshire in eastern England (Lockwood, 1974). The nursery of the German Bight spawners lies to the north on the Danish coast. In 1963, many larvae were drifted further north than the Texel Gate (Beverton and Lee, 1964;

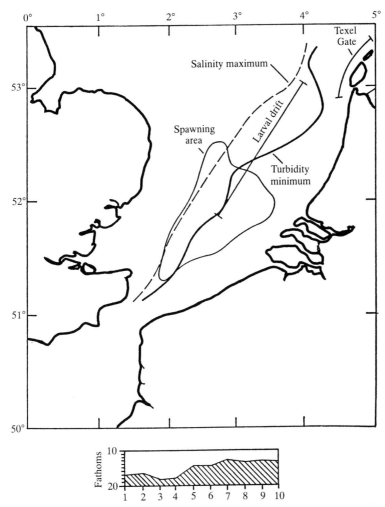

Figure 4.5 The larval drift of plaice from the spawning ground to nursery ground (after Cushing, 1972). The diagram below shows depths along the line of the larval drift.

Harding and Talbot, 1973). An interesting point is that the numbers of larvae of plaice and flounder entering the Wadden Sea are correlated with subsequent recruitment (of the whole group of southern North Sea plaice), which implies that part of the regulatory process occurs during the larval drift, which will be discussed more fully in Chapters 6 and 7. Subsequently, they grow at maximal rate (Zijlstra *et al.*, 1982). On the nursery ground the little plaice are eaten by shrimps at settlement and

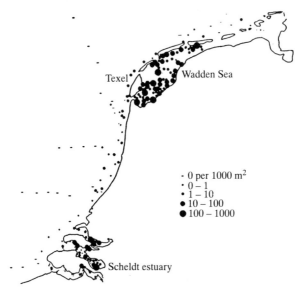

Figure 4.6 The nursery grounds of the plaice in the estuary of the Scheldt, on the open Dutch coast and in the Wadden Sea (after Zijlstra, 1972).

their mortality is density dependent (Figure 7.2; van der Veer, 1986).

An interesting feature of Figure 4.6 is that a patch of O-group plaice is found in the estuary of the Scheldt (Rijnsdorp and van Stralen, 1982), but individuals are also found on the open beaches. It is likely that the Scheldt nursery derives from the eastern Channel spawning group (Houghton and Harding, 1976), but the spread of individuals along the beaches or the spread of larvae to the north from time to time implies a low leakage from spawning group to spawning group within this unit of management, the southern North Sea (German Bight, Flamborough Off Ground and Southern Bight) and eastern English Channel. (It is implicit in the continuous distribution of stage 1 eggs, shown in Figure 4.2.). When the fish leave the nursery ground they spread away from the Wadden Sea to the north and west. A line of trawl stations was worked each month in the thirties from the Haaks Light Vessel off Texel Is. to the Leman Bank off the English coast. Beverton and Holt (1957) used this material to describe the movement north westward as a diffusion. Presumably the juveniles continue to spread until they mature and join the adult stock.

The initial work on migration was executed by tagging experiments. Harden Jones (1968) analysed earlier work and established that mature

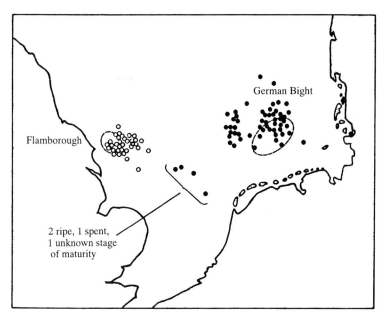

Figure 4.7 Recoveries of plaice tagged on their spawning grounds in the German Bight (●) and on the Flamborough Off Ground (○) (after de Veen, 1962).

and spent plaice migrated in the western side of the Southern Bight; Rauck (1977) showed that the first-time spawners came into the Southern Bight from the north east. But the most interesting result came from the work of de Veen (1962) who tagged plaice on the spawning grounds in the German Bight and off the Yorkshire coast of England. Of recoveries on these grounds in the following year, only 1% or 2% strayed to another ground and none to the Southern Bight (Figure 4.7). Rauck (1977) showed that of the plaice tagged on the German Bight spawning ground in the first quarter of the year, 1 out of 50 were subsequently recovered in the same quarter, but in the last quarter the proportion increased to 1 in 10. So the proportion of adults that stray between spawning groups is relatively low.

De Veen (1978) charted the distribution of tag recoveries on the feeding grounds from each of the three spawning grounds, Southern Bight, German Bight and Flamborough (Figure 4.8). Those from the Southern Bight do not migrate north of the Dogger. From Flamborough they move as far north as Aberdeen and from the German Bight the plaice migrate towards the Skagerak. Because the spawning grounds lie

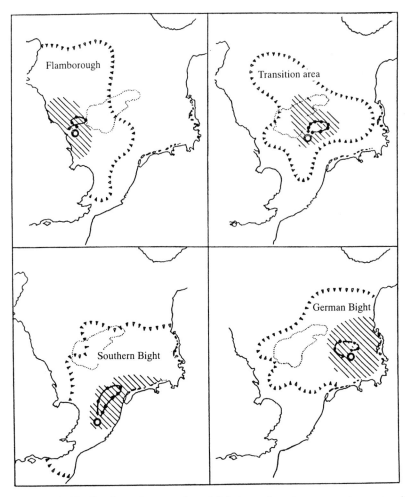

Figure 4.8 Distribution of recoveries of fish tagged on four spawning grounds, effectively the distribution of feeding grounds (after de Veen, 1978). Most tags were in the shaded areas; dotted lines mark boundaries to distribution of tags.

in different places, the feeding grounds are also distinct to some degree, although there is some overlap.

During the spawning season, de Veen (1978) displaced tagged fish from the German Bight spawning ground to the Flamborough ground off the Yorkshire coast in Northern England. After four to six months, the recovered fish showed a distinct tendency to return towards the German Bight. After a year, during the spawning season, the centre of distribution of recovered fish had shifted half way across the North Sea towards the German Bight. This is slight evidence for a return to the ground of first spawning which is distinct from homing, the return to the native ground. But, as will be shown below, any open sea homing must differ profoundly from that of the Pacific salmon.

Summarizing the evidence so far, the plaice spawn on the same ground at the same time each year. The larvae are drifted to the north-east into shallow and productive water as the spring outburst develops. They pass through the Texel Gate into the Wadden Sea and the magnitude of recruitment is established to some degree just before settlement. At settlement they are eaten by shrimps and their mortality is density dependent. They grow on the nursery ground at maximal rate until they diffuse away to join the adult stock (see Chapter 7). There must be a leak of larvae to other nursery grounds, which is probably low (compare the number of individuals on the open coast with the numbers on the nursery grounds in Figure 4.6) and a low leak of adults to other spawning grounds. The proportions are low and may not even bias the vital parameters of the stock or population if the numbers of immigrants and emigrants are about equal.

Spawning migration

The plaice migrate to their spawning ground and return from it by selective tidal transport (Harden Jones *et al.*, 1978; Harden Jones, 1980*a,b*, 1984). With this mechanism fish stay on the seabed during one tide and move up into the midwater at slack water, so they can travel on the next tide for some distance quite quickly in tidal waters. Selective tidal transport by elvers was discovered by Creutzberg (1958) and by Creutzberg *et al.* (1978) in plaice larvae. Harden Jones and his colleagues used a sector scanning sonar; the sonar transmits on a wide 30° beam at 300 kHz and receives on a narrow beam of 0.33°; the range resolution is 7.5 cm. The returned signals are displayed on a B-scan of angle on range; pictures of the seabed reveal sand ripples (Figure 4.9) and the mines for which it was designed. The transducer can be rotated quickly into the vertical plane so that the vertical movements of the fish can be recorded. With transponding tags (Mitson and Storeton-West, 1971) on the migrating fish, their vertical and horizontal movements can be recorded for long periods; indeed, one fish was recently followed for more than six days (Dr G.P. Arnold, personal communication).

Figure 4.10 shows a typical series of observations (Greer-Walker *et al.*, 1978). For three successive tides the depth distribution of a plaice, with

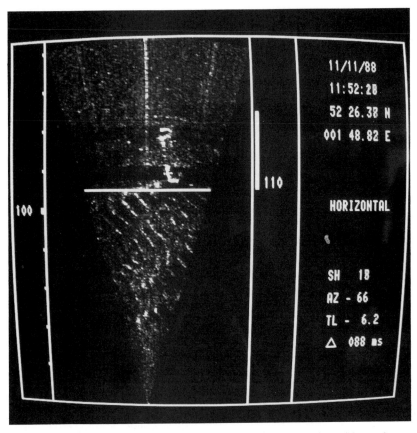

Figure 4.9 Sector scanning picture. This is a horizontal display with sandunes shown up to a range of 100 yd (91 m).

an acoustic tag, off Lowestoft (England) was recorded. The fish swam in midwater on the northbound tide and stayed on the seabed on the southbound one. For the period of two-and-a-half tides the fish moved 25 miles to the north. More generally, the maturing fish were shown to migrate on southbound tides and the spent plaice on northbound ones. The two forms of migration were confirmed with midwater trawl hauls on consecutive tides, mature fish being caught on southbound tides and spent ones on the northbound ones (Arnold and Cook, 1984). On the feeding grounds to the north the plaice rise from the seabed at night and return in the morning but they do not use selective tidal transport (Arnold, 1981). If the fish migrate south to spawn in the western half of

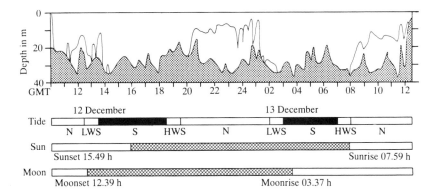

Figure 4.10 Migration of plaice with selective tidal stream transport as demonstrated by an acoustically tagged fish transponding to a sector scanning sonar (after Greer Walker *et al.*, 1978). The track of the fish is shown as a continuous line above the seabed.

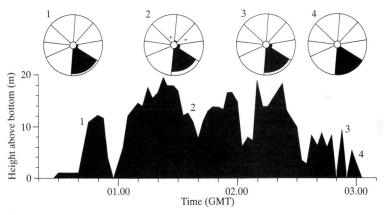

Figure 4.11 The heading of plaice to the south-east in midwater as shown by compass tag records (after Harden Jones, 1984).

the Southern Bight they do so on the flood and the spents move north on the ebb (Harden Jones *et al.*, 1979).

A compass tag was made (Mitson *et al.*, 1982) which resolved direction in eight sectors. Figure 4.11 shows the average heading for 2.5 h to the south-east. But more generally, the plaice do not maintain a heading for such a long period without periodic contact with the seabed; indeed there was some evidence that the plaice oriented to the tidal current before they 'took off' from the seabed (Arnold *et al.*, 1989). The

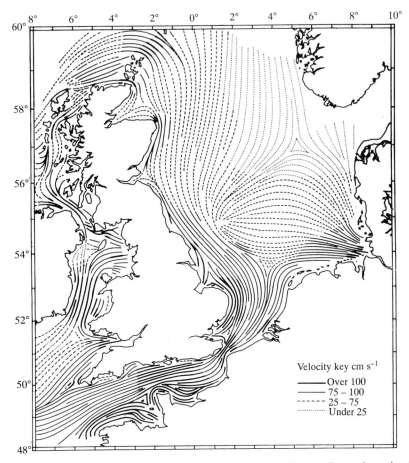

Figure 4.12 Tidal stream paths made by connecting adjacent lines of maximal tidal flow (after Harden Jones *et al.*, 1984).

method by which this might be maintained is unknown. Recently, Metcalfe *et al.* (1993) have shown that the plaice in midwater maintain their direction more rigorously at night than in the daytime; further, they swim downtide faster than the tide itself.

Figure 4.12 shows the tidal stream paths made by connecting adjacent lines of maximal tidal flow (Harden Jones *et al.*, 1984). The tidal structures of the Southern Bight and the German Bight are clearly distinct and the spawning migrations would be expected to be quite different. Figure 4.13 (Arnold and Cook, 1984) shows simulated migration tracks in the Southern Bight made by starting migrations from given

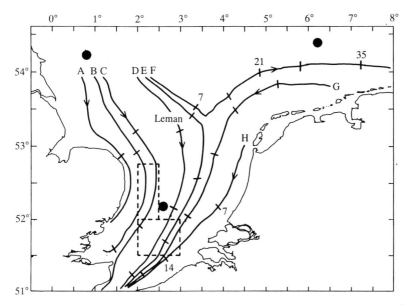

Figure 4.13 Simulated plaice migration tracks with selective tidal transport in the Southern Bight of the North Sea (after Arnold and Cook, 1984). The spawning area of the plaice is shown as the two dotted rectangles. The three amphidromic points are shown.

positions by selective tidal transport and linking them to make a continuous track. The three centres of spawning are marked as black circles. Tracks C and D show the southerly movement on the flood in the western part of the Southern Bight. Track G shows the direction from which the first-time spawners come moving south on the ebb on the eastern side of the Southern Bight, as suggested by Rauck (1977). Track F displays a migration into the German Bight; Tracks F and G end in quite different places despite the fact that they start very close to each other. This arrangement may be one of the reasons why the German and Southern Bight groups tend to remain distinct from one another.

Two questions remain: How does migration start and stop? How is the spawning ground detected? Harden Jones (1980*a*) suggested that the southbound migration of maturing plaice started when low water slack and moonrise occurred at sunset. They might start the northbound migration when high water slack occurs at sunrise. Arnold and Cook (1984) thought that slack water might be detected by reduced turbulence or by reduced noise. Both questions remain unanswered, although Harden Jones (1984) suggested that freshwater springs contained distinctive molecules to which the plaice might return, which implies homing (but see below).

The spawning migration and its complement, the return of spent fish, is the mechanism by which the spawning group retains its identity. If the spawning groups in the unit of management, Eastern Channel, Southern Bight, German Bight and Flamborough, are compared, the larval drifts, the nursery grounds, the feeding grounds and the spawning grounds are geographically distinct. Each spawning group must have its particular larval drift and its particular nursery ground. That for the Southern Bight has been described in detail because it is the best known. The immature fish diffuse away from their nursery grounds to their particular feeding grounds and they probably join the adult stocks on the spawning migration. The positions of the four spawning grounds have arisen in the course of evolution so that the best combination of larval drift, nursery ground and feeding ground has been obtained. There are four distinct migration circuits. The major difference is between the Southern Bight group and that in the German Bight in that the migration circuits are arranged about the two amphidromic points and, as noted above, the tidal structure keeps them apart.

In the North Sea, plaice spawn also in the Firth of Forth and in the Moray Firth on the coast of Scotland. In the English Channel, they spawn also in the west which is part of a distinct unit of management. The stocks in the Irish Sea are genetically distinct from those in the North Sea (Purdom and Wyatt, 1969). Larvae might be drifted south from the Moray Firth or the Firth of Forth or eastward from the western Channel, but they would settle to the seabed long before they reached a nursery ground characteristic of the southern North Sea plaice. Similarly, adults from the southern North Sea do not find themselves very often on the spawning grounds in the western Channel or on those off the east coast of Scotland.

A stock is contained in its migration circuit which is based on spawning ground, nursery ground and feeding ground. The plaice of the Southern Bight of the North Sea use selective tidal transport for the migration of the maturing and spent fish and their larvae, half way down the water column, are drifted to their nursery ground in a wind-driven residual current. The adult migrations are relatively short, two or three hundred kms, and they happen quite quickly (*c* 20 days); what determines the distance of migration is not known. However each spawning group in the North Sea has evolved a distinct pattern.

Such mechanisms are enough to isolate the spawning groups to some degree. Within the four groups in the southern North Sea, some larvae may be lost because of change in the wind-driven currents. Most probably end on a nursery or an open coast as in Figure 4.6, but some may well end on a nursery downstream of that normally expected. In a similar way, some adults may appear on different spawning grounds. The proportion is small, as shown above, but it is probably large enough to vitiate genetic identity between spawning grounds in the southern North Sea. The arguments apply also to the more distant grounds. The newly developed methods of model tracking of larvae (Bartsch *et al.*, 1989) should estimate the stray of larvae; few tagged adults have appeared on the distant spawning ground despite the large numbers released.

There is a mosaic of spawning grounds around the British Isles. Between them there is a potential stray of adults and larvae which might constitute a gene flow. From the arguments given above, the gene flow from one spawning ground to another must decrease with distance. Where it has become reduced by distance to very low levels, genetic differences might be expected, as between the North Sea and Irish Sea plaice, as noted above. Larvae from the Southern Bight could not reach the Irish Sea before settling and the distance is probably too far for the adults.

For a fisheries biologist, the vital parameters of the population studied should be homogeneous. In their study of the unity of the Downs stock of herring, Cushing and Bridger (1966) showed that mortality rates within the four fisheries in the Southern Bight and eastern English Channel were the same. Cushing (1992) showed that the recruitments to the Downs and Bank stocks of herring were not correlated at all for a long period, 1924–89. Differences in the growth rates between cod stocks in the North Atlantic are considerable (Garrod, 1988). Any stock/recruitment relationship must be based on the fact that the recruitment arises from that stock.

The errors of estimation of the vital parameters are relatively low. In a well-known and well-sampled stock, like that of the plaice of the southern North Sea, the coefficient of variation of a vital parameter is low. Stray from spawning group to spawning group is expected because the strength and direction of wind-driven currents are both variable, but the stray from the assemblage of spawning groups to others in the northern North Sea or western English Channel is obviously less. All that is needed is that the stray from the unit of management should be less, perhaps considerably less, than the error of estimation. Implicit in this statement is the assumption that within the period of examination, emigration and immigration are about equal.

Homing

Because plaice, cod and herring, amongst many others, spawn on the same ground each year at the same time, it was commonly believed that they return to their native grounds to spawn and that they home to them. The return of the sockeye salmon (*Oncorhynchus nerka*) to its native stream was shown by Foerster (1937), who fin-clipped many thousands of smolts leaving Cultus Lake, on the Fraser River in British Columbia. The mature fish returned to the lake to spawn four years later. Hasler and his colleagues have investigated the problem since the fifties (Hasler, 1966). Hasler and Scholz (1980) suggested that 'juvenile salmon become imprinted to the unique chemical odour of their natal stream during the smolt stage and subsequently use this cue to locate their stream during the spawning migration'. Kleerekorper (1969) listed the low threshold levels for detection of various chemicals, for example 0.000001 ppm for morpholine (Wisby and Hasler, 1958).

Differences in stream chemistry exist and the juvenile salmon learn or imprint the odour of the home stream in a very short time (hours) and they retain the imprint for some years, when they return on their spawning migration. Experiments on the choice between home waters and non-home waters showed that home preferences were established for four salmonids. By blocking the nostrils, the need for chemical information for homing was established, and this was shown in electro-physiological responses. Streams were artificially odoured with morpholine or phenyl ethyl alcohol and fish imprinted in such streams returned later to spawn in them. Fish exposed to an odour will return to another stream artificially carrying that odour and such results were supported by electrophysiological information. This brief account was taken from the article by Smith (1985).

The salmonids do not migrate downstream towards the sea until smoltification starts. In the open sea, homing would require that the eggs or larvae became imprinted on the spawning ground; so homing of the salmonid form is peculiar to the rivers in which they live. If fish in the sea home to their native grounds they do so in a different way. The question must arise, whether the little brain of the fish larva is competent to acquire an imprint, and in any case the larvae are hatched from pelagic eggs at some distance from the spawning ground. For rather similar reasons it is difficult to show that in the open sea fish return to their native grounds to spawn.

The circuits of migration of other groups

The hydrographic containment of the plaice in the Southern Bight of the North Sea has been described in detail, because of the very large amount of work that has been carried out there during this century. A short account of some other circuits of migration follows, first that of the Pacific halibut. Thompson and van Cleve (1936), on the basis of egg distributions, distinguished two groups of spawners, one off Cape Spencer at the southern end of Queen Charlotte Is., in British Columbia, and another off Yakutak, in Alaska. Tagging experiments (Thompson and Herrington, 1930) showed that the spread of adults from the two regions, which are far apart, appeared to be low. On the basis of these observations, two units of management were established. Figure 4.14(a) shows the circulation in the North East Pacific. However, St Pierre (1984) showed that spawning grounds were in fact distributed from Cape Spencer westwards along the shelf of the Alaska gyral as far as Unimak Is. in the Aleutian chain and in the Bering Sea (Figure 4.14(b)). Skud (1977) re-examined the tag recoveries of adult halibut off Yakutat and showed that they spread east and west from the Aleutian Is. to Washington State and eventually to northern California, but Skud's most important conclusion was that juveniles from the Bering Sea to Cape Spencer all moved eastwards presumably in a countercurrent (Figure 4.14(c)). Thus there is one continuous spawning ground from Cape Spencer to the Bering Sea; the larvae are drifted westwards in the Alaska current until they settle and the juveniles are carried back to the east. On the basis of these observations, Skud (1977) established a single unit of management for the Pacific halibut.

Bailey *et al.* (1982) described the migration circuit of the Pacific hake (*Merluccius productus*). The adult fish migrate south in winter in the California Current over the shelf. They spawn off southern California in the region from Cape Mendocino to Punta San Eugenio and the larvae are found as far as 300 km offshore and just below the mixed layer in depth . As a consequence, they are taken inshore and the juveniles are found inshore of the 200 fm line off central and southern California; high recruitment is associated with downwelling (Bailey, 1981). During the summer the adults move north, presumably in the California Undercurrent, a summer flow. The circuit is illustrated in Figure 4.14(d);

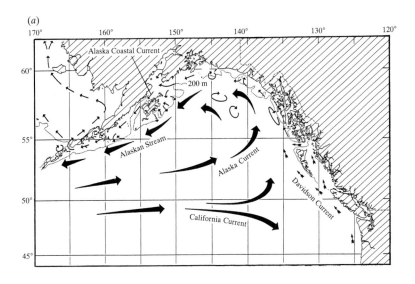

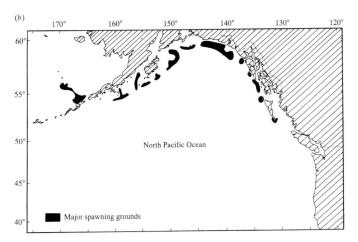

Figure 4.14 Some circuits of migration.
(*a*) Circulation in the North East Pacific (after St Pierre, 1984).
(*b*) Spawning ground of the Pacific halibut (after St Pierre, 1984).
(*c*) Migration of juvenile Pacific halibut (Skud, 1977). Distributions of recoveries
from distant tagging positions.
(*d*) Migration of the Pacific whiting (hake) (after Bailey *et al.*, 1982).

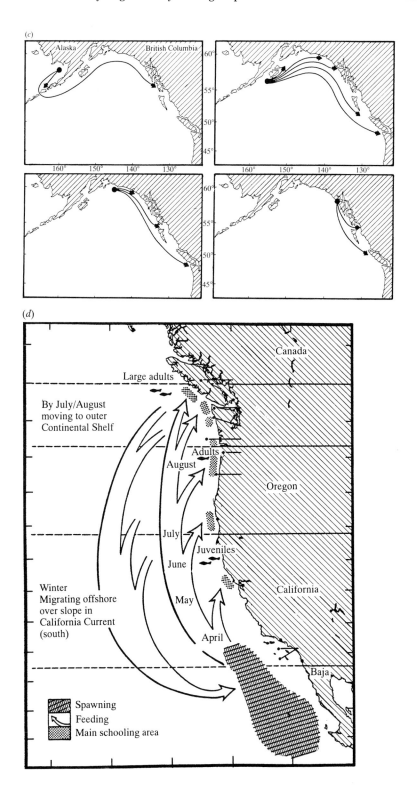

(c)

(d)

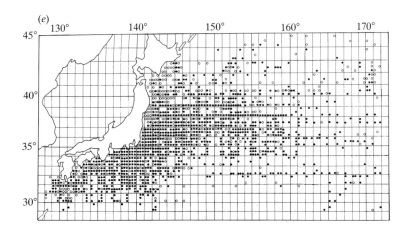

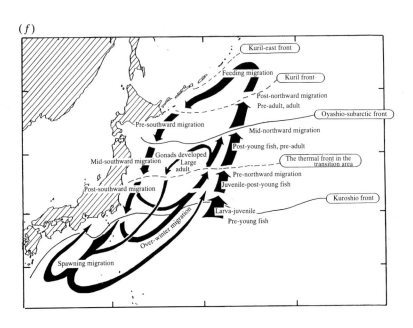

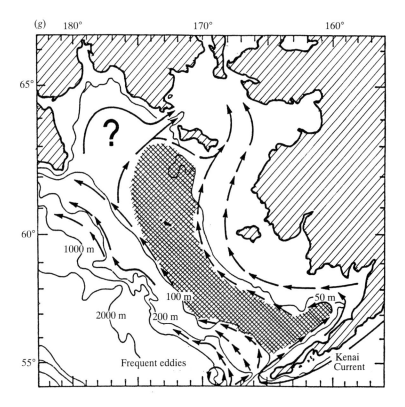

Figure 4.14 (*cont.*)
(*e*) The distribution of larval production of the Japanese saury (Watanabe and Lo, 1989). Solid circles indicate positive records.
(*f*) The circuit of migration of the Japanese saury (after Kosoka, 1986).
(*g*) Circulation in the Bering Sea (after Incze and Schumacher, 1986); the hatched area indicates the spawning ground.

(*h*)

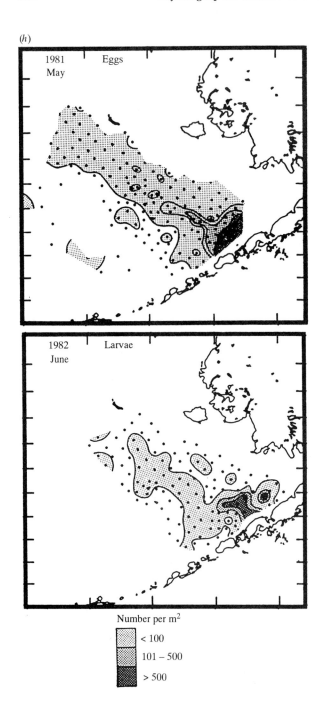

(i)

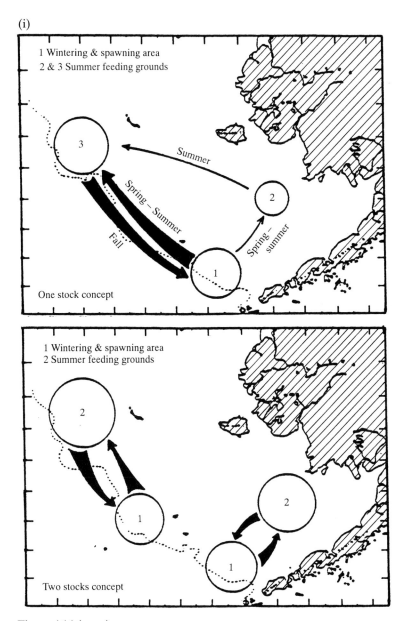

Figure 4.14 (*cont.*)
(*h*) Distribution of eggs of the Bering Sea pollock in May 1981 and of larvae in June 1982 (after Dunn and Matarese, 1987).
(*i*) Two possible migration circuits of the pollock in the Bering Sea (after Bailey *et al.*, 1986).

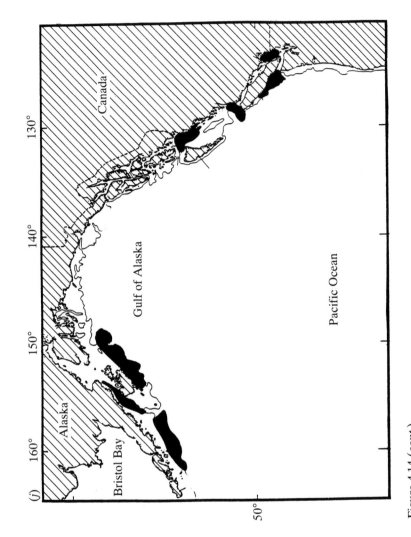

Figure 4.14 (cont.)
(j) Spawning grounds of the Alaska pollock.
(k) Migration circuit of the bluefish (after Sherman, 1986).

(k)

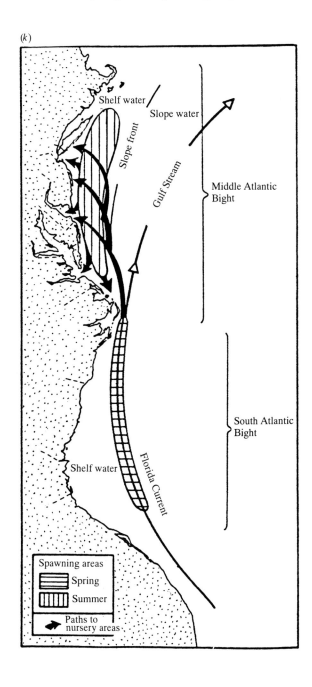

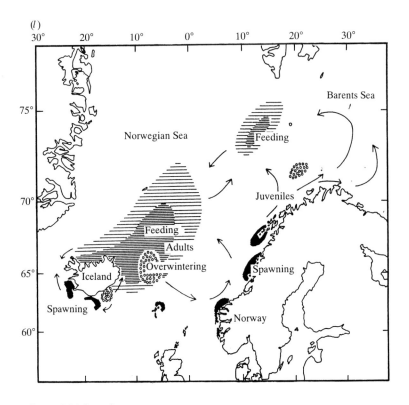

Figure 4.14 (*cont.*)
(*l*) The migration circuit of the Atlanto-Scandian herring (after Bakken, 1983);
the spawning grounds of the Icelandic spawners are also shown.

there are small distinct stocks in Puget Sound, in the Strait of Georgia and a dwarf one off Bahia California.

Watanabe and Lo (1989) investigated the larval production of the Pacific saury across an area of 2.5 million square miles to the east of Japan (Figure 4.14(*e*)). The survey was made with surface nets at all times of the day for 15 years. The general migration circuit off the east coast of Japan is shown in Figure 4.14(*f*).

The Bering Sea pollock (*Theragra chalcogramma*) yields a very large fishery. Figure 4.14(*g*) shows the circulation pattern in the region (Incze and Schumacher, 1986). In Figure 4.14(*h*) is given the distribution of eggs in May 1981. There are two possible migration circuits, for the stock structure has not yet been decided Figure 4.14(*i*) (Bailey *et al.*, 1986).

During the eighties much work was done on the stock of pollock which spawns in the Shelikov Strait between Kodiak Is. and the Alaska Peninsula. Between 1.3 and 6.0 billion fish spawn near Cape Kekurnoi at about the same time each year from late March to mid-April; ten batches of eggs are laid at a depth of 150 m, but the larvae migrate to the upper 50 m, where they grow at about 2 mm d^{-1}. There is a large patch of eggs or larvae off Semidi Is. by the end of May and off the Shumagin Is. by June or July, with lengths of 20–30 mm (Figure 4.14(j)). Later, the juveniles are found in the bays along the Alaska Peninsula.

The migration circuit of the bluefish off the Eastern Seaboard of the United States is displayed in Figure 4.14(k) (Sherman, 1986). Figure 4.14(l) shows the migration circuit of the Atlanto-Scandian herring. There are three major spawning grounds, each separated from the other by some hundreds of km and the feeding area stretches from Iceland to Bear Is. in the Barents Sea. The stock depends on the Atlantic stream as it flows north off the coast of Norway and upon the cyclonic gyre in the Norwegian Sea. Further, the fish feed in summer all along the polar front from Iceland to Jan Mayen.

From this brief study of migration circuits, it is seen that they are common to the large populations of exploited fish stocks. Fishes migrate over very great distances, apparently in an orderly manner, because the ocean currents are regular in their appearance as, of course, are the tides. The isolation of such populations from each other depends on slight stray or gene flow from one group to another. In the example of the Southern Bight plaice examined above the boundaries of the population could be delineated, as being where the chance of larva or adult reaching it from the spawning ground is very low. One of the questions that arises is: how do the fish migrate where the tidal streams are not as strong as in the North Sea? Obviously they could use a selective current transport, and then the further question is: how do they find the right current at the right time?

Genetic studies

If fish did return to their natal grounds to spawn and if there were no strays or gene flow from one mosaic of grounds to another, genetic differences between them might be expected. The plaice of the southern North Sea stray as larvae and as adults. At some level of examination, the stray should be the same as the gene flow. During the fifties and early sixties, immunological methods were used to estimate allele frequencies in different groups of fishes. In the mid-sixties this somewhat laborious method was replaced by electrophoresis with which allele frequency could be measured much more easily and more frequently; enzymes such as lactate dehydrogenase and proteins such as haemoglobin or transferrin were used. From family studies of cattle, Jamieson (1966) showed that the transferrins are genetically based, so distributions of transferrins were effectively distributions of alleles. The allele frequencies are said to be in equilibrium if they are distributed by the Hardy–Weinberg Law; with homozygotes, AA and aa, and heterozygotes, Aa, alleles are distributed in binomial form, with probabilities p and q, $p^2 AA + 2pq aA + q^2 aa$. Originally the equilibrium was tested in 2×2 tables with chi^2, but Levene's (1947) correction should have been used for small samples (see also Fairbairn and Roff, 1980). A maximum likelihood method, the log-linear G statistic, has been used in recent work. Further, a deficit of heterozygotes is always of interest for it indicates heterogeneity of one form or another.

Differences in allele frequencies between groups are due to selection, drift or isolation. In the early seventies, Kimura and his colleagues (Kimura, 1968) suggested that proteins and enzymes were selectively neutral in their expression and discussion continues on this point (see, for example, Cain, 1983). As most fish populations studied in the open sea are large, (say 10^{10}), differences between groups have been assumed to be due to isolation. But some populations of the Pacific salmon are quite small and differences between them in allele frequencies may be the result of drift. The simplest conclusion for the large stocks of fish is that differences in allele frequencies are due to isolation or to selection; indeed, isolation could be due to selection (particularly in stocks which are heavily exploited, sometimes restricting the area of a spawning ground) and provided that the selective processes persist, the mechanism need not be specified.

Present sampling requirements were given by Ihssen *et al.* (1981), who summarized the proceedings of a conference on stock identification: each sample should comprise more than 50 individuals at each of at least 20 loci (a locus is the position of an allele or gene on a chromosome): samples should be taken from the same breeding unit, which in many fish stocks are often well defined, as shown above. Confidence intervals should be estimated for the allele frequencies. In addition, there should be reliable estimates of heterozygosity.

Genetic studies have developed a little differently in the various groups, salmon, cod and herring; to some degree this leads to differences between regions. Now follows a short history of fish stock genetics by region.

The Pacific North West

Extensive work on electrophoretic genetics has been carried out in the Pacific North West because of the difficult stock problems of the trout and the salmon. The salmonids return to their native streams to spawn but many spawning groups have been exploited in common by the Japanese in the North Pacific. The early work was a little disappointing because differences tended to appear only across great distances; Utter *et al.* (1980) found that there were three populations of sockeye (*Oncorhynchus nerka*) in Bristol Bay, the Skeena and the Columbia River. Similarly, there were three populations of chum salmon (*Oncorhynchus keta*): Kamchatka, British Columbia and the Columbia River. Utter *et al.* (1980) studied the stocks of sockeye in Cook Inlet, Alaska. They used 13 enzymes at 26 loci and equilibrium was established with the G test; a dendrogram was made with Nei's genetic distance, D,

$$[D = -\ln((\Sigma x_i y_i)/(\Sigma x_i^2 \Sigma y_i^2)^{1/2}].$$

Over considerable geographical distances the differences were low.

Aspinwall (1974) examined three proteins in 32 populations (each of about 2500 individuals) of pink salmon (*Oncorhynchus gorbuscha*). (Recall that the pink salmon lives for only two years.) Three groups were established: Kodiak Is., northern British Columbia and Vancouver Is.; but the odd-year populations were genetically distinct from the even-year ones, presumably because the isolated populations have diverged. Beacham *et al.* (1985) examined 14 enzymes in stocks of pink salmon from four rivers in 1982 and 21 in 1983. A dendrogram showed small differences between Puget Sound, the Fraser River and the Straits of San Juan de Fuca (all of which are rather close together), but again the greatest difference was found between the odd- and even-year broods with different allele frequencies ($D = 0.008$). The chance of mixture between the odd- and even-year broods of the short-lived pink salmon must be very low. Isolation may go back to the last glaciation and exploitation may have reduced the stray.

Shaklee *et al.* (1991) studied the stock structure of the odd-year pink salmon in Washington State and British Columbia. They used samples of 100 to 600 fish from 26 putative stocks from 52 sites. With G tests, there were no significant deviations from the Hardy–Weinberg law and the distribution of alleles in time was steady. A dendrogram in genetic distance (Cavalli-Sforza and Edwards, 1967) revealed three major groups in the Hood Canal, in Puget Sound and the Fraser River, in southern British Columbia and in northern British Columbia. In the Hood Canal there were five stocks, each of which was distinct. The improvement in resolution was attributed to more information and to the sampling by stocks, already established (that is, by streams).

Pella and Milner (1987) examined the chinook salmon (*Oncorhyncus tsawytscha*) in the Columbia River in 14 hatchery stocks with enzymes at nine loci. A dendrogram in D was made, in which differences in all enzymes were low. With a simulation model, they estimated the mixture of hatchery stocks in the Columbia River. The differences are trivial but the model might be useful in analysing the mixture of stocks, which is of considerable importance in many fisheries throughout the world.

The results on the Pacific salmon were a little disappointing. As the fish return to their native streams there was the hope of revealing a complex genetic structure in the wild stocks, particularly where they mix at sea. With the exception of the difference between the odd and even years of the pink salmon, some of those found are really the consequence of the last glaciation. But more recently, Utter *et al.* (1989) examined the populations of chinook salmon from British Columbia to California. They used 25 loci from 86 places and tested for deviations from the Hardy–Weinberg law with the G statistic. Nine groups were distinguished by a dendrogram based on D and by four principal components of the allelic variability. Figure 4.15 shows the nine groups and the dendrogram; glaciation reached the Fraser but not the Columbia rivers. There are some transplanted hatcheries on the lower Columbia river. Each of the nine groups comprises a river system with many tributaries or a group of rivers, between each of which there must be some gene flow. Analysis of the genetic diversity revealed that most of it was from differences between populations. The advance made by Utter *et al.* (1989) lay in the concentration of samples by rivers and their tributaries; hardly a stream was not sampled.

Grant and Utter (1984) sampled Pacific herring (*Clupea pallasi*) from Asia, the North East Bering Sea, the South East Bering Sea, the Gulf of Alaska and the eastern North Pacific for differences in 40 enzymes from eye, muscle and liver. G tests were used to establish equilibrium and to assess the geographical differences between allele frequencies. A cladogram based on D showed that there were two groups, East and West

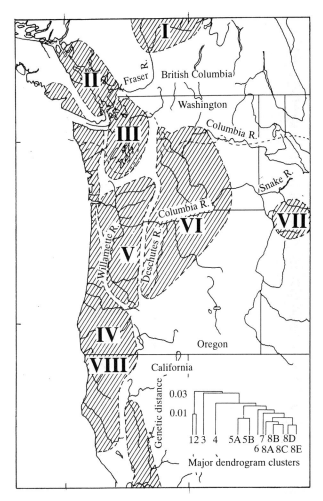

Figure 4.15 The groups of chinook salmon (*Oncorhynchus tsawytscha*) in British Columbia, Washington State and northern California (after Utter *et al.*, 1989).

Pacific, probably separated at the last glaciation. This result is less encouraging than that on the salmonid species in the North Pacific.

In some contrast, Grant and Utter (1980) found differences in the walleye pollock (*Theragra chalcogramma*) on a somewhat shorter scale of geographical distance. They used 20 enzymes at 28 loci and showed equilibrium with G tests. A dendrogram-based D showed differences between the eastern and western Pacific, but also small but detectable differences between the South East Bering Sea and the Gulf of Alaska. Another interesting difference is that between populations of the Pacific hake (*Merluccius productus*), established with two enzymes and two proteins, one population in the eastern Pacific from British Columbia to Mexico and the other in Puget Sound, an inlet in Washington State.

Australasia

One of the most remarkable differentiations has been found in the stocks of barramundi (*Lates calcarifer*) in northern Australia by Salini and Shaklee (1988). It is a large perch that spawns outside or near river mouths, and it is found in eight rivers across a distance of more than 1000 km. Eleven enzymes and one protein were used at 12 loci. Seven groups were distinguished using chi^2 with Levene's correction for small samples. Wright's index of fixation (a measure of isolation) showed little or no exchange. The index is:

$$F_{st} = (4N_e m + 1)^{-1}$$

where F_{st} is the index of fixation, N_e is the effective population size, here derived from tagging experiments, and m is the rate of immigration estimated from tagging data.

When F_{st} is high, there is little immigration; the values ranged from zero to 0.627 with an average of 0.087. With a one-dimensional stepping-stone model, Salini and Shaklee estimated 2.6 movements per year between populations. The tagging experiments showed that 8/278 fish returned had moved between two rivers less than 100 km apart and one river at 120 km. In two rivers which were fairly close to each other, the indices of fixation were low and the heterozygosity was also low, indicating a mixture. But the general conclusion from this work was that rivers more than 150 km apart contained fish that were genetically distinct, which has obvious and important relevance for management. A most important point is the association of genetic data with those from tagging experiments. From the point of view of stock identity, the established isolation is remarkable, in view of the fact that the larvae live off the river mouths. It would be of the greatest interest to investigate how they and the juveniles find their way back to the river.

The North Atlantic

Salmon

The Atlantic salmon (*Salmo salar*) spawns in the rivers of northern Europe from northern Spain to Norway and in North America from the rivers of Long Island Sound to Ungava Bay in Canada. Payne *et al.* (1971) distinguished a Celtic group from a Boreal group in the rivers of the British Isles. The Celtic group is distributed in South Wales and in south-west England and the Boreal group is found in the rest of the British Isles (Figure 4.16). A remarkable detail was that two Irish rivers,

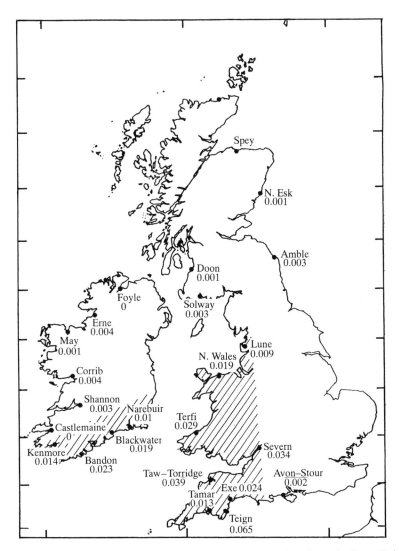

Figure 4.16 Allele frequencies in populations of Atlantic salmon in Great Britain and Ireland; the shaded area shows the distribution of the Celtic group (after Payne *et al.*, 1971).

the Blackwater and the Bandon in south-east Ireland, differed in the allele frequencies of two enzyme systems (Cross and Payne, 1978). A real difference distinguished salmon from North American and European rivers in that the North American fish have two transferrins not found in European fish (Tf^3 and Tf^4) and a third transferrin appears in the European (Tf^2) but not in the North American fish (Thorpe and Mitchell, 1981). Both groups are exploited as they return to their native rivers, but in addition both groups have been exploited as immatures off west Greenland. In this mixed fishery it would be possible and desirable to estimate the proportions of the stocks from each side of the Atlantic. Payne (1980) attempted this; the proportion of North American salmon off west Greenland is

$$A = (Tf^1/Tf^4)/[2q(Tf^4)(1 - qTf^4)]$$

They found that the North American proportion varied as follows: 23% in 1970, 53% in 1971 and 20% in 1972. There is one remarkable difference between North American and European salmon and it is that the North American salmon carries 72 chromosomes and the European 74 (Hartley and Horne, 1984). This would be the simplest way of differentiating individual fish.

Stahl (1988) examined 19 enzymes in the salmon of 23 rivers in the East and West Atlantic and the Baltic; the estimated heterozygosity was low. Equilibrium was tested with an extended chi^2 and the homogeneity of allele frequencies were confirmed with the G-statistic. From a dendrogram derived from four estimates of D, four groups were distinguished: East Atlantic, West Atlantic, Baltic and a landlocked group; the same conclusion emerged from a principal component analysis. McElligot and Cross (1991), using eight loci, examined salmon from one river in Iceland, two Canadian rivers and six Irish rivers, including nine tributaries of the Blackwater. Although the distances between them were low, there were significant genetic differences between stocks in rivers and tributaries. Again, the increase in resolution follows from an increase in information and by sampling the putative stocks (that is in the rivers and their tributaries).

To investigate the effect of the west Greenland fishery, more than a million smolts were tagged between 1959 and 1971 in North America and in Europe. Table 4.1 shows the results in recoveries per thousand fish tagged.

From the British Isles, France and North America, 0.284% of fish tagged were recovered, compared with 0.4% taken in their home

Table 4.1. *Recoveries of salmon tagged in Europe and North America: Recoveries per thousand tagged*

Tagged at	West Greenland	Norwegian Sea
North America	2.3	
Scotland	2.6	0.1
England, Wales	2.0	0.0
Ireland	2.5	0.0
France	4.8	0.0
Iceland	0.0	0.0
Norway	0.2	0.6
Sweden	0.7	0.0
USSR	0.0	0.0

Source: Ruggles and Ritter (1980); Swain (1980).

streams. The ratio of recoveries at west Greenland to those in the home streams was 0.71. In contrast, from Iceland, Norway, Sweden and the former USSR, only 0.0225% of fish tagged at west Greenland were recovered; but 1.3% of fish tagged appeared in the home streams. The ratio of recoveries at west Greenland to those in the home streams was 0.0173, more than 40 times less than those from the British Isles, France and North America. Hardly any fish were recovered from the Norwegian Sea and the Faroes. Thus, the west Greenland fishery exploited stocks from France, the British Isles and North America.

In 1972, 62% of fish taken came from North America and 37% from Europe. The tagging experiment was a very large one involving many vessels at sea and the fact that it yielded a greater North American proportion than expected from the distribution of genetic markers (20%) need not cause concern, because the statistical power of the tagging experiment was much greater. However, the tagging experiment yielded information on the distribution of recovered tags which would not have appeared from genetic analysis. A combination of both methods might have been instructive.

Herring

The Atlantic herring spawn on narrow distinct grounds at the same time each year. This fact has led people to believe that the fish home to their native grounds, but there is so far no evidence for this. The long collection of morphometric measurements and of meristic characters made since the time of Heincke (1898) listed differences of environmentally determined characteristics. Many believed that such differences might also have had a genetic basis. There is no reason why environmentally determined differences should not be used for stock distinction, but it would require a constant environment, improbable in the sea.

Andersson *et al.* (1981) using 25 loci, compared Kattegat spring spawners and Baltic herring and unexpectedly found very little difference. Kornfeld *et al.* (1982), using 29 enzymes at 42 loci, examined spring and autumn spawners in the Gulf of Maine and in the Gulf of St Lawrence, but the Wahlund effect (that is, the gain or loss from emigration or immigration) is not very sensitive. The lack of heterozygote deficit in the spring spawners suggested little gene flow between the spring and fall spawners. Ryman *et al.* (1984) studied the differences between Baltic herring, Kattegat and Skagerak herring, Norwegian and North Sea spawners. Seventeen enzymes at 17 loci were used. Equilibrium was established with an extended chi^2 test. The differences between allele frequencies were established with the *G* statistic and a dendrogram of *D* displayed differences, but they were low ($D = 0.001$). With meristic characters or with morphometric measurements the five groups would have been considered as quite different, but there are no genetic differences. The stocks examined are obviously different merely in the distinction of the migration circuits. King *et al.* (1987) examined the population genetics of herring from places as diverse as the Baltic, the Blackwater (a river that flows into the Thames estuary) and the west coast of the British Isles including Ireland. They used 16 enzymes at 30 loci (12 of which were polymorphic) at ten localities. Each sample comprised about 100 individuals and equilibrium was established after using Levene's correction for small sample size. The allele frequencies were common to eight year classes and the variability within each year class was low. The average genetic distance was 0.0004 and the index of fixation was low, 0.002. There was a difference between the coastal and the shelf stocks, but it was low. Jørstad *et al.* (1991) studied samples of Atlantic herring from 28 positions in the Baltic, Norwegian Sea, Irish Sea, the Thames estuary, off Newfoundland and off New Brunswick. There were no differences in the oceanic and pelagic stocks across the North Atlantic, but significant ones in the Norwegian fjords. An interesting point about some of the results of King *et al.* is that some of the samples lay outside the region of ice cover in the last glaciation.

The three studies on the herring on each side of the Atlantic reach the same conclusion, that genetic differences do not really exist between spawning groups. The most remarkable result is the lack of difference between the Baltic herring and the rest; the taxonomists have classified the Baltic herring as *Clupea harengus membras*, perhaps needlessly. Despite the long history of racial studies on the herring, there is no evidence of genetic differentiation, presumably because there is persistent gene flow between the narrow and distinct spawning grounds.

About a decade after collapse, the Georges' Bank herring recovered on its original spawning ground. There had been no spread to it from other grounds as might have been expected from recolonization. The age structure was quite different (effectively the 1983 year class, i.e. spawned in 1983) from that on Jeffery's Ledge and south west Nova Scotia, the nearest neighbour stocks. The distribution of alleles and of heterozygosity resembled those of other herring stocks in the north west Atlantic, but one isozyme, PGI^{35}, was not observed elsewhere in the Gulf of Maine. Hence the stock is recovering on its original ground and not by recolonization from neighbours (Stephenson and Kornfeld, 1990).

The differences observed of meristic characters and morphometric measurements are of course determined environmentally. There are three major stocks amongst those groups in the North Atlantic sampled for genetic difference: the Baltic herring, the Norwegian and the North Sea herring. The spawning grounds are distinct and far apart and the migration circuits occupy different seas (Figure 4.17). I have distinguished two groups of autumn-spawning herring in the North Sea, the Downs stock and the Bank stock (Cushing, 1992). The distinction is based on the consistent differences in fecundity (and egg weight) and the lack of correlation between recruitments. The stray from one group to another is low; from a large tagging experiment made on the Norwegian stock, none were recovered on the southern North Sea spawning grounds, but from North Sea taggings a few Bank spawners were recovered on the spawning grounds in the southern North Sea. The question of stock identity was decided in the location of the migration circuits and in the differences in the vital parameters of the populations. Where the study of genetics has failed, there is no other choice but to rely on environmental characterization.

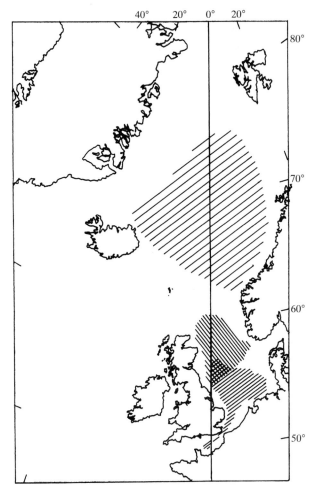

Figure 4.17 The migration circuits of three stocks of herring in the North East Atlantic.

The Atlantic cod

The Atlantic cod (*Gadus morhua L.*) lives in the North Atlantic and in the Baltic. It has been exploited for a very long time, since the twelfth century in the Vestfjord in Northern Norway (Rollefsen, 1956) and since the sixteenth century on the Grand Bank and off Iceland (Cushing, 1988*a*). Figure 4.18 shows the major stocks, with their spawning grounds and feeding grounds; the recognized stocks are Baltic, North Sea (with three or four spawning grounds), Faroe Is., Faroe Bank, Arcto-

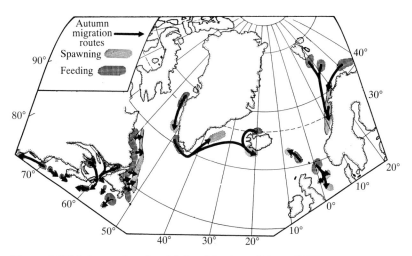

Figure 4.18 Major stocks of cod (after Martin and Jean, 1964).

Norwegian (with the main spawning in the Vestfjord, but some occurring north of the Lofoten Is. and some to the south towards Møre), Iceland, east Greenland, west Greenland, Labrador and north-east Newfoundland, Southern Grand Bank, Flemish Cap, north-east Gulf of St Lawrence, St Pierre, the Magdalenian shallows of the Gulf of St Lawrence, Banquereau, Browns Bank, Georges Bank and off Nantucket Is. (Martin and Jean, 1964). The migration circuit of the Arcto-Norwegian stock has been described (Cushing, 1968) and others might be proposed, such as those on the isolated banks like Flemish Cap, the stock of which might be retained in a Taylor column; but most, such as that in the North Sea and those on the coast of North America, have not been described. The main reason for this is that the eggs of cod and haddock are hard to distinguish and so the spawning grounds are often not well delineated.

The genetic studies started in the early sixties. The present state is as follows. de Ligny (1969) summarized Jamieson's work on haemoglobins and transferrins, showing marked differences across the ocean which supported the conclusion that many of the stocks listed above were genetically distinct. Jamieson and Turner (1978) examined the North American stocks with 13 transferrin alleles at 18 positions on the shelf. Differences were found between groups. However, Gauldie (1984) showed that there were differences between years (ideally, they should be differences between year classes). For the five alleles quoted by

Gauldie the following differences remain between West Greenland and the Grand Bank (and Rittu Bank):

	Tf A-2	Tf B-2	Tf C-1	Tf C-3	Tf D-1
West Greenland	0.086–0.160	0.140–0.240	0.007–0.020	0.570–0.720	0.040–0.098
Grand Bank	0.003–0.005	0.360–0.480	0.100–0.210	0.202–0.320	0.097–0.120

Thus there is a difference between west Greenland and the Grand Bank, but the work needs repeating to show the allele frequencies in a series of year classes with confidence limits.

The use of haemoglobin as a genetic marker has been doubted by Mork and his colleagues (Mork *et al.*, 1983, 1984, 1985), who worked on a coastal cod stock in the Trondheimsfjord in Norway. Three alleles differed in their haematocrits and there may have been differences in general physiological characteristics. There was a little evidence of potential gene flow, but Mork *et al.* believed that some selection was taking place. Jamiesen and Birley (1989) have summarized the earlier work on haemoglobin using material from many year classes; their results were as follows.

	Allele frequencies
Baltic	0.35
British Isles	0.57–0.67
Faroe Plateau	0.06
Faroe Bank	0.19
Iceland	0.09–0.61
West Greenland	0.05–0.13
North America	0.00–0.09

The differences between groups are significant as estimated by chi^2 with Levene's correction. The considerable range at Iceland is due to the Wahlund effect (the mixture of alleles from different groups) because the Iceland spawners include some from west Greenland which return to Iceland to spawn. Jamieson and Birley concluded that the differences were due to selective adaptation.

Mork *et al.* (1985) examined 19 loci at nine positions in the Atlantic with 12 enzymes; equilibrium was established with the method of Vithayasai (1969) and allele frequencies were tested with the G statistic. A dendrogram of D showed differences between the Baltic and the rest (also shown in the transferrins) and between the east and west North Atlantic; D was low, 0.004 to 0.008. Cross and Payne (1978) had studied the North American cod stocks with six transferrin alleles and six enzyme alleles. With Rogers' (1972) estimate of genetic distance (which is based on Pythagoras), they found a sharp difference between Flemish Cap and the rest and a lesser difference north and south of the Laurentian Channel (which was not recorded by the enzymes).

Jamieson and Birley (1989) studied the stocks of haddock (*Melanogrammus aeglefinus*) in the North Sea, off the west coast of Scotland, at Rockall and off the Faroe Is. With 19 transferrin alleles they examined 15 year classes in age groups 1 to 13. Allele frequencies were analysed by age, sex and brood with the use of the G statistic. Three races were distinguished: Rockall, west of Scotland/western North Sea and Faroe Is./eastern North Sea. At first, the division of the North Sea by the Greenwich Meridian appears difficult to understand. However, there are three streams that enters the North Sea from the north: the Fair Isle current, the Shetland current and the Tampen Bank current (Figure 4.19). The latter flows through the Faroe–Shetland Channel subsurface and turns back across the northern North Sea at the shelf edge and reaches the surface sometimes at the edge of the Rinne, the Norwegian deep water. The Fair Isle current flows from off the west coast of Scotland and turns into the North Sea south of the Faroe–Shetland Channel. Jamieson and Birley believe that there are either barriers to gene flow or that intense selection maintains the difference. It is possible that the separation by currents is sufficient, but that remains to be shown. This successful use of transferrins reinforces the need to re-examine the stocks off the coast of North America.

The North Atlantic cod has been tagged since 1898 (Schroeder, 1930) off North America. Templeman (1979), from tagging experiments, distinguished five stocks (see Figure 4.18) between St Pierre and northern Labrador: the Labrador/north-east Newfoundland stock, the Northern cod, spawns offshore in March and April, the eggs and larvae move south in the Labrador Current and the nursery ground lies in the bays of southern Labrador and eastern Newfoundland. A second stock is that of the Northern Gulf (of St Lawrence): the cod spawn in April and May and the nursery ground is in northern Newfoundland. There is

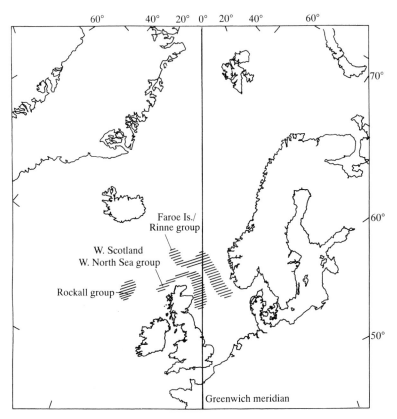

Figure 4.19 The three stocks of haddock in the North Sea, west of Scotland, Faroe Is. and Rockall (adapted from Jamieson and Birley, 1989).

a third stock on Flemish Cap. The Avalon stock spawns between St Pierre and the Southern Grand Bank and the eggs and larvae are carried to the nursery grounds in the region of the Avalon peninsula. The Southern Grand Bank stock spawns mainly in May and the fish are retained in that region. From the 13 main positions of tagging, the average distance of recapture was 151 miles, which implies that the stocks are rather localized in this very extensive region (the distance from St Pierre to northern Labrador is about 1100 miles, at sea).

Thompson (1943) tagged cod in five regions north of the Laurentian Channel: Avalon, Fortune, Raleigh, Gulf of St Lawrence and Labrador. Figure 4.20(*a*) shows the results for the Fortune area as distance in miles against number of days that the fish were free. There are four yearly groups as the cod tend to return to their grounds of first spawning. But

(a)

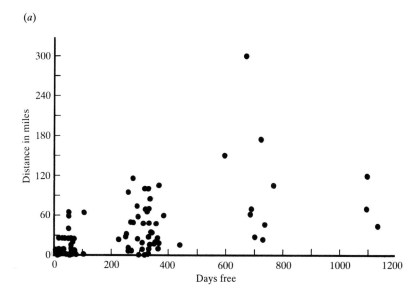

(b)

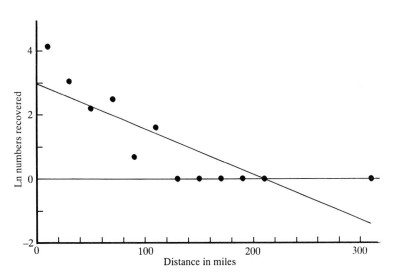

Figure 4.20 Cod on the Canadian Shelf. (a) The tagged cod tend to return to their grounds of first spawning, but there is also a distinct emigration. (b) The dependence of ln numbers recaptured on distance (after Thompson, 1943).

five tags were recovered at more than 300 miles from the spawning ground, an emigration rate (distant recaptures/all recaptures) of 0.0255. For the five groups the emigration rates were:

Avalon	Fortune	Raleigh	Gulf of St Lawrence	Labrador
0.0263	0.0255	0.0385	0.159	0.0198

The emigration rate from the Gulf of St Lawrence is much higher than the rest and the average was 0.0375.

Another way of analysing the recaptures is to plot ln numbers recaptured against distance (Figure 4.20(*b*)). The furthest distance is that at which numbers recaptured were reduced to one from mortality and emigration:

Avalon	Fortune	Raleigh	Gulf of St Lawrence	Labrador
210	420	320	295	280

The mean furthest distance of the five groups is 305 miles (twice Templeman's average distance). Thus the groups of cod live in rather small groups of about 300 miles across in the region north of the Laurentian Channel (about 1100 miles). Cod tend to return to their ground of first spawning, but they do emigrate at an average rate of 0.0375, which would promote a relatively high rate of gene flow between each group.

McCracken (1959) and Martin and Jean (1964) made an analogous study on the Scotian Shelf. Martin and Jean tagged fish on the Emerald Bank, on Banqereau and on the southern edge of the Laurentian Channel; McCracken tagged cod on the Magdalenic Shelf. Very few cod crossed the Laurentian Channel or to the western banks of the Scotian Shelf. Those tagged on the southern Laurentian Channel migrated into the Gulf and back again and those tagged on the offshore banks tended to remain there. Thus, south of the Laurentian Channel there is a mosaic of stocks which do not migrate very far from their

Table 4.2. *Estimates of emigration rates from the distant recaptures as proportion of numbers recovered*

Position	Number tagged	Number recovered	Distant recaptures	Emigration rate
Barents Sea				
1	62286	3884	14	
2	70150	1600	0	
3	40044	4729	20	0.0033
North Sea	4000		1	
Faroe Is.	6183	1475	2	0.0014
Faroe Bank	585	87	2	0.0230
Iceland	4939	466	23	0.0494
W. Greenland	8500	1067	258	0.2418
Labrador	18186	473	6	0.0127
Newfoundland				
1	8651	897	1	
2	18822	4234	8	
3	47560	14219	26	0.0018
Scotian Shelf				
1	8774	727	4	
2	2459	481	0	
3	21750	2200	0	0.0012
Nantucket	27247	2900	0	0.0000
	Sum	39439	365	0.0093
	Sum (excluding Iceland and W. Greenland)	37906	84	0.0022

Sources: Strubberg, 1922; Graham, 1924; Schroeder, 1930; McKenzie, 1934, 1956; Tåning, 1937; Thompson, 1943; McCracken, 1959; Gulland and Williamson, 1962; Martin and Jean, 1964; Postolakii, 1966; Maslov, 1972; Templeman, 1974, 1979; Rauck, 1977; Lebed *et al.*, 1983; Godo, 1984; Warnes, 1989.

spawning grounds. It would be of great interest to examine the genetic structure of this mosaic in more detail.

Another way of looking at the problem is to list the distant recaptures from the major tagging experiments, such as the transatlantic passage of one fish from the North Sea to the Grand Bank (Gulland and Williamson, 1962) or the recovery in the Faroe Is. of two cod tagged on Faroe Bank (Strubberg, 1916). Table 4.2 shows the positions of tagging (of the major tagging experiments) and the numbers of fishes tagged and recaptured, also the numbers of distant recaptures and the emigration rates (distant recaptures/numbers recaptured). If the emigration rates are averaged with the zero distant recoveries the average emigration rate is 0.00052.

There are three types of distant recapture: (a) the colonization of west Greenland from Iceland in the twenties and thirties and the return of mature fish to Iceland, the proportions of which are high, (b) the low-level recapture from distant grounds, and (c) none, for example the Scotian Shelf, 1959 to 1962 and the Barents Sea, 1934 to 1939. On both grounds there were recaptures in other series of tagging experiments.

The more interesting information arises from the emigration rates, distant recaptures as a proportion of all recaptures from a particular ground. Very roughly, the currents move from east to west; the West Spitzbergen Current could carry fish into the Arctic into the east Greenland Current to west Greenland and thence to the Labrador Current and the tagging experiments are arranged in that order. Then the expected distant recoveries might distribute themselves about the diagonal (which they do roughly, $r^2 = 0.58$). The average emigration rate was 0.00052, which is low. As might be expected this is much less than that observed from Thompson's experiments (Thompson, 1943) within the Newfoundland region. Recently, Dickson and Brander (1993) have shown that with the 1957 year class, there was a distinct migration from west Greenland to Labrador.

If one assumes equilibrium between mutation and migration (although migration must always be more effective than mutation), an index of genetic diversity, G_{st} is given by:

$$G_{st} = 1/(1 + 4N_e m),$$

where N_e is the effective population size, in numbers, and m is the migration rate, here that of emigration. $N_e m$ is the gene flow and if $N_e m \gg 1$, there is little genetic diversity. If $N_e m \ll 1$, there is the possibility of genetic diversity. Suppose that there are 10^5 individuals in any of the Atlantic cod populations, with an emigration rate of 0.00052, then G_{st} is very low indeed, but the gene flow is 52, and so there may be little genetic differentiation.

Smith *et al.* (1990) reviewed the idea of the unit stock. They criticized it on two grounds, that the stock is not always genetically homogeneous and that repeated samples showed that electromorph frequencies are not always stable. They concluded that a stock must be defined in genetic terms, which for the Atlantic herring would be disastrous. Yet Stephenson and Kornfeld (1990) have shown that the population of the

Georges Bank herring recovered not by colonization but by resurgence. The evidence of the discreteness of this stock was genetic. The 'stock' of the fisheries biologist requires only that the vital parameters are homogenous, that emigration/immigration rates are low enough to prevent bias and that recruitment is generated by spawning stock biomass.

Conclusion

Evidence for a unit stock is often based on a good description of the migration circuit as it is linked by a succession of fisheries, as for example in the Arcto-Norwegian cod (see Cushing, 1968). This requires knowledge of the spawning grounds and nursery grounds together with good evidence of the migrations. In this account the migration circuit of the plaice in the Southern Bight of the North Sea is described in as much detail as is known and as always the detailed study reveals further questions. The spawning group in the Southern Bight is part of a mosaic with spawning grounds all around the British Isles. If the chances of larval stray, from the path from spawning ground to nursery ground, and of adult stray by spreading are considered, it becomes possible to set the bounds of a unit of management and those of a genetic unit free of gene flow. Ideally, the use of genetics in this subject should be delimited in these terms.

The study of genetics as an aid to defining the unit stock has not yet been fully successful. There are two reasons for this: first, variability between position, year and year class was not fully described and second, expectations were too high. After apparent failure with the stocks of Pacific salmon by the earlier workers, Utter *et al.* (1989) started to discriminate stocks of chinook salmon. The work of Shaklee and Salini (1985) on the barramundi, of Jamieson and Birley (1989) on the North Sea haddock and of Utter *et al.* suggest that the genetic approach should be continued as greater resolution is revealed with more detailed sampling and more elaborate techniques.

It has taken a long time for the genetic study of stock structure to become established. One reason is that in the herring, differences may not exist at all and that the gene flow in the Pacific salmon is higher than expected from fish that return to their native streams to spawn. Results of tagging experiments on salmon and cod have been presented and in each case they revealed information that might not emerge from genetic studies alone. Both techniques require sampling from research vessels and so could well be used together. Then the spread of tags in space and of gene flow between spawning groups could perhaps be combined. The problem of stock structure could then be investigated in a more active manner.

The investigation of the spawning group of plaice in the Southern Bight of the North Sea has revealed how the fish sustain their circuit of migration, although some interesting questions remain unanswered. At least once, larvae were blown beyond their usual nursery ground and others drifted on to open beaches away from that ground. The emigration of adult cod as recaptures distant from the major spawning groups has been estimated and it was low, but on the Laurentian shelf the stray was distinctly greater. It should not be forgotten that the west Greenland cod stock was probably established by stray from Iceland during the twenties and thirties (see Cushing, 1982). With tagging experiments conducted on appropriate scales, a stock could be defined as that from which the stray is low, significantly lower than variability of the vital parameters.

Genetic studies on stock structure have developed much more slowly than hoped and indeed they are of little use to a herring biologist, with the one startling exception on George's Bank cited above. In contrast, for the Australian barramundi, the North American chinook salmon and the North Sea haddock, quite new evidence has been produced. Such stocks would be defined as those from which gene flow is very low.

The requirement of the fisheries biologists is that the vital parameters of the group sampled should not be biased by stray or gene flow. In the herring stocks, differences in recruitment across the North Atlantic are very great. The gene flow or stray between them is enough to prevent the establishment of genetically based stocks, but is probably sufficiently low to obviate bias in the estimation of the vital parameters.

5

Climate and fisheries

Introduction

Fish stocks respond to climatic change in a number of ways. Patterns of migration may alter, as, for example, the invasion of the Svalbard Shelf by the Arcto-Norwegian cod stock in the late twenties or the colonization of east and west Greenland by the cod stocks at Iceland (see Cushing, 1982). The commonest effect of climatic factors is to augment or diminish the magnitude of recruitment profoundly over a period of time. In this chapter some remarkable changes to the fish stocks are recounted; indeed, it now appears that the populations are most sensitive indicators of climatic change.

There are three sequences of climatic events known to have affected fish stocks. The first is the North Atlantic Oscillation, the pressure difference in the atmosphere between Iceland and the Azores; linked to it, but distinct, is the evolution of the Greenland High. A long-term consequence of changes in the position of the Greenland High was the second sequence of events, the great salinity anomaly of the seventies, a large cool slug of water that moved from Iceland to the Grand Bank and thence across the Atlantic to the Barents Sea and once more to Iceland over a period of more than 14 years (the Great Slug). The third sequence of events is the El Niño/Southern Oscillation (ENSO), which recurs every five years or so in the subtropical Pacific, but with possible links to the North Atlantic.

The link between recruitment and climatic factors

First, it is desirable to establish that the magnitude of recruitment does indeed depend on climatic factors. Shepherd *et al.* (1984) studied the recruitments to 18 stocks in the North East Atlantic between 1962 and

Table 5.1. (A) *The correlations between the first principal component of the ln recruitments and the three principal components of sea surface temperature (T₁, T₂, T₃) and (B) the correlations between the ln recruitments of nine stocks and the three principal components of sea surface temperature in the North Sea and off the west coast of Scotland* T_1, T_2 *and* T_3

	T_1	T_2	T_3
A First principal component of recruitments	−0.80	−0.33	−0.23
B Recruitments			
Sole	−0.45	0.35	0.25
Plaice	−0.55	0.45	−0.26
Cod	−0.65	−0.26	0.25
Haddock	0.60	0.13	−0.26
Whiting	0.43	0.17	−0.13
Whiting (west of Scotland)	0.65	0.07	−0.24
Saithe	0.64	0.06	0.13
Saithe (west of Scotland)	−0.06	−0.19	−0.13
Herring	−0.56	−0.14	−0.26

1976. These authors executed a principal component analysis on recruitments (in ln numbers) by regions and the first two eigenvectors were significant, which implies common sources of variation within regions. They carried out a second principal component analysis on sea surface temperatures in February, March and April in three areas of the North Sea and another off the west of Scotland, four in all. The rationale for this procedure was that most of the fish were spring spawners and temperature may be a convenient proxy for the processes (as yet unknown) that govern the generation of the spring outburst. The larvae might well depend on the food produced at that time. Three components were significantly correlated with those derived from the recruitments.

Table 5.1 shows the correlations between the principal components of recruitment and those of the sea surface temperature and also correlations between ln recruitments of nine stocks and the principal components of sea surface temperature. Some are negative and some positive, which means that, if real, different processes are linked to common proxies, as for example the time of onset of the spring outburst. The development rates of eggs and larvae depend inversely on temperature and this process must play a part, if a minor one. Because recruitment can be predicted from the abundances of the 0-group fish and even in the late larval stages, it is likely that major processes determining recruitment take place during larval life (see Chapter 6); indeed, this is really the only justification for any dependence of recruitment on climatic factors. This is because wind, irradiance and temperature affect primary production and hence the processes of growth and mortality during larval life. The paper of Shepherd *et al.* (1984) established the clearest link between recruitment and climatic factors.

Hollowed *et al.* (1987) examined the recruitments to 59 stocks of 28 species in five regions in the North East Pacific, California, Washington–Oregon, Canada, Gulf of Alaska and the Bering Sea. They distinguished interannual changes from the trends. Because of their interest in extreme year classes, each recruitment was expressed as a proportionate deviation from a five-year running median; although recruitment is sometimes considered to be log-normally distributed (Hennemuth *et al.*, 1980; but see the book by Rothschild, 1986, for a discussion of this point), Hollowed *et al.* considered that a log transformation tended to confound the interannual differences with those due to the trends and in this instance it was not used.

Hollowed *et al.* (1987) classified the year classes by quartiles and the extremes lay in the upper or lower ones. Figure 5.1(*a*) shows the distributions of relative year class strengths by groups, with extreme year classes indicated; there were three groups of high year classes in 1951 and four in the period 1961 to 1963. Figure 5.1(*b*) shows a more extensive time series for the North East Pacific groundfish stocks (Hollowed and Wooster, 1992). The years 1961 and 1962 coincide with the build up phase of the Great Slug in the North Atlantic and to the start of the gadoid outburst in the North Sea; those of 1958, 1959 and 1982 occurred during ENSO events.

Because only 23% of the time series were normally distributed (as shown with a Kolmogorov–Smirnov test), a large number of rank correlations were used for the interannual material and Pearson correlations for the trends. A plot of the number of probabilities $>p$, against $(1-p)$, (Schweder and Spjøtvoll, 1982) showed that the great majority of probabilities were ordered as expected; then the large number of correlations could be used without the Bonferroni correction. It was shown that, within regions, year classes were positively correlated, but between them they were not, which confirms the result of Shepherd *et al.* (1984). There is a most interesting contrast, that year classes are correlated positively within regions but not between them, although the extreme year classes were common to three regions. Hence, there are two distinct scales of processes.

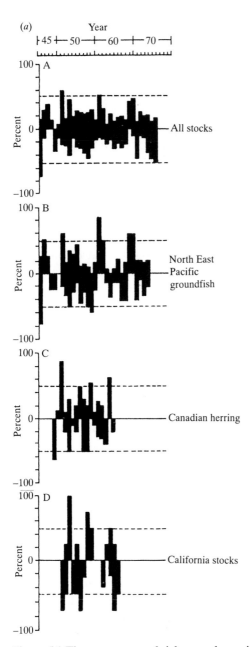

Figure 5.1 The occurrence of rich year classes in the North East Pacific: (*a*) the common strong year classes, in all stocks, North East Pacific groundfish, Canadian herring and Californian stocks (after Hollowed *et al.*, 1987); (*b*) a more extensive distribution of rich year classes in the North East Pacific Ocean (after Hollowed and Wooster, 1992).

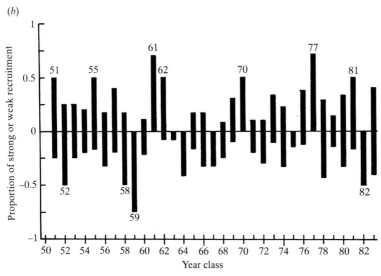

Figure 5.1 (*cont.*)

Figure 5.2 shows the recruitments to five cod and four haddock stocks in the North West Atlantic between west Greenland and the Gulf of Maine for periods between 1950 and 1980 (Koslow *et al.*, 1987). All stocks fluctuated in about the same way throughout the period and most were correlated with each other. Autocorrelative methods suggested a half period of about 10 to 20 years. Various environmental factors were linked with recruitments in a multiple stepwise regression: local winds, salinity, river run off and the principal components of sea surface temperature and of atmospheric pressure. None reveals the actual processes but a link with climatic factors was established. The half period may be associated with the North Atlantic Oscillation and the lowest observations in the time series with the passage of the great salinity anomaly of the seventies (see below).

Myers (1991) examined the recruitments of cod, haddock and herring stocks separately in the east and the west North Atlantic. The standard deviations of the logarithmically transformed recruitments of 53 stocks were plotted on latitude between 40° N and 69° N with a quadratic curve. Except for the western Atlantic herring, the differences from a median latitude were significant and the constants were significantly different from zero. The same result appeared when detrended observations were used, which eliminates the possibility that trends in recruit-

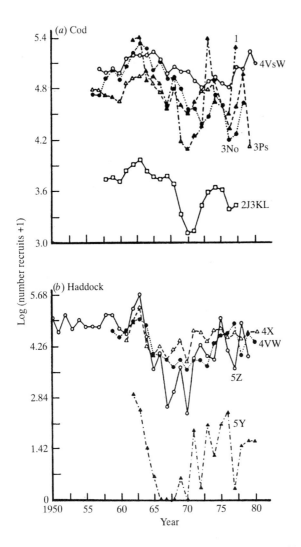

Figure 5.2 The recruitments to five cod and four haddock stocks in the North West Atlantic between 1950 and 1980 (Koslow *et al.*, 1987).

ment were linked to latitude. Recruitment was more variable at the northern and southern ends of the range, which implicates climatic factors.

These four papers establish the link between recruitment and climatic factors. Variability tends to be common within regions and not between them, where each region is about the size of the average depression or anticyclone. Yet, a very few extreme year classes were common to somewhat broader regions, which means that some more extensive influence was at work. In contrast, the gadoid year classes in the North West Atlantic showed a 'periodicity' of 20 years or more that must reside in the memory of the ocean. As is shown below, some of the above observations have been associated with the major climatic changes since the fifties: ENSO events, the Great Slug and the North Atlantic Oscillation.

The North Atlantic Oscillation

The difference in surface air pressure between the Azores and Iceland indicates the strength and duration of the Atlantic air circulation. It was named by Sir Gilbert Walker at about the same time as he established the existence of the Southern Oscillation (the alternation between the subtropical Pacific high and the Indonesian low, Walker and Bliss, 1932). van Loon and Rogers (1978) published data on the trend in the North Atlantic Oscillation between 1895 and 1983 (Figure 5.3); the upper time series shows the difference in surface pressure between the Azores and Iceland and the lower represents the pressure anomaly over an area of southern Greenland. The dotted lines in the lower figure give the data smoothed with a low pass filter. A high or low index may persist for a number of years; for example, in 1904 to 1911, the index was high and the Iceland low dominated the North Atlantic, so south-westerly winds blew in the Norwegian Sea when the very strong 1904 year class of Norwegian herring was hatched.

Cushing (1982) described the rise of the west Greenland cod fishery from the Iceland stock as the Greenland pressure anomaly slowly rose. The cod were probably absent from west Greenland from about 1850 until 1908 to 1909 when a few were found there by the Tjalfe expedition. Cod were caught on the west Greenland banks from 1912 onwards; with the year classes 1917, 1922, 1924, followed by those of 1934, 1936, 1942 and 1945, the stock increased hugely, as the Greenland high intensified, until annual catches of nearly half a million tonnes were being taken in the fifties (see Figure 5.4). Stock and fishery collapsed in the late sixties as the Greenland high disappeared. Figure 5.4(*a*) illustrates the catches of cod and haddock on Fylla Bank off west Greenland between 1925 and 1985 (Dickson and Brander, 1993), Figure 5.4(*b*) shows the year class strengths of the west Greenland cod stock between 1953 and 1984,

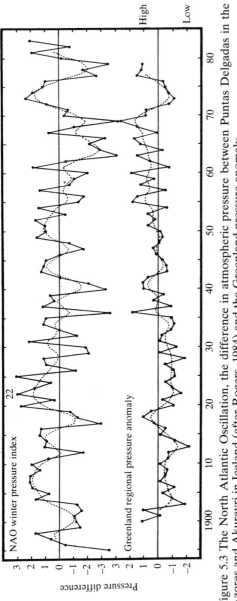

Figure 5.3 The North Atlantic Oscillation, the difference in atmospheric pressure between Puntas Delgadas in the Azores and Akureyri in Iceland (after Rogers, 1984) and the Greenland pressure anomaly.

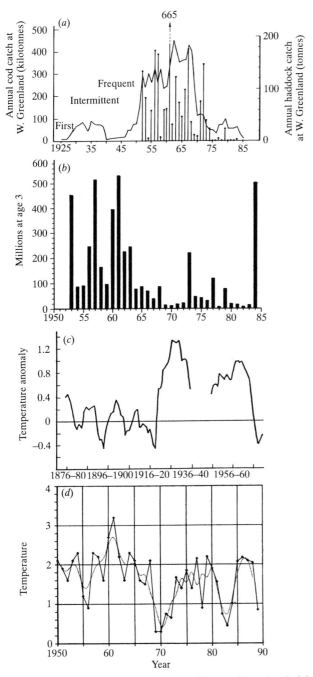

Figure 5.4 Cod and temperature off west Greenland: (*a*) catches of cod (and haddock) at west Greenland, 1925 to 1985 (after Dickson and Brander, 1993); (*b*) recruitment to the west Greenland cod stock (1953 to 1984); (*c*) temperatures at west Greenland from 1876 to 1970; (*d*) temperatures on Fylla Bank off west Greenland, 1950 to 1989 (Hovgård and Buch, 1990).

Figure 5.4(*c*) shows the temperatures from 1876 to 1970, Figure 5.4(*d*) displays the temperatures on Fylla Bank, 1950 to 1989 (Hovgård and Buch, 1986, 1990). The relatively warm period in the fifties is obvious. Since 1963, most of the year classes have been poor, but the larger ones were hatched in waters of 1 to 2 °C; from 1965 onwards there were two cold periods, 1965 to 1972, and 1981 to 1982, with no good year classes. From 1973 to 1980 the water off west Greenland was fairly warm with three moderate year classes. The best recent year class, that of 1983, was hatched in cool water of 1 °C. If the temperature on Fylla bank indicates the presence of a high over west Greenland, there is a slight indication that recovery might depend upon it. The recruitments of 1956, 1961 and 1963 were the last good ones before the collapse. The stock increased steadily and so did the catches as the Greenland high pressure anomaly increased from the twenties to the fifties (Figure 5.3). Then both stock and fishery collapsed as the North Atlantic Oscillation became negative in the sixties. The possible mechanism of decline of that stock is now examined.

The west Greenland cod stock was colonized from Iceland as larvae and juveniles were carried across the Denmark Strait by the Irminger Current and the east Greenland Current, which mix as they round Cape Farewell (Hansen, 1968). The little fish survived if the flow persisted across the Denmark Strait and if the water off west Greenland was relatively warm, about 2 °C (Figure 5.4). The colonization was described in a large tagging experiment at both Iceland and Greenland from 1924 onward (Hansen *et al.*, 1935, Tåning, 1937). Dickson (cited in Cushing and Dickson, 1976) described the collapse of the west Greenland cod stock in an atmospheric scenario. Figure 5.5 shows the increment in pressure at sea level between 1900 to 1939 and 1956 to 1965; a high appeared over west Greenland in winter in the fifties and sixties and northerly winds increased in the Norwegian Sea and in the North Sea. The change in winter air pressure between 1956 to 1965 and 1966 to 1970 is shown in Figure 5.6 (Dickson and Lamb, 1976); the high had changed its shape and northerly winds increased over the east Greenland Current converting it into a polar stream (this was also the time when the great salinity anomaly of the seventies, or the Great Slug, was formed, see below). Figure 5.7 shows the change in winter sea level pressure between 1966 to 1970 and 1971 to 1974. The Greenland high was replaced by low pressure with cold water off west Greenland under the northerly wind there and the southerly wind in the Denmark Strait inhibited the transport of larvae and juveniles from Iceland (this is

Winter pressure change
1900–39 → 1956–65

Figure 5.5 The change in mean winter sea level pressure over the North Atlantic from the period 1900 to 1939 to that of 1956 to 1965 (after Cushing and Dickson, 1976).

shown in the North Atlantic Oscillation and in the collapse of the Greenland anomaly), so by 1971 the conditions for good survival of the west Greenland cod stock no longer held. Figure 5.3 shows that the Greenland anomaly built up gradually from the twenties as the stock grew slowly. The anomaly reached a high positive level from 1955 to 1971 when the Greenland high developed and this was the period of high catches off west Greenland.

A more complete explanation of the phenomenon is given by Dickson (1992). Figure 5.8 shows the change in winter mean sea level pressure and in winds as cold succeeds warm conditions or vice versa: (*a*)

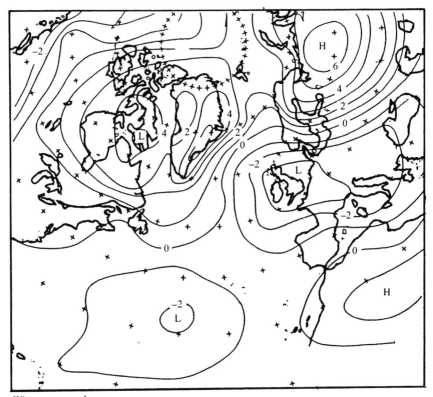

Winter pressure change
1956–69 → 1966–70

Figure 5.6 The change in mean winter sea level pressure over the North Atlantic from the period 1956 to 1965 to that of 1966 to 1970 (after Dickson and Lamb, 1972).

compare 1900 to 1914 (cold) with 1925 to 1939 (warm), (*b*) compare 1970 to 1984 (cold) with the earlier warm period, 1950 to 1964. In both changes, from cold to warm, easterly winds increase across the Denmark Strait above the Irminger Current; as the cold period appeared in the late sixties, the winds across the Denmark Strait slackened. This is a sufficient explanation in atmospheric terms of the rise and fall of the west Greenland cod stock.

The colonization of the west Greenland cod stock from Iceland depended on the transport of larvae and juveniles (in the Irminger Current) across the Denmark Strait under strong easterly winds during

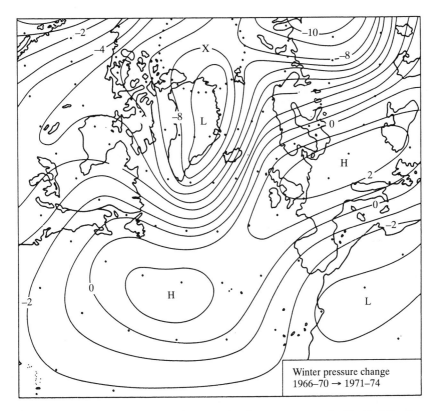

Figure 5.7 The change in mean winter sea level pressure over the North Atlantic between 1966 to 1970 and 1971 to 1974 (after Dickson and Lamb, 1972).

the warm periods, when the juveniles survived in the relatively warm waters off west Greenland. The total catch between 1950 and 1970 amounted to nearly 6 million tonnes, worth about £8 bn at present prices. In the long term, this event was associated with the presence of the Greenland high, which was developed as the North Atlantic Oscillation reached a minimum.

Dickson and Namias (1976) analysed the changes between 1956 and 1971 over North America and Greenland. They found that the Greenland high was associated with cold winters in the south-east United States. During winters of extreme cold there, the zone of peak winter storm frequency is drawn to the south-west and this decreases cyclonic activity over Iceland and Greenland. Because the ocean retains heat, the baroclinic fields intensify. The low over the south-east United States is

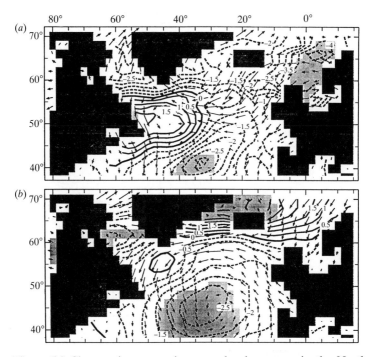

Figure 5.8 Changes in mean winter sea level pressure in the North Atlantic between different periods: (*a*) 1900 to 1914 (cold) and 1925 to 1939 (warm); (*b*) 1970 to 1984 (cold) and 1950 to 1964 (warm). The change from cold to warm conditions leads to an increase in easterly winds across the Denmark Strait and vice versa (after Dickson and Brander, 1993).

associated with a high over the Rockies and cold air over the North Pacific. This sequence of low and high pressures reflects the changes in the path of the jet stream. Figure 5.9 (Horel and Wallace, 1981) shows such a sequence of events with warm air over the subtropical Pacific. This structure forms the basis of the sequence and the teleconnections between the Southern Oscillation and the North Atlantic Oscillation (see van Loon and Rogers, 1978, 1981; Egger *et al.*, 1981; van Loon and Madden, 1981). Horel and Wallace correlated indices of the Southern Oscillation with the atmospheric parameters of northern hemisphere winters.

The changes in pressure difference in time between the Azores and Iceland express the changes from a system dominated by the Greenland high to one dominated by the Iceland low. It was this type of event that

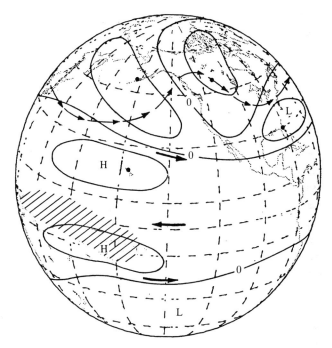

Figure 5.9 The train of anticyclone and depression following an ENSO event, atmospheric high (H) over the subtropical Pacific, low (L) over the Alaska gyral, high over the Canadian Rockies and low over the south-east United States. 0 indicates the first eigenvector of sea surface temperature (after Horel and Wallace, 1981).

led to the major changes in the North Atlantic in recent decades. Four major phenomena occurred between the fifties and the seventies: the collapse of the west Greenland cod stock, the changes in the plankton in the waters in the North Sea and the North East Atlantic, the gadoid outburst in the North Sea and the passage of the great salinity anomaly of the seventies around the North Atlantic for more than 15 years.

Changes in the North Sea plankton

The following account of the changes in the North Sea plankton is taken from the article by Dickson *et al.* (1988*a*). Between the fifties and the seventies, the zooplankton and the phytoplankton in the North Sea and west of the British Isles declined, to recover in the early eighties

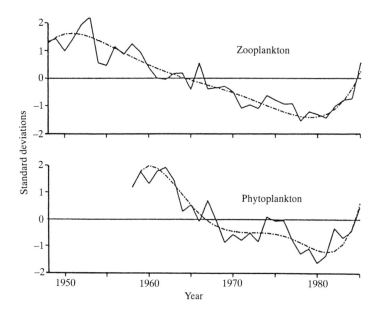

Figure 5.10 The decline in zooplankton and phytoplankton and their recovery between 1948 and 1983, based on 12 areas and 24 species of both algae and animals (after Colebrook *et al.*, 1984).

(Colebrook *et al.*, 1984) (Figure 5.10); *Pseudocalanus* declined by a factor of 5 between the fifties and the seventies (Glover, 1979). Figure 5.11 shows the changes in spring species of phytoplankton by season and year from 1958 to 1985 in the central western North Sea and in the central eastern North Sea (areas C_2 and C_1 in the continuous plankton recorder network). The decline in abundance from the fifties to the seventies and the recovery in the eighties is common to both areas. But in the west, production was delayed by about a month or more in the sixties and the seventies as compared with the fifties. In the east, however, there was no delay, although the stocks of phytoplankton were reduced. It is possible that in the west the delay started in 1962, when the gadoid outburst began (see below).

A northerly index was calculated in the air pressure difference between the fifties and the seventies in the rectangle, 10° E to 20° W by 35° N to 65° N. By the seventies, the northerly index had increased in November, March and April as compared with the fifties. Because the latter months are those of the spring outburst, the mean hemispheric difference in surface pressure was plotted for the seventies less that of

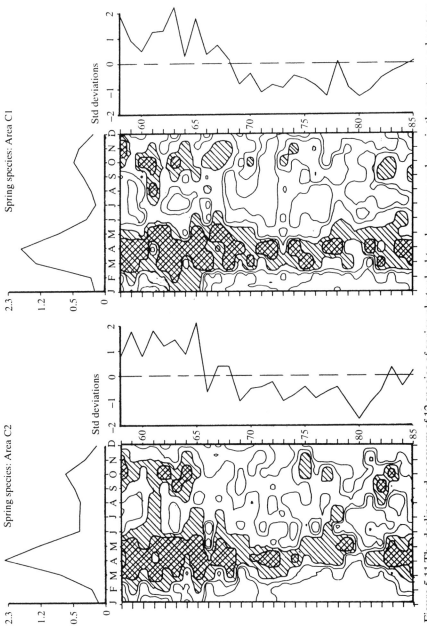

Figure 5.11 The decline and recovery of 12 species of spring phytoplankton by season and year in the western and eastern North Sea (after Dickson *et al.*, 1988*a*); the mean seasonal trends in each are shown as is the interdecadal trend.

Figure 5.12 A surface pressure difference diagram for the months March to April for the seventies less those of the fifties in the northern hemisphere; the point of interest is in the North Atlantic (after Dickson *et al.*, 1988*a*) where a ridge appears.

the fifties, that is across both ends of the trend in the plankton (Figure 5.12). An intense ridge of pressure difference became established over the eastern North Atlantic from the Faroe Is. to North West Africa. So northerly winds intensified in the North Sea and there were stronger north-easterly winds off Morocco. It was shown that northerly winds did

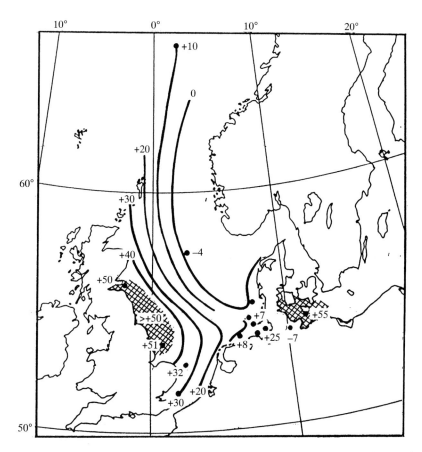

Figure 5.13 The increased incidence of gales at selected light vessels in the North Sea from the fifties to the seventies; the major increment occurred in the western North Sea (after Dickson *et al.*, 1988*a*). Contours are 0, +10%, +20%, +30%, +40% from east to west. Shaded areas are 50% increased incidence.

increase in the North Sea between the two periods, but there was also an increment in the incidence of gales. Figure 5.13 shows the percentage increment in gale force winds per decade between the fifties and the seventies; there was a sharp increase in the western North Sea, but little change in the east.

The decline in the stock of phytoplankton by the seventies is common to both the western and eastern North Sea, but the delay in primary production occurs only in the west central North Sea and this is probably associated with the greater incidence of gales there. If one looks at Figure 5.14 (Sverdrup, 1953), on the development of the critical depth, it

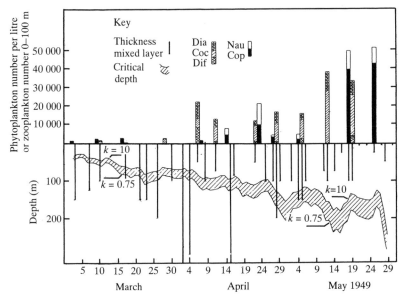

Figure 5.14 The development of critical depth, mixed layer depth, phytoplankton and zooplankton at Ocean Weather Ship M in the Norwegian Sea, in spring 1949 (after Sverdrup, 1953); the development of the spring bloom was hesitant and interrupted. Zooplankton number is numbers per haul, 100 m to the surface. Dia = diatoms, Coc = coccolithophorids, Dif = dinoflagellates, Cop = copepods, Nau = nauplii. k is the attenuation coefficient of irradiance in sea water.

is obvious that at least two storms delayed the development of the spring bloom at Ocean Weather Ship M in 1949. The incidence of gales in the Sverdrup mechanism has not been stressed before in the discussion of the development of the critical depth. The decrease in phytoplankton in both the western and the eastern North Sea was due to the increment in windspeed which accompanied the increase in the northerly winds, but the delay in the west was probably due to the incidence of gales.

The changes in the plankton between the fifties, the seventies and the eighties were probably the consequence of an increase in the Azores high in spring with an increment in Northerly winds and gales in the North Sea followed by a decline in the eighties. The period 1956 to 1971 is clearly shown in the time series of the North Atlantic Oscillation, a period of the Greenland high with warm air over west Greenland but cold northerly winds over east Greenland. In contrast, during the seventies, pressure was low over Greenland from 1970 to 1976, which implies that the North Atlantic Oscillation had intensified (Figure 5.3). There were two sources of northerly wind in the North Atlantic: (a) that associated with the Greenland high which ultimately impacted the west Greenland cod fishery, and (b) that associated with a secular ridge which developed over the North East Atlantic between the fifties and the seventies and the decline (and recovery) of the plankton in the North Sea and to the west of the British Isles.

The gadoid outburst

The 1962 year class of haddock in the North Sea was 25 fold larger than the average between 1918 and 1961 (Jones and Hislop, 1978). Subsequently, there were also high year classes of cod, whiting and saithe (and possibly Norway pout) during the sixties and seventies; the increments in stock of the five species between 1959 and 1970 were highly correlated (Cushing, 1980). During the eighties there were signs that the recruitments to the gadoid stocks were declining, that the so-called gadoid outburst is over (although the most recent year classes of haddock are high). Larval cod, haddock and saithe in the Gulf of Maine feed on the naupliar and copepodite stages of *Calanus* (Marak, 1974). Jones (1973) suggested that cohorts of larval haddock grew alongside those of *Calanus*. With continuous plankton recorder material, Cushing (1982) showed that *Calanus* was most abundant in the north-eastern North Sea. Although *Calanus helgolandicus* is present in the North Sea, *Calanus finmarchicus* predominates. Colebrook (1978) wrote that before 1950 the North Sea was dominated by small copepods (*Pseudocalanus* and *Paracalanus*), but after that date *Calanus* became most abundant. This event can be seen in the material from the Flamborough Line, which sampled a limited section between the Dogger Bank and the coast of England between 1933 and 1963. *Calanus* is common in the Norwegian Sea (Oceanographic Laboratory, Edinburgh, 1973). Tait (1957) and Tait and Martin (1965) had found that part of the water flowing through the Faroe Shetland Channel to the Tampen Bank Current that entered the North Sea by the Rinne originated in the Norwegian Sea. Tait also characterized the flow in 1949 to 1950 as 'arctic water' that had not then been noticed since 1927. Riepma (1980) wrote that the Tampen Bank Current flowed south-easterly under all wind conditions. *Calanus finmarchicus* does not overwinter in the North Sea and its populations are sustained by inflow. Backhaus *et al.* (1994) have created a model of the passage of *Calanus finmarchicus* from 600 m in the Norwegian Sea by the shelf counter current to the North Sea, by the Fair Isle current and the Tampen Bank current; the input from the Fair Isle current means that the input is not confined to the north-eastern North Sea.

Cod probably spawn in March in the central North Sea (Daan, 1978) and the larvae may start feeding up to a month later. Dickson *et al.* (1973) published an inverse correlation of cod recruitment between 1954 and 1968 against sea surface temperature in the central North Sea. Indeed, Cushing (1982) listed a number of correlations between cod and haddock and numbers of *Calanus*. Cushing (1984) compared the average monthly abundance of *Calanus* in the periods 1948 to 1961 and 1962 to 1978 in the north-eastern North Sea and found that the production of the stock was delayed by a month or six weeks, which was also shown in the plankton more generally, as described in the last section (but not known in 1984). In time series (Figure 5.15) the delay follows the pattern of decline and recovery of zooplankton in the North Sea in that it was least in the fifties, greatest in the seventies and was low again in the eighties. Delay in the production of the *Calanus* stock each year can be

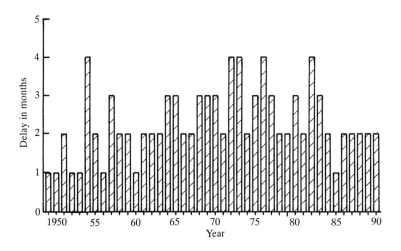

Figure 5.15 The changes in delay of the production of the stock of *Calanus finmarchicus* from 1948 to 1990 as shown in the continuous plankton recorder material (with permission of the Director of the Sir Alister Hardy Foundation for Ocean Science).

Table 5.2. *Multiple regression of North Sea cod recruitment on* Calanus *production, delay in production and March temperature*

Calanus abundance	−0.0013*
Delay in production	0.4537**
March temperature	0.1496**
Calanus abundance × delay	0.0007*
Temperature × delay	0.0854**

Note:
*$p < 0.05$, **$p < 0.01$, $r^2 = 0.75$.

expressed as the number of months from March onwards in which the peak stock occurred (including July when the little cod start to feed on small fish (Cushing, 1980)). Cushing (1984) calculated a multiple regression of cod recruitment on the abundance of *Calanus*, the delay in months and the sea temperature in the central North Sea in March between 1954 and 1977 (that is across the dates at which the gadoid outburst increased) (Table 5.2).

Brander (1992) re-examined the regression for the period 1962 to 1986 (excluding some missing observations). The distributions of *Calanus* from the continuous plankton recorder (CPR) material in the North Sea were divided into four quadrants. *Calanus* was most abundant in the central North Sea and in the north-east North Sea and there was little trend in abundance during that period. A multiple regression of cod recruitment against delay and *Calanus* abundance was not significant.

Brander reproduced Henderson's (1961) figure of the distribution of cod larvae in the North Sea from the CPR material from 1948 to 1956; there are two areas, one in the western North Sea off Scotland and the other lying between the Southern Bight and the Skagerak (Figure 5.16(*a*)). The first spawning ground corresponds to the area in the central North Sea where the production of stock was delayed between the fifties and the seventies, and the second ground lies in the region where the production of stock declined but was not delayed, as discussed in the previous section. There is no evidence yet that the spawning distributions have changed; indeed the pelagic 0-groups are found in the north-eastern North Sea (Brander, 1992). Riepma (1980) showed that,

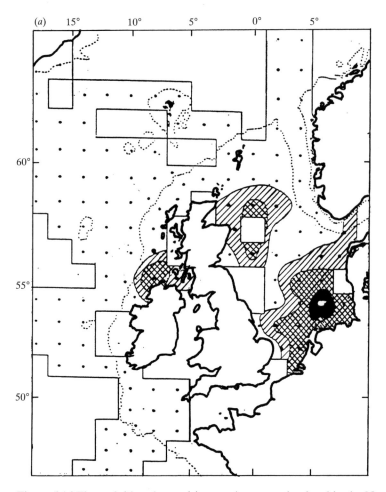

Figure 5.16 The gadoid outburst: (*a*) spawning grounds of cod in the North Sea (Henderson, 1961); (*b*) dependence of cod recruitment on the delay in production of *Calanus* in the months March to June for the period 1955 to 1971; (*c*) the changes in the correlation from 1955–1971 to 1955–1980.

under a northerly wind, waters off the Scottish coast would move north-easterly. It is possible that the western group of larvae move to the north-east and that the eastern group migrate to the north. Then the western group may have suffered from the delay in production but the eastern group did not.

I have re-examined the problem using material from 1955 onwards and from the months March, April, May and June (because the original

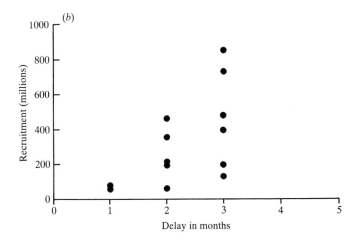

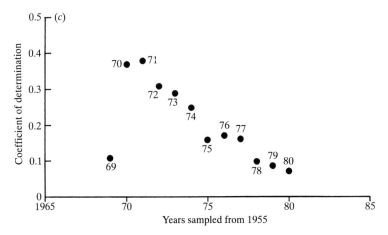

Figure 5.16 (*cont.*)

delay was observed from April to June (Cushing, 1982)). The object was to diagnose the failure. Figure 5.16(*b*) shows the dependence of cod recruitment in delay in the north-western North Sea from 1955 to 1971 ($r^2 = 0.38$; $p < 0.01$). Figure 5.16(*c*) shows the trend in the coefficient of determination from (1955 to 1971) to (1955 to 1980). Between (1955 to 1970) and (1955 to 1974), the correlations are significantly different from zero ($p < 0.01$ to $p < 0.03$) and subsequently they are not; at the most, a little more than one-third of the variance was accounted for. Thus, the original thesis that the gadoid *outburst* depended on the delay in production is sustained, but the more interesting point is that the relationship decayed in the early seventies.

There are two forms of explanation for the decline in correlation with time. The first is that the main variable was the increment in recruitment in the early sixties, which became less prominent as the years passed by. The second is that, with the increased stock, a density-dependent effect appeared which modified the processes of earlier years.

The salinity anomaly of the seventies

The high over Greenland appeared in the late fifties and it intensified in the period 1965 to 1971. Northerly winds increased over the western Norwegian Sea. The east Greenland and east Icelandic currents became cooler and less saline as the proportion of polar water increased. Indeed, between 1964 and 1971 they became polar currents carrying ice. Figure 5.17 shows the anomalies of temperature and salinity north of Iceland (Dickson *et al.*, 1985). Because the surface layers were relatively fresh they did not mix with the saltier water below, so ice could form. A large

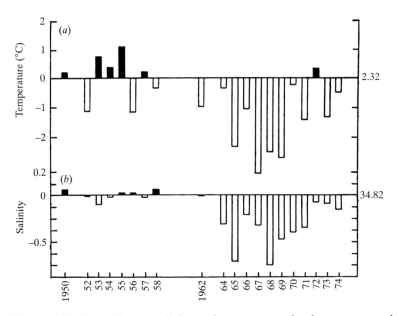

Figure 5.17 Anomalies of salinity and temperature in the waters north of Iceland, 1950 to 1974 (after Dickson *et al.*, 1985).

body of cool fresh water was formed down to 700 m (it was reduced to 500 m after its passage through the Faroe–Shetland Channel) and it was called the great salinity anomaly of the seventies, the Great Slug (Dickson *et al.*, 1988*b*).

The Great Slug was traced from east Greenland all around the North Atlantic from 1961 to 1982. From north of Iceland it moved to west Greenland and then across the Labrador Sea towards the Grand Bank, whence it drifted across the North Atlantic to a region west of the British Isles and back to Iceland, through the Faroe–Shetland Channel and through the Norwegian Sea to the North Cape. During this period, the Slug became mixed to some degree; it started with a 'salt deficit' of 72×10^9 tonnes and ended with one of 47×10^9 tonnes and it travelled more than 10 000 km. It took 7 years to pass through the area north of Iceland and 6 years to pass the North Cape, 14 years later.

North of Iceland, primary productivity was reduced from 2.6 mg C m^{-3} h^{-1} (1958 to 1964) to 0.7 mg C. m^{-3} h^{-1} (1965 to 1971) (Thoradottir, 1977) and zooplankton was reduced by a factor of 3 (Astthorsson *et al.*, 1983). On the Grand Bank the quantities of phytoplankton, copepods and euphausiids were reduced by a factor of 3 in 1972 as compared with the previous 13 years, as sampled by the continuous plankton recorder network (Robinson *et al.*, 1975). This reduction is of the same order as that off Peru in 1972 to 1973 during El Niño, when the recruitment of anchoveta was reduced by an order of magnitude (Guillen, 1976).

Before the Great Slug appeared off Iceland, the herring fishery north of the island occurred at the polar front (Jakobsson, 1980), and the herring of the Atlanto-Scandian stock lay just offshore, feeding on *Calanus finmarchicus*; but in May 1962 the drift ice lay close to the north-west coast of Iceland. By 1966 and 1968 the herring shoals were 100 to 400 miles north and east of Iceland (Malmberg *et al.*, 1967, 1968; Malmberg and Vilhjamsson, 1968) presumably because herring did not cross the gradient to the fresher water of the Great Slug and *Calanus* did not live there either (Figure 5.18). In the Barents Sea the echo distributions of herring were shifted 10° to the west (Middtun *et al.*, 1981) and those of the capelin to the south and south-east (Loeng *et al.*, 1983) when the mass of cool fresh water passed through the region. In the English Channel, whiting and Norway pout appeared between 1976 and 1979 (Southward and Mattacola, 1980; Blacker, 1982) at the time when the Great Slug passed offshore.

Fifteen 'deep water' stocks were examined and the year classes

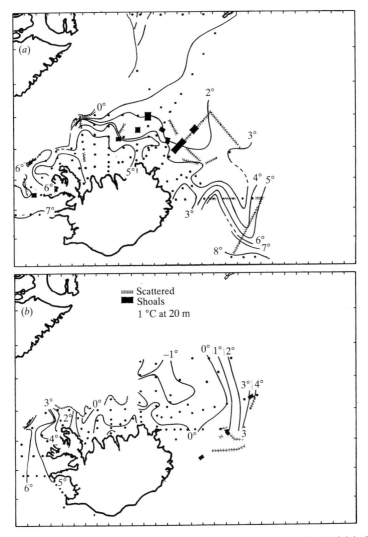

Figure 5.18 The distribution of herring at 20 m north of Iceland: (*a*) in May 1962 before the passage of the Great Slug; (*b*) in May 1965 in the cool, freshish waters of the Great Slug (after Jakobsson, 1992).

hatched during the passage of the Great Slug were compared with those in other periods (see Table 5.3) (Cushing, 1988*b*).

The years of the Great Slug were identified from the evidence given by Dickson *et al.* (1988*b*). The differences in recruitment in numbers in years of anomaly were examined with a Wilcoxon rank test. Significant differences were established in 11 stocks out of 15.

Table 5.3. *Reduction in recruitment in deep water stocks during the passage of the Great Slug* (anomalous years taken from Dickson *et al.*, 1988*b*)

Stock	Anomalous years	Number of year classes	*p*
Icelandic summer herring	1965–71	36	0.01
Icelandic spring herring	1962–71	45	0.01
E. Greenland cod	1965–71	13	0.05
W. Greenland cod	1969–72	15	0.01
N. Grand Bank cod	1971–73	20	n.s.
S. Grand Bank cod	1971–73	22	0.05
W. Scotland saithe	1974–78	16	0.01
North Sea saithe	1975–77	18	0.05
Faroe saithe	1975–77	18	0.01
Faroe plateau cod	1975–77	18	n.s.
Faroe plateau haddock	1975–77	18	n.s.
N.E. Arctic saithe	1978–81	19	n.s.
N.E. Arctic cod	1978–81	21	0.01
N.E. Arctic cod 0-groups	1978–81	15	n.s.
N.E. Arctic haddock	1978–81	22	0.01
N.E. Arctic haddock 0-group	1978–81	16	n.s.
Blue whiting	1978–81	6	0.01

All anomalous recruitments were low. Figure 5.19 shows some of the time series of recruitment; the Iceland spring spawners did not recover and have not been caught since, presumably because the Great Slug passed at a time when the stock was heavily exploited and it collapsed through recruitment overfishing. In other stocks, for example the Iceland summer spawning herring, the west Greenland cod stock, the Grand Bank cod, the west of Scotland saithe, the north-east Arctic haddock, the north-east Arctic cod and the North Sea saithe, the anomalous years stand out clearly.

It can be shown that a reduction of about 1 °C might lead to a delay of about a month in the seasonal development of surface temperature (as sampled at the Ocean Weather Ships), which represents a delay in the generation of buoyancy and hence a delay in the onset of the spring bloom. If cod spawn at a fixed season (Cushing, 1969), then a delay in the time of onset of the production cycle may starve the larvae.

Changes in subtropical pelagic fish stocks

Since the fifties, the pelagic stocks of fish, herring, sardine and anchovy

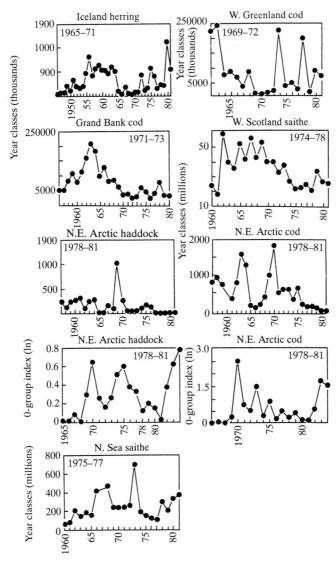

Figure 5.19 Time series of recruitments to some deep water stocks in the North Atlantic; the reduced recruitments during the anomalous years of the Great Slug are shown, with dates of anomalous years indicated in each plot (after Cushing, 1988*b*).

have suffered very great changes in abundance. The typical sequence was an expansion of catch followed by a collapse due to over-exploitation. But such stocks are also vulnerable to environmental change, as, for example, the dramatic increment in stock with the 1904 year class of Norwegian herring. Here the changes in the subtropical stocks, sardine and anchovy, are examined.

The habitat of upwelling

Bakun and his colleagues have made a comparative study of the physical environment in which sardines and anchovies live in the four major upwelling areas (California, Peru, Canary and Benguela). Bakun and Parrish (1980) showed that there was a group of half-a-dozen species common to all four regions (anchovy, sardine, horse mackerel, hake, Spanish mackerel and bonito). They also listed relationships between the recruitments of eight species and various indices of upwelling. Bakun and Parrish (1982) made use of the cube of the wind speed as an index of turbulent energy. They plotted it against Ekman transport, offshore and onshore (Figure 5.20); the sardines and anchovies, save off Peru, tended to live and spawn at low positive Ekman transport and low turbulence. They suggested that the fishes spawn where the offshore flow by Ekman transport is low.

Cushing (1971) noted that the greatest production occurs at low rather than high rates of upwelling. With Steele and Menzel's (1962) relationship of production against depth, production in the rising water (without grazing) was calculated. Then the total production was estimated as an inverse function of upwelling velocity. The reason for this is that if the water rises quickly, it reaches the surface before much production has taken place and it might then become vulnerable to grazing or loss due to turbulence or advection. In the slowly rising water such factors do not come into play and the stock of algae develops its full potential.

Bakun (1985) plotted the late spring temperature anomalies (compared with temperatures some distance offshore) in the four major regions. Sardines and anchovies tended to spawn where the cold anomalies were less intense, another index of low offshore Ekman transport. But he found that the anchovies off Peru spawned at higher offshore Ekman transport than those off California. The apparent conflict between the two regions was resolved when it was found that the mixed layer depth off Peru was shallower. Ekman transport per unit

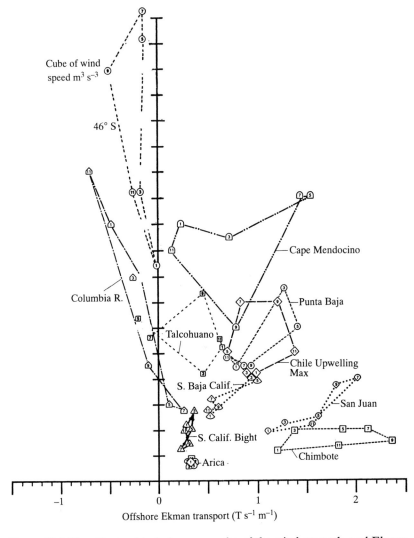

Figure 5.20 The effects of turbulence, as cube of the wind strength, and Ekman transport (after Bakun and Parrish, 1982).

mixed layer depth indicates the offshore velocity of the surface mixed layer and Bakun found that the anchovies off California, off Baja California and off Peru all appeared to minimize their offshore movement. In the major upwelling areas the anchovies and sardines spawn where the water at the coast rises relatively slowly and where the offshore transport is slow. There are two advantages, first that the larvae are not drifted offshore and second, that the production of food is greater.

Bakun and Parrish (1990) extended their observations on sardines in the eastern boundary currents to a population of *Sardinella aurita* that lives in a coastal bight inshore of the Brazil current, a western boundary current. They estimated wind stresses, cube of the wind stress, cloud cover, solar radiation and Ekman transport for a broad area off southern Brazil. They found that the transports were one-third to one-half of those in the major upwellings in the eastern boundary currents. But there is strong upwelling from Cabo Frio and Cabo Sao Tomé. Bakun and Parrish showed that *Sardinella* spawns between December and January in the centre of the coastal bight during a period of least wind stress (and hence least turbulent mixing) and low Ekman transport. The spawning grounds lie downstream from the strong upwelling from Cabo Frio and so planktonic food should be available. Bakun and Parrish (1982) had pointed out that in the eastern boundary currents the sardines and anchovies spawned in coastal bights downstream from major upwellings.

Bakun and Parrish (1991) extended this study to the anchovy (*Engraulis anchoita*) in the South West Atlantic. These spawn in both winter and summer between Cabo Frio and Cabo Tres Puntas, some 2500 km to the south. The same physical parameters were estimated as in the *Sardinella* study. The anchovies spawn in winter in the coastal bight, south and downstream of the upwelling off Cabo Frio. Between the coastal bight and Mar del Plata, spawning in winter depends on the upwelling at the shelf break. Further south, in summer, the anchovies spawn in a region of productive tidal fronts. It is of great interest that over a broad range, and at all seasons, the anchovy manages to exploit three different productive systems.

Cury and Roy (1989) used a method that estimates optimal transformations of multiple regressions, essentially a form of non-linear fitting. They related the recruitment of the Pacific sardine to stock and to turbulence (as cube of the wind speed). Although recruitment and stock were positively correlated, recruitment depended on turbulence in a dome-shaped manner; above 5 to 6 m s^{-1}, recruitment was reduced by turbulence. They proposed that there was an environmental window within which the sardines were able to survive (Figure 5.21). The argument was extended to other upwelling areas with the same result, for Peruvian anchoveta, Californian sardine, Senegalese sardinella and Moroccan sardine. Large-scale turbulence would not affect the offshore

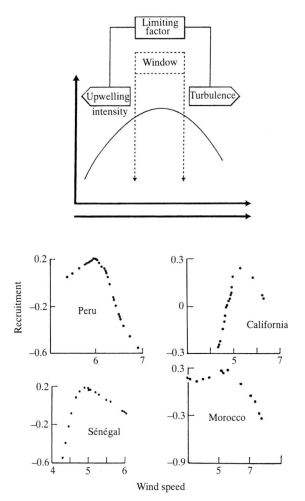

Figure 5.21 The window of upwelling in which optimal conditions occur at moderate intensity of upwelling and at moderate turbulence; the dependence of recruitment on wind speed is shown for four stocks: Peruvian anchoveta, Californian sardine, Senegalese sardinella and Moroccan sardine (after Cury, 1991).

transport but it would affect the production of food. The importance of this concept is that the non-linear nature of many of the essential processes has been revealed.

Apparent switches between pelagic stocks

Figure 5.22 shows the apparent switches from sardine to anchovy off

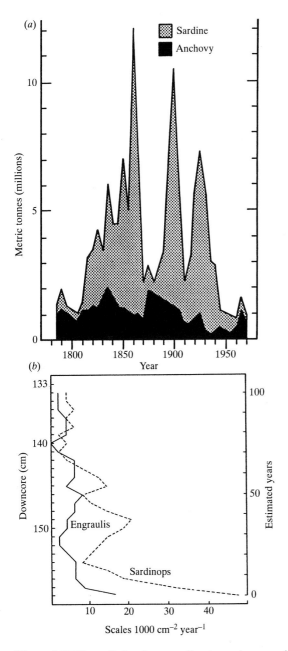

Figure 5.22 The switches from sardine to anchovy as shown by the abundance of scales in sediments (*a*) off California (after Smith and Moser, 1988); (*b*) off South Africa (Shackleton, 1988). (*c*) A more recent analysis of the scale deposits off California (after Baumgartner *et al.*, 1992).

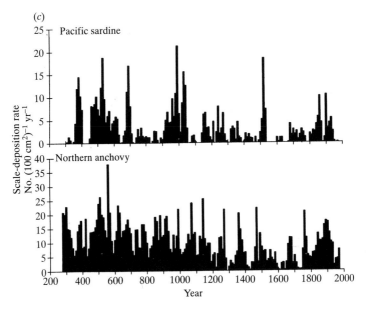

Figure 5.22 (*cont.*)

Southern California for the last 200 years (Smith and Moser, 1988; Figure 5.22(*a*)). The long-term changes were derived from the scales deposited in anoxic sediments off southern California from 1785 to 1970, in millions of tonnes. The sardine stock was more abundant and there is some evidence of alternation with the stock of northern anchovies. Smith and Moser analysed the material with an integrated periodogram and found periods in five species ranging from 30 to 65 years. Baumgartner *et al.* (1992) made a fuller study with more information and Figure 5.22(*c*) shows the scale deposition rate for the sardine and anchovy off California. The sardine stock is more variable with more pronounced periodic changes. Shackleton (1988) has found changes in the sardine and anchovy stocks in the Benguela region with the use of the same method (Figure 5.22(*b*)) and again the stocks appear to alternate. The changes occurred long before exploitation in the present century and they were most profound.

Kawasaki (1983) published the concomitant trend in catches of the Far Eastern sardine, the Californian sardine and the Chilean sardine between 1900 and 1981 (the dates are those of the Far Eastern sardine; the large catches of the Chilean sardine appeared only in the seventies) and noted that 'the fluctuations were in phase with each other' (Figure

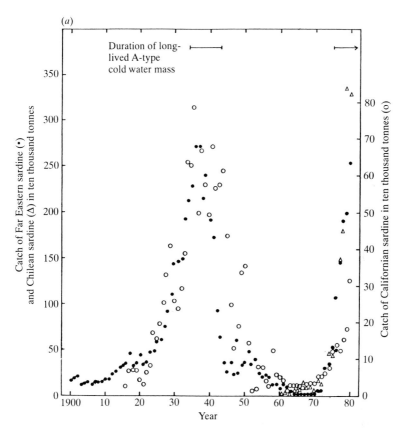

Figure 5.23 The Kawasaki diagram: (*a*) catches of the Californian sardine (from Baja, California), the Far Eastern sardine and Chilean sardine from 1900 to 1981; (*b*) the solar radiation, the northern hemisphere air temperature and sea surface temperatures in the North Atlantic and the North Pacific (after Lluch-Belda *et al.*, 1989). The sardine and anchovy catches are shown in more detail: (*c*) off Japan (Kondo, 1988), together with the long-term northerly winds off California (Ware and Thomsen, 1991); (*d*) off South America in the Humboldt current; (*e*) in the Benguela current (after Lluch-Belda *et al.*, 1989).

5.23(*a*)); note the difference in scale of the catches. Kawasaki's diagram is the most remarkable presentation of a pan-Pacific phenomenon with concomitant catches, up to 40 years apart. He noted that the peak catches of the Far Eastern sardine occurred when the Kuroshio meander south of Honshu spread to the west and south in cold water. Kawasaki and Omori (1988) showed that the catches of each of the three stocks were highly correlated with each other. With air temperature anomalies between 64.2° N and 90.0° W (Hansen *et al.*, 1983) they found that

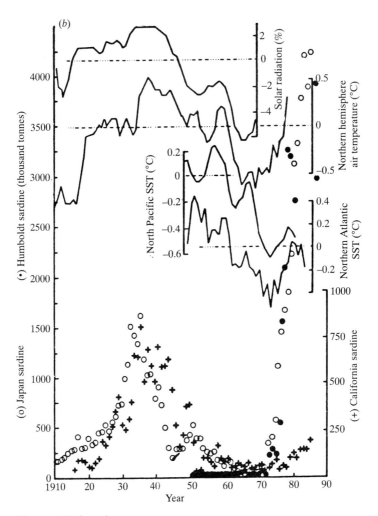

Figure 5.23 (*cont.*)

catches in each stock increased as the temperature anomalies were increasing and decreased as they decreased; in other words the high catches occurred during warm periods and they fell as the temperatures declined. Lluch-Belda *et al.* (1989) (Figure 5.23(*b*)) related the decline of sardine stocks in the Pacific to a 4% decrease in solar radiation, a fall in northern hemisphere air temperature of about 0.5 °C and in sea surface temperature (SST) in both the North Atlantic and in the North Pacific of about 0.8 °C. The SSTs and air temperature recovered in the early

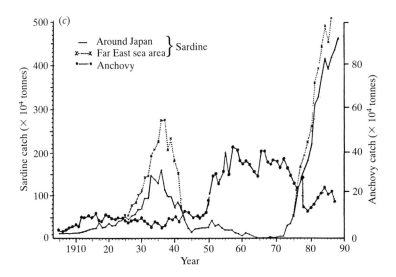

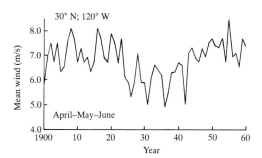

Figure 5.23 (*cont.*)

seventies at the time when the sardine stocks in the Pacific recovered; they may be proxies for the non-linear dependence of recruitment on food in an upwelling system. The important point is that the 'period' is of the order of 40 years. In the North Atlantic, a half 'period' of about 20 years has been described in the North West Atlantic cod stocks. Such periods, if they exist, can only reside in the memory of the ocean.

Kondo (1988) reported earlier periods of abundance in the Far Eastern sardine (1640 to 1660, 1680 to 1730, 1820 to 1840, 1860 to 1880, 1920 to 1940, 1970 onwards). The present period started with the 1972 year class and subsequently the 1980 year class predominated. In the

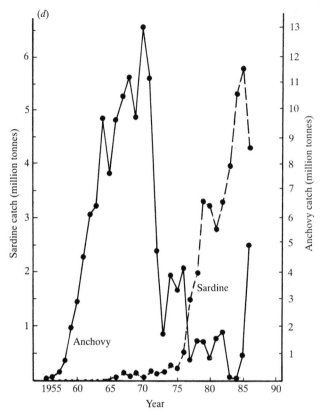

Figure 5.23 (*cont.*)

summer of 1980, cold water over the spawning ground (south of Honshu) was replaced by warm water, that is when the Kuroshio replaced the Oyashio. There will always be a front offshore of the spawning ground, on the shoreward edge of the major current, but with the Oyashio there is downwelling there instead of upwelling. Upwelling would be of advantage to the larval fish

Figure 5.23(*c*)–(*e*) (from Lluch-Belda *et al.*, 1989) shows the changes from sardine to anchovy or anchovy to sardine for long periods, 1905 to 1987, in the four major upwellings and inshore of the Kuroshio off Japan, an extension of the Kawasaki diagram. The catches off Japan are shown in Figure 5.23(*c*) (top) (Kondo, 1988). The sardine catch reached a peak in 1935, after which it declined until a major recovery took place in the seventies and eighties. The catches of anchovies (on a scale one-fifth of that of the sardines) declined during the thirties, but recovered in

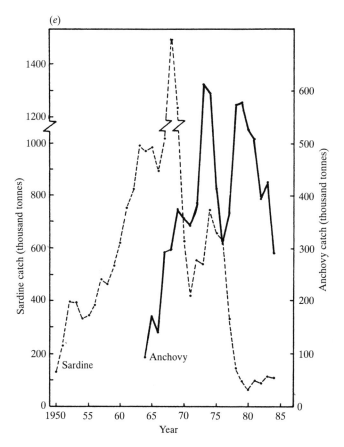

Figure 5.23 (*cont.*)

the fifties and sixties, to fall again somewhat in the seventies as the catches of sardines advanced. The lower diagram in Figure 5.23(*c*) shows the period of decreased northerly wind off California during the sardine periods of 1925 to 1945 (Ware and Thomsen, 1991). Aligning the two plots on the same time scale, it can be seen that the sardine period was one of low upwelling off California. Upwelling is a function of along-shore wind stress, itself an effect of the intensity of the North Pacific high. It is possible that the pan-Pacific changes observed in the sardine catches depend upon reduced wind strength across the whole Pacific, with reduced upwelling; but how does this affect the processes inshore of the Kuroshio? In the Humboldt current (Figure 5.23(*d*)), the catches of sardines succeeded the anchovies, but most of the latter were taken off Peru with a smaller fishery south of Arica in northern Chile. The sardines off Chile (note: the scale of anchovy catches is twice that of the sardines) were caught in the same region as the catches of anchovy declined, but a later fishery for anchovy developed off Talcahuano, much further south. Figure 5.23(*e*) shows the changeover from sardine to anchovy in the Benguela system. Recruitment to the South African and the South West African pilchards were correlated, high in the fifties and lower in later decades; anchovy recruitment increased during the seventies (Crawford and Shannon, 1988).

The switches in catches (or biomasses or recruitments) from sardine to anchovy and vice versa are not at all precise, but upwelling varies in time on a broad scale. The wind stresses may very well be the same for great distances in the major upwelling areas, but differences in direction could change the magnitudes of Ekman transport. Hence it is not surprising that the recruitments to the South African and South West African pilchard stocks march in the same way. Figure 5.23(*c*) shows the changes in northerly wind off Southern California (30° N; 120° W) from 1920 to 1960 (Ware and Thomsen, 1991). The period 1925 to 1945 was one of weak northerly wind and hence, low upwelling. The subsequent anchovy period was one of stronger upwelling. The implication is that the sardines survived well in periods of weak upwelling and the anchovies in periods of strong upwelling.

From the thirties to the fifties and sixties, the wind stresses across the Pacific may have increased (as indeed they did in the North Atlantic), but from the sixties onwards they probably rose again. Brodeur and Ware (1992) examined the summer stocks of zooplankton in the Alaska gyral and found that they had doubled between 1956 to 1962 and 1980 to 1989. Wind stress augmented between the two periods; there was a positive dependence of zooplankton on Ekman transport, as divergence spread across the gyral (with increased catches of salmon). At Ocean Weather Station P, there was no such positive increment during the period, possibly because a single station represents a poor sample of the gyre. Venrick *et al.* (1987) found an increment in the stock of the deep chlorophyll maximum in the centre of the North Pacific gyre as wind stress increased across the whole North Pacific. Of the greatest interest is the long-term change in wind stress across the North Pacific with effects on the deep chlorophyll maximum in the centre of the North Pacific gyre, the zooplankton and the salmon catches in the Alaska gyral.

Figure 5.23 shows two remarkable events: (1) the pan-Pacific changes in sardine catches with an apparent period of about 40 years; (2) the switch to catches of anchovies between 1950 and 1975. The first event must reflect changes in upwelling, perhaps in intensity off Peru, Chile and California, and in the nature of the front shoreward of the Kuroshio off Japan. The catches involved are very large, more than 20 million tonnes. Off Japan the peak catch of sardines in the thirties reached nearly three million tonnes and in the eighties more than five million tonnes; those of anchovies between the fifties and the seventies amounted to about two million tonnes.

The decline of the zooplankton in the waters around the British Isles was correlated with changes in the upwelling indices off Portugal; upwelling increased between the fifties and the seventies and declined in the eighties. This followed from the pressure difference ridge over the British Isles and the increase in northerly winds in the North Sea. There was a similar increase in north-easterly winds off the Iberian peninsula. Catches of sardines were inversely related to the indices of upwelling, or low upwelling favours the recruitment of sardines, already familiar from the work of Bakun and his colleagues.

Between April and September, a ridge of pressure difference appeared off California, in 33° N, between the fifties and the seventies. Ahlstrom's (1966) material on the seasonal distribution of sardine and anchovy larvae reveals that the spawning period changed during the fifties from March/May to February and July/September as the indices of upwelling increased (Figure 5.24). It was shown that the sardines prefer 100 to 150 tonnes s^{-1} per 100 m of coastline (Bakun's (1973), indices of upwelling). As the upwelling increased from year to year, the sardines shifted their spawning season to months of slower upwelling. But the seasonal distribution of spawning of the northern anchovy remained the same (January to June) and they tolerated a greater rate of upwelling, 124 to 170 tonnes s^{-1} per 100 m of coastline (Dickson *et al.*, 1988*a*).

During the first 11 days after hatching, sardines grow faster than anchovies. Sardine larvae grow in length at a rate of 1 mm d^{-1} (Kimura, 1970; Butler and de Mendiola, 1985) whereas the anchovy larvae grow in length 0.2 to 0.5 mm d^{-1} (Kramer and Zweifel, 1970; Methot and Kramer, 1979). The anchovy larva is slow to learn to feed (Hunter, 1972*b*) and reaches metamorphosis later in life than the sardine (Theilacker and Dorsey, 1980). The sardine may grow in an upwelling plume for it needs the good supply of food expected from a moderate rate of upwelling. The anchovy grows more slowly and can presumably tolerate both intermittent periods of food lack and somewhat more

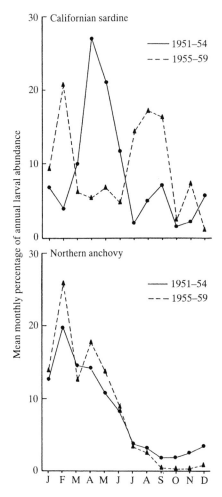

Figure 5.24 The changes in spawning season of the sardines off California during the fifties and the lack of change in the spawning season of the northern anchovy (after Dickson *et al.*, 1988*a*).

turbulent conditions characteristic of stronger upwelling. The difference in growth rate between the sardine and the anchovy may express quite different survival strategies, one in the plume of continuous and moderate upwelling and the other in intermittent and stronger upwelling. (This account is taken from that by Dickson *et al.* (1988*a*).) Is this difference in growth rates enough to create the enormous changes in sardine and anchovy catches during the present century?

The decline of the Pacific sardine catches in the forties and recovery in the late seventies was linked by Lluch-Belda *et al.* (1989) to the decline of SSTs in the North Atlantic and North Pacific in the fifties and their recovery in the seventies and eighties. The physical change was world-wide, with the decline and recovery of northern hemisphere air temperatures leading the SSTs by a decade or so. This is also the change that occurred in the North Atlantic during the cooling between the fifties and the seventies. During this period the changes in the plankton in the waters around the British Isles and the increment of gadoid catches in the North Sea, sometimes called the gadoid outburst, the passage of the Great Slug from one end of the North Atlantic to the other for a period of 14 years, were all linked to the pressure difference ridge in the North East Atlantic generated between 1965 and 1971. The same form of event occurs in the apparent half 'period' in the gadoid recruitments in the North West Atlantic: the full 'period', extrapolated from this thinnest base of 40 years, is exactly that observed.

The major events in the subtropical ocean were the switches from sardine to anchovy and back again, which may be due to changes in the upwelling systems. Table 5.4 shows the major sardine and anchovy periods off Japan, off Peru, in the Benguela and off California.

Each period is defined somewhat roughly, but there are three distinct ones evinced in the four systems. The sardine period of the twenties and thirties is related to weakened northerly winds and hence weakened upwelling off Southern California. Three are the classical upwelling areas; the Kuroshio system is not usually classed as an upwelling area, but there must be upwelling inshore of the current, possibly geostrophic, that is, function of the wind stress across the whole Pacific. It is surprising that the same events can be detected in the Benguela upwelling region. Furthermore, the anchovy period is the same as that of the northerly winds in the North Atlantic. Lluch-Belda *et al.* (1989) give anomalies in global air temperatures from 1910 to 1986. There was a

Table 5.4. *The major sardine and anchovy periods in three upwelling regions and off Japan*

	Sardine	Anchovy	Sardine
Japan	1920–1940	1950–1975	1975 onward
Peru		1955–1973	1975 onward
Benguela	1950–1970	1965–1975	1975 onward
California	1930–1950	1965–1982	1975 onward

cool period from 1960 to 1975, a period of relatively strong upwelling, hence an anchovy period (for at least part of the time). From 1925 to 1960 and from 1975 onwards, there were relatively warm periods when the sardines flourished and presumably the upwelling winds were weaker.

ENSO (El Niño and the Southern Oscillation)

El Niño was the name given by the Peruvians to a warm current from the north at Christmas in the austral summer when the upwelled water became warm and unproductive and the Guanay birds emigrated to Ecuador. It recurs every three to five years. From the events in 1957 to 1958 and 1965 to 1966, Bjerknes (1966, 1972) elaborated a pan-Pacific explanation based on an atmospheric circulation in the plane of the equator, which he named the Walker circulation. This is part of the Southern Oscillation. In the equatorial Pacific, the thermocline slopes upward from west to east. When the trades slacken, in the western Pacific, this slope is changed, the thermocline shallowing in the west and deepening in the east. The equatorial system comprises the north and south equatorial currents flowing to the west. At 5° to 9° N, the equatorial countercurrent flows to the east and at the equator the Equatorial Undercurrent flows to the east in the thermocline at about 100 m. This complex structure has been called the equatorial waveguide. When the trades slacken, a Kelvin wave is generated in the layer above the thermocline. In this very long planetary wave, water is transported at 250 cm s^{-1} across the Pacific within the waveguide for a period of two or three months. In this way, the eastern tropical Pacific becomes flooded with warm water, part of which flows off Peru just before midsummer, at Christmas. The Kelvin wave splits at the American Shelf and travels as a coastally trapped wave both north and south.

Bjerknes (1972) showed that when the ENSO cycle matures, two anticyclones appear north and south of the equator, which he called the 'Hadley anticyclones'. To the north appears a short train of cyclone and anticyclone, over Alaska, the Rockies and over the south-east United States (see Figure 5.7). Consequently, there is a potential for tele-connections between atmospheric systems across the northern hemisphere.

The whole ENSO event lasts for about 12 to 18 months. After the eastern tropical Pacific has been flooded with warm water, the processes reverse and the trades blow strongly, once again stacking up water in the western Pacific. The essential process is the alternation between this stack and the flow of this water to the east.

Outside the equatorial waveguide, when the wind stress slackens, baroclinic Rossby waves are generated in the layer above the thermocline by Ekman pumping, a downward flow as Ekman transports converge. They are also planetary waves, which are slower than the Kelvin waves and they cross the Pacific towards the west in about nine months on average. At the continental shelf in the western Pacific, the Rossby waves transform into Kelvin waves that at the equator travel the waveguide. Figure 5.25(*a*) (Enfield, 1989) illustrates the standard or canonical El Niño as sea levels at Truk (in the western equatorial Pacific) and at Callao (in Peru) for a period of three years. As the sea level rises in the east, it falls in the west. The 18-month period of El Niño is shown with the double peak at Callao. In 1982 to 1983, one of the strongest El Niños for a century took place, but it was six months late, in the austral winter. Sea temperatures off Callao were up to 7 °C warmer than usual. Two or three Kelvin waves appeared in the North Pacific and the South Equatorial Current disappeared; by the end of 1982 the Equatorial Undercurrent had vanished and the eastern tropical Pacific flooded with warm water. There were terrible consequences with heavy flash floods in Ecuador and Peru, heavy rains in Chile, storms in California, a record drought in Australia and heavy rains in the southeast United States and northern Caribbean. Figure 5.25(*b*) shows the sea surface temperature at Puerto Chicama, the Southern Oscillation index and the Cd/Ca ratio (an index of temperature) in the Galapagos coral; it is really an index of upwelling from the Equatorial undercurrent. (Some of this account is taken from Enfield (1989).)

Graham and White (1988) have suggested that this system is a 'coupled oscillator'. The warm water which appears in the eastern tropical Pacific changes the wind stress field across the tropical ocean. Such changes propagate the two forms of wave as described above to generate warm water once again in the eastern tropical Pacific. If the proposed oscillator were not more than a metaphor, an explanation would emerge of the recurrence of ENSO events every three or five years without an unspecified trigger, when the trades slacken. The model studies by Charnock and Philander (1989) indicated a number of possible complex interactions. Barnett (1991) found that the ENSO signal comprised two components, a quasi-biennial one and a low frequency one (of three to seven years). The latter is a standing oscillation and the former propagates energy to the east.

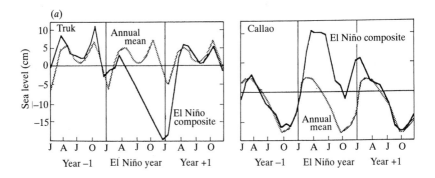

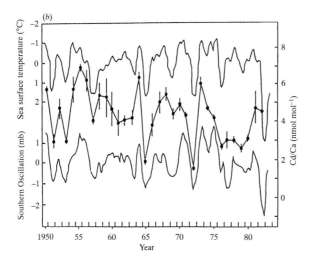

Figure 5.25 El Niño: (*a*) the standard or canonical El Niño (*b*) sea surface temperatures of Puerto Chicama, the Southern Oscillation and Cd/Ca ratio on the reefs off the Galapagos Is. (after Enfield, 1989).

The appearance in the waters off southern California of the pelagic crab, *Pleuroncodes planipes* (Longhurst, 1967), in 1957 to 1958 was probably an effect of a Kelvin wave during an ENSO event. Radovich (1961) recorded the appearance off California, during the El Niño of 1957 to 1959, of fishes that lived further south, a pomfret (*Taractes asper*), a pelagic ray (*Dasyates violocea*), requiem shark (*Carcharinus improvisus*), smooth stargazer (*Kathetostoma averruruncus*), bigeye (*Parathunnus sibi*), moonfish (*Vomer declivifrons*), roosterfish (*Nematistius pectoralis*), a halfbeak (*Hemirhamphus saltator*), two pompanos

(*Trachinotus rhodopus* and *T. paitensis*) and the mobula (*Mobula japonicus*). In the very strong ENSO event of 1982 to 1983, exotic animals were found far to the north of Vancouver Is. (Mysak, 1986); they were the oceanic sunfish (*Mola mola*), chub mackerel (*Scomber japonicus*), bonito (*Sarda chilensis*), pomfret (*Brama japonica*), swordfish (*Xiphias gladuis*), Pacific sardine (*Sardinops sagax*) and the leatherback turtle (*Dermochelys schlegelli*).

Mysak *et al.* (1982) detected a five to six-year and a three-year oscillation in data on sea level, temperature and salinity off Vancouver Is. and off Ketchikan further north. The first was attributed to the northward propagation of a coastally trapped baroclinic Kelvin wave and the second, possibly to the westward propagation of baroclinic Rossby waves. Pearcy (1983) drew attention to the fact that the year classes of herring off south-east Alaska and in the eastern Bering Sea were high in the El Niño years from 1926 to 1961. Sinclair *et al.* (1985) established that the survival index (ln recruitment per stock of Pacific mackerel) between 1928 and 1965 was positively correlated with sea level averaged between San Francisco and San Diego. High sea level indicates flow from the south during an ENSO episode. Smith (1985) showed that during the El Niño of 1957 to 1959, the quantities of copepods, euphausiids and Thaliacea were all reduced. Bailey and Incze (1985) recorded high year classes in 1957 to 1959 in the stocks of sardine, jack mackerel and the Petrale sole, but there were many stocks which evinced no such increments.

Mysack's evidence links the higher herring year classes with the appearance of the Kelvin waves. The higher sea level might improve survival of the herring, which spawns on the beach, by reducing the chance of desiccation. But that special case does not apply to the other stocks. The Pacific mackerel may profit by the upwelling of the warm water from the south, but a general mechanism remains obscure.

From this brief summary, the range of the biological effects of ENSO may be briefly sketched. Changes in the nature of upwelling, in the Kelvin wave transport, in the equatorial upwelling, off the Americas and even in the Alaska gyral, will have biological effects over very broad areas. The most profound effect of ENSO occurred off Peru where the stock of anchoveta survives in a region of restrained offshore velocity, as noted above. The stock collapsed in 1971 to 1973; recruitment failed in 1971 and again during the ENSO event of 1972 to 1973. In the two latter years, primary production was reduced by a factor of 3 and recruitment by an order of magnitude (Guillen, 1976).

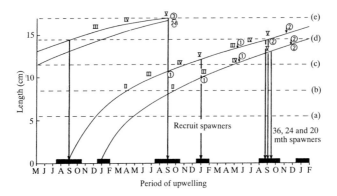

Figure 5.26 The spawning and growth of the anchoveta off Peru. Ages are given in roman numerals and spawnings in arabic.

The main period of upwelling lasts from May to November during the southern winter and spring. Figure 5.26 shows the timing of growth and spawning. The adult fish spawn towards the end of the main period of upwelling, when zooplankton is available for the larvae. The little fish start to mature at 8.5 cm in length at the age of about six months, but they do not spawn until they are about 15 months old. They live up to five years, as indicated by the tagged survivors of the 1969 year class. The recruits spawn first in high summer in December to February in a period of weak upwelling; then they spawn again in September and October. The anchoveta grows very quickly and spawns early in its life, converting energy quickly. The catches were high in the largest fishery in the world (up to 13 M tonnes yr^{-1}), which collapsed in the period 1971 to 1973.

Figure 5.27 (Muck, 1989) displays events in the Peruvian upwelling system from the fifties to the early eighties. It is the result of the remarkable analysis of the large quantities of data by Pauly and Tsukuyama (1987, 1989). In 1972 to 1973, the anchoveta biomass fell from about 15 M tonnes to about 2 M tonnes and for 17 years there has been no recovery. The sardine stock increased from about 0.3 M tonnes to 2 to 5 M tonnes. The stock of guano birds declined in three stages, during the El Niños of 1957–8, 1965–6 and 1972–3, as they suffered in competition with the fishermen (because the young fish do not feed in the years of El Niño). The stocks of bonito declined in two steps, 1971 and 1976 (a later El Niño). The stock of hake increased from the early seventies, but dropped during the strong El Niño of 1982–3. There was

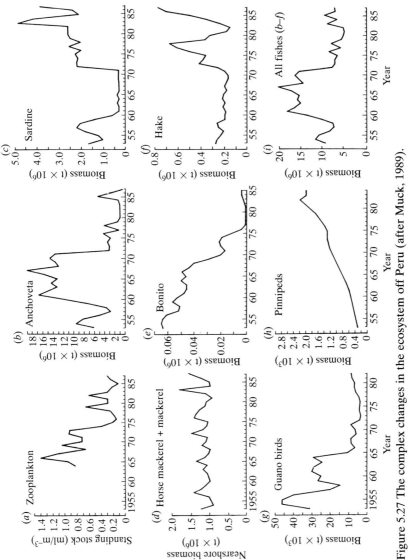

Figure 5.27 The complex changes in the ecosystem off Peru (after Muck, 1989).

no change in the stock of mackerels and horse mackerels and the pinnipeds increased slowly across the decades.

An interesting point is the decline of zooplankton in the seventies, which cannot be accounted for in the other events with a superficial examination. The events of 1971 to 1973 had a profound and lasting effect on the ecosystem of the Peruvian upwelling area. The clearest change was the transfer from the anchoveta to the sardine. The recruitment failure in 1971 is not well understood, but may have been an effect directly or indirectly, of fishing. The failures in the recruitments of 1972 and 1973 were due to the ENSO events which generated apparently irreversible changes. The full multispecies analysis by Pauly and his colleagues will be awaited with some interest.

Understanding of ENSO events has slowly increased in the last two decades and it has extended from an eastern tropical Pacific phenomenon to a world-wide one. The coastally trapped Kelvin waves on the western coasts of the Americas affect not only events off Peru but those as far north as the Alaska gyral. The sequence of events is as shown in Figure 5.7, from the North Pacific high, to that over the Rockies, and the low over the south-eastern United States. Thence, there is a link to the systems in the North Atlantic. Is it possible that the great El Niño of 1982–3 played a part in generating a number of outstanding year classes in the North Atlantic in 1983?

Conclusion

In recent years it has been established that fish stocks respond to those environmental factors which change the climate. Year class strengths vary in somewhat the same way within fairly large regions, but not between them; however, some extreme year classes are common between them. The major climatic factors that modify the production cycle may possibly govern life in the larval stages.

There are different scales of temporal change of interest to a fisheries biologist, from year-to-year differences to decadal changes. In the next chapter the year-to-year responses of recruitment to environmental factors will be discussed in more detail. Here, accounts have been given of changes lasting from 5 to 40 years.

Table 5.4 gives the sardine and anchovy periods in the major upwelling areas and off Japan. The anchovies tended to flourish from the mid-fifties to the mid-seventies; the sardines were abundant from the twenties to the late forties and from 1975 onwards (the pattern in the Benguela Current does not fit well, although the 1975 return to sardines is shown there). Lluch-Belda *et al.* (1989) showed that the anchovy period was a cooler one, from trends in solar radiation, northern hemisphere air temperature, North Pacific sea surface temperature and North Atlantic sea surface temperature. Ware and Thompson (1991) showed that the first warm period or sardine period was one of slacker winds off California, increasing in strength as the anchovy period supervened.

The most astonishing event was the association made by Kawasaki of the common trends in catches of the sardines in the three major fisheries in the Pacific. It was shown that the great changes were associated with switches from sardine to anchovy and back again and that the two species respond differently in their growth patterns to changes in the rate of upwelling. Indeed the period of the Californian high sardine catches, from the mid-twenties to the mid-forties, is associated with a period of reduced northerly winds, or reduced upwelling. It is likely that similar events occurred off Chile and off Japan. This is a long-term atmospheric event which is perhaps not yet well understood.

The cool anchovy period of the subtropical seas corresponds to the four prominent events in the North Atlantic. The west Greenland cod stock flourished from the year classes of 1942 and 1945 to that of 1963, when the high stood over Greenland in winter. The Great Slug was generated between 1962 to 1965 and 1971 when the Greenland high collapsed. Between the fifties and the seventies, the plankton in the North East Atlantic declined, to recover in the eighties, when the sardines flourished again in the subtropical seas.

The events in the North Atlantic depend on two events, the build up and collapse of the Greenland high (Figure 5.3) and the north-eastward extension of the Azores high between the fifties and the seventies (Figure 5.12). The first was associated with the rise and fall of the west Greenland cod stock and the generation of the Great Slug. The decline and recovery of the plankton in the North Sea and the North East Atlantic is associated with the gadoid outburst in the North Sea as the Azores high advanced. Both were events during the cooler anchovy period following the warm one in the twenties and forties.

The North Atlantic Oscillation diminished in the early sixties and the northerly wind over east Greenland generated a polar current which developed into the Great Slug, the great salinity anomaly of the seventies. The Slug found its way all around the North Atlantic for 14 years and reduced the recruitments to 15 'deep water' stocks during its passage.

A third event in the North Atlantic was the pressure difference ridge in the North Sea and the North East Atlantic with increased northerly winds, which may have generated decline in the zooplankton between the fifties and the seventies. The ridge might have been described as an extension of the Azores high. It is likely that the gadoid outburst started during this period as the production of *Calanus* was delayed. Thus a number of biological events in the North Atlantic were linked to changes in the atmospheric distributions over long periods of time.

6
Recruitment

Introduction

In the last chapter, some of the dramatic and extensive changes to stocks of fish were described. They followed from considerable and sustained changes in the magnitude of recruitment, but the possible causes were not discussed. There are two potential causes of differences in recruitment from year to year, in a closed population: first, changes in egg production and second, changes in survival. There are two main sources of density-dependent mortality: first, that caused by the aggregation of predators, and second, predation as growth is reduced by food lack, causing the period of predation to be extended, the Ricker–Foerster thesis. Both are somewhat hard to establish from field observations. In an open population, there are gains and losses by diffusion and advection, which could potentially vary from year to year, but which today can be estimated by physical oceanographers with some degree of accuracy. Temperature is a pervasive factor for two reasons: (a) the development of eggs and larvae is an inverse power function of temperature and so modifies growth rate and the incidence of predation, (b) it is a proxy for other changes in physical factors on the climatic scene, for example differences in wind strength and direction: a drop in temperature can indicate an increment in wind strength. As shown in the last chapter, fish stocks are profoundly modified by climatic change and a special case is the match/mismatch hypothesis which proposes that a source of variation in the magnitude of recruitment lies in differences from year to year in the time of onset of primary production in spring outburst or in upwelling area.

Three technical problems

There are three technical problems that should be considered: (a) the effect of temperature on growth, (b) the study of the daily rings on the otoliths, and (c) the sampling of fish larvae.

The effect of temperature on growth

Gray (1929) and Medawar (1945) showed that there was no general growth function. That is, growth cannot be described by some function with specific parameters of precise biological meaning. Ricker (1979) came to the same conclusion, but noted that within limited ranges or data sets, particular equations may be useful. Adults experience a broad range of temperature during the course of each year, but the effects of temperature on growth have often been ignored because fisheries biologists are often only interested in the growth of adults from year to year. But for larvae and juveniles, the differences in temperature from year to year may have considerable effects. Growth rate, ingestion and the standard rate of metabolism depend on temperature and the same must be true of predation, because it depends on the difference between the attack velocity of the predator and the escape velocity of the prey.

The link between growth and temperature appears simple because enzyme action increases with rising temperature and with it, respiration and the demands of maintenance. There have been many attempts to rationalize the processes but the best description remains that from the experiments of Brett *et al.* (1969). Sockeye salmon of 13 g wet weight were fed *ad libitum* and on smaller rations at a range of temperatures. Figure 6.1 summarizes the results of the experimental work. At R_{max}, the maximal ration when the salmon were fed *ad libitum*, growth rate (in % wt. d^{-1}) between 0 °C and 16 °C increased by more than an order of magnitude in a near-linear manner. The greatest growth was found at 15 °C, but it was only slightly greater than that at 10 °C. At temperatures above the optimal at 15 °C, growth rate declines to zero at 24 °C.

The standard metabolic rate increases with temperature in a quasi-linear manner; it may be imagined as a negative growth rate in Figure 6.1. It obviously plays a part in the decline of growth rate above 15 °C, but it is not sufficient. Brett *et al.* (1969) suggested that oxygen uptake might be limiting at high temperature and at the highest temperatures some sockeye refused to feed.

With rations less than maximal, 1.5% d^{-1} to 6% d^{-1}, the optimal

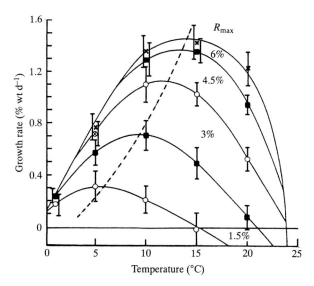

Figure 6.1 The dependence of growth rate on temperature at different rations for 13 g sockeye salmon (after Brett *et al.*, 1969).

growth rate occurs at lower temperatures at lower standard metabolic rates. At the optimal temperature of 15 °C, growth rate is a linear function of temperature but at temperatures both lower and higher, the function is curvilinear, which reflects the different parts played by the standard metabolic rate and the unspecified process above the optimal temperature.

The life history of fishes divides itself into three stages: larvae, juveniles and adults. Each tends to live in a distinct environment. For example, the larval plaice live in the midwater, the juveniles live on the flats on the continental coast of Europe between France and Denmark and the adults live on the seabed of the open North Sea, the bigger ones in deeper water. Similar distinctions can be established for most other migratory species, for example, cod, herring and salmon. Fish larvae live for perhaps two or three months at a relatively low range of temperatures in any given year, but which could differ considerably from year to year. The juvenile fish in temperate waters tend to live in shallow water near the coast where the summer temperatures are high and the winter temperatures are very low. Adult fish are faced with a lesser span of temperature within seasons and from year to year because they tend to live in deeper water.

Much of the study of recruitment processes takes place during rather short periods of time; life as eggs and larvae lasts for two or three months in temperate seas and much less in the tropics. Life as 0-groups may last for some months. Temperatures in such short periods may differ considerably from year to year and growth must be compared at standard temperatures.

The growth of fish larvae and the use of daily rings on the otoliths

Blaxter (1988) summarized much of the recent information, particularly on terminology and on events during development (for example, the northern anchovy). He also showed that reared fish larvae grow more quickly than do wild ones (in dry weight, condition factor and percentage fat); indeed, some larvae starved in the laboratory grow more quickly than do the wild ones. This most important observation implies that *the growth of fish larvae is nearly always food limited.*

Kramer and Zweifel (1970) grew the larvae of the northern anchovy in the laboratory at 17 °C and at 22 °C. They found that the Laird–Gompertz equation fitted the data for 35 days at 17 °C more effectively than did a simple exponential. The equation is:

$$W_t = W_0 \exp\{(A_o/a)\ (1 - \exp - at)\},$$

where W_0 is the initial weight and W_t that at time t, A_0 is the initial growth rate, a is the rate at which the growth rate declines with age and the rate itself is expressed by the constant (A_0/a).

The rationale for using this equation is that growth rate declines towards metamorphosis. Pannella (1971) discovered that the otoliths of fish larvae bore daily rings and that, in principle, the growth of fish larvae could be perhaps estimated at sea (Figure 6.2). Brothers *et al.* (1976) showed that there were daily rings on the otolith of the northern anchovy, the growth of which can be studied in the laboratory. With the Pacific sardine, Zweifel and Lasker (1976) used the Laird–Gompertz equation to describe the development of eggs and early larvae; they applied the method to observations on other species. Struhsacker and Uchiyama (1976), from daily rings on the otoliths, found differences in the growth of the nehu (*Stolephorus purpureus*) between Pearl Harbor and Kaneohe Bay, on Oahu in the Hawaiian Is. Hunter (1976) grew the northern anchovy in the laboratory under different feeding conditions; the Laird–Gompertz equation was fitted to the data, showing differences in growth rate. Methot and Kramer (1979) examined the growth of anchovy larvae at sea; the growth rate varied from 0.341 to

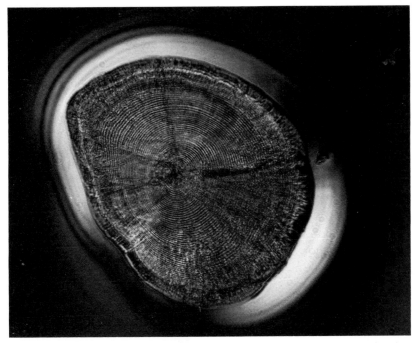

Figure 6.2 The daily rings on the otolith of *Clupea harengus*, the herring (courtesy of John Nicholls).

0.552 mm d^{-1}. There was no relationship between growth rate and temperature, so the differences were probably due to food.

Townsend and Graham (1981) studied the growth and age of larval herring in the Sheepscot River estuary in Maine. They distinguished two populations, one hatched before October and the other hatched after that month (Figure 6.3). This was the first use of the daily rings to establish the dates of hatching from larvae caught much later in their lives. With the lampfish and the northern anchovy, Methot (1981) found little variation in growth rate, for (A_0/a) in the Laird–Gompertz equation was assumed to be a function of temperature, possible but not shown. Growth rate is a function of temperature, but one ought also to establish the ratio of growth rate to maximal growth rate at a given temperature. Checkley's (1984) work on herring larvae showed that the growth rates of well fed and starved larvae were quite different. With the data of Lough *et al.* (1980), Jones (1985) estimated growth rates and distributions of hatching dates of larval herring; she found that those hatched early grew more quickly than those hatched late and that the

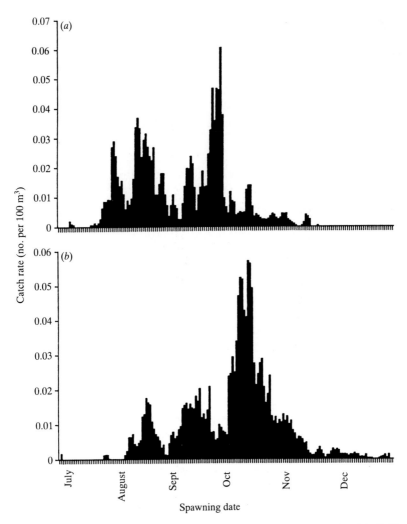

Figure 6.3 The distribution of hatch dates of herring larvae in (*a*) Sullivan Harbor and (*b*) the Sheepscot River in the fall and winter of 1986–7.

difference was the same in two years. Penney and Evans (1985) established differences in growth rate of redfish between two years.

Until 1985 progress in the use of the daily rings on the otoliths of fish larvae was slow. A notable review by Campana and Nielson (1985) summarized the reasons. There are technical difficulties in removing the calcium overburden on the otoliths by sectioning, grinding, polishing and acid etching. Further, the number of rings is sometimes less than the number of days observed in experimental arrangements. The intervals between rings were sometimes so narrow that the rings could not be resolved with a light microscope. With a scanning electron microscope, however, they can be resolved. The important point is that there is a circadian rhythm of calcium deposition under endocrine control (Simkiss, 1974) that might be entrained by photoperiod as a zeitgeber. So the rings are probably laid down each day and the problem is to count them.

Since 1985, work has continued. Victor (1986) found that metamorphosis of a coral reef fish (*Thalassosoma bifasciatus*) was delayed with reduced larval growth. The delay of metamorphosis as function of reduced growth is of much importance. Bailey and Stehr (1986, 1988) reared the larvae of the walleye pollock, *Theragra chalcogramma*, and with the daily rings found differences in growth rate when fed well and starved. Kendall *et al.* (1987), with the same method, revealed differences in growth rate at sea between years in the walleye pollock, 0.09 mm d^{-1} in 1981 and 0.21 mm d^{-1} in 1983. Walline (1983) used the daily rings on the otoliths to establish differences in growth rate between species: *Engraulis encrasicholus*, *Sardina pilchardus* and *Sardinella aurita*. Warlen (1988) made an extensive study of the growth at sea of larval menhaden, *Brevoortia patronus*; of eleven monthly samples between December 1979 and February 1982, there were significant differences between months and years in the constants of the Laird–Gompertz equation, that is differences in growth rate.

Campana *et al.* (1987) showed a linear relationship between number of increments and age, but the residuals were not randomly distributed. They thought that the trouble was that the rings were not resolved by the light microscope; indeed they suggested that the so-called 'hatch check' might comprise a number of narrow increments.

The larvae of *Macruronus novaezelandiae* were examined by Thresher *et al.* (1988); there were significant differences between years in growth rate, together with significant differences between regions. Further, they found that larvae closer to the spawning ground grew faster. Haldorson *et al.* (1989) found significant differences between years in the growth rate of the walleye pollock, but not in the flathead sole (*Hippoglossoides elassodon*). Hovenkamp (1989) examined the growth rate of larval plaice with the Laird–Gompertz equation and there were significant differences between years. Chambers *et al.* (1988), with laboratory reared larvae of the winter flounder (*Pseudopleuronectes americanus*), found that the age at metamorphosis varied between 48 and 60 days (see Chapter 7 for a discussion on work by Cushing and Horwood, 1994). Campana and Hurley (1989) examined the growth of larval cod and haddock in the Gulf of Maine. They wrote that there were no differences in specific growth rates between cruises or regions. But, on Browns Bank in April, there were significant differences between years and in 1985 there were significant differences between banks; in April 1985 there were significant differences between cod and haddock. Hovenkamp (1990) found growth differences in larval plaice with daily growth rings on the otoliths, which were confirmed by RNA/DNA ratios. Karahiri *et al.* (1989) found significant differences in 0-group plaice between years and between the months of May and July. They also established that the first increment was laid down 4 to 6 days before hatching. With the use of the Sr/Ca ratio in herring otoliths, Townsend *et al.* (1989) traced the early temperature history of the larvae and found their time of entry into the Gulf of Maine. They showed that the autumn spawners died off and were replaced by later spawners from offshore. Yoklavich and Bailey (1990) found significant differences between regions in the walleye pollock, but none between years; they also published distributions of hatch dates, some of which are bimodal.

The use of daily rings on the otoliths of fish larvae has not advanced as quickly as might have been hoped. There are technical reasons, for example in the preparation of the otoliths and in the inability, from time to time, to resolve the narrow increments with a light microscope; but there appears to be a belief that the growth rate of fish larvae is constant despite evidence to the contrary. Differences in (A_0/a) in the Laird–Gompertz equation are sometimes attributed to temperature, but are really differences in growth rate. A brief survey of the literature shows that differences in growth rate do occur between regions, between years and between species. If that is so then it would be desirable to increase the sampling rate considerably. It is, however, difficult to increase the sampling rate with the scanning electron microscope because of the cost.

The growth of the adult fish population is well known and samples are stratified by region, season and by length, using long-established methods. The samples of fish larvae are often small, and sporadically arranged in season and in region. If the larvae and early juveniles live for three months, samples might be taken in every week of that period, stratified by region and time of spawning. Then the number of larvae taken would rise from one or two hundred to thousands, and differences in growth rate in the life of the larvae at sea might become properly established.

The distributions of hatch dates have revealed much of interest, particularly on the origin of herring larvae in the bays of the Gulf of Maine. If the sampling effort was increased considerably the distribution of hatch dates as sampled each week would be of great interest, whether they remain the same or whether they shift in time. The comparison of such distributions with those of spawning would be instructive in finding the conditions which encourage survival.

The sampling of fish larvae

To study the growth and death of larvae in the sea, they should be sampled properly at all sizes; for example, Bückmann (1942) sampled herring larvae in the southern North Sea with a Petersen Young fish trawl and caught them up to lengths at metamorphosis, 40 to 45 mm, (as compared with maximum lengths of up to 20 mm with many other gears) but that net was not quantitative. Morse (1989) estimated night/day ratios for many larval fishes from the Marmap surveys on which 61 cm Bongo nets were used; he calculated correction factors for time of day and for length. Somerton and Kobayashi (1989) used analogous methods and found, as might be expected, that the variance increased quite sharply with the night/day ratio. Brander and Thompson (1989) compared three different nets (Longhurst Hardy plankton recorder, Mocness and a pump) in vertical hauls with a night/day ratio; of these, the LHPR was the most successful, catching herring larvae at night at a mean length of about 15 mm. From the International Herring Larval Surveys in the North Sea, Brander and Thompson estimated the night/day ratio from high speed tow nets (Figure 6.4). Not only were one-third caught in daytime at 20 mm, but none were caught larger than 25 mm (the larger herring larvae are now caught in the North Sea with larger nets). Beverton and Tungate (1967) showed that the night/day ratio of plaice larvae remained at unity for all sizes; indeed they grow only to about 12 mm before they metamorphose, so from Figure 6.4 a high speed net might be expected to sample them properly.

Sissenwine *et al.* (1984) listed larval abundances from six or seven years of four species (herring, cod, haddock and silver hake) and showed that the largest year classes were not generated by the largest larval abundances. The larvae were sampled with a Bongo net which tends to catch the smaller animals. The lack of relationship implies that there is

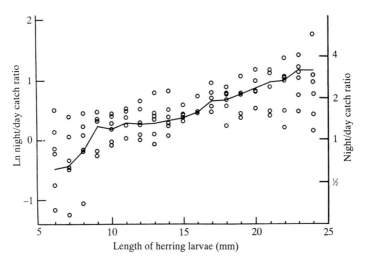

Figure 6.4 Night/day ratios of catches of herring larvae (after Brander and Thompson, 1989).

no relationship between the numbers of early larvae and subsequent recruitment. The numbers of later larvae have not always been well sampled.

The work of Leggett and his colleagues

A small exposed beach called Bryant's Cove in Conception Bay, Newfoundland, is the site where Leggett and his colleagues worked from 1978 onwards. Capelin spawn there on the ebb of a spring tide in summertime and lay their eggs in the sediments. In 1978 and 1979, samples were taken twice a week in daylight at low tide for a period of nearly three months. Figure 6.5(*a*) shows time series of numbers of larvae in the sediments and in the water and Figure 6.5(*b*) gives the concentration of larvae as function of hours since the last onshore wind. Figure 6.5(*c*) illustrates the strength and duration of the onshore wind in 1978 and 1979 (Frank and Leggett, 1983). The concentration of the capelin larvae is correlated with air temperature, irradiance and the onset of onshore winds. The latter disturb the beach sediments and the larvae emerge in the relatively warm water. Frank and Leggett (1981*b*) found that the time to hatching varied with temperature and with position on the beach.

Frank and Leggett (1982) extended Templeman's (1948) thesis that

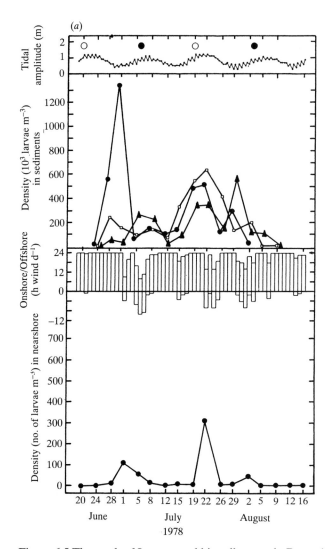

Figure 6.5 The work of Leggett and his colleagues in Bryant's Cove:
(a) time series of catches of capelin larvae at high (●), mid (○) and low (▲) tide
in the sediments and in the water (after Frank and Leggett, 1981a);

the capelin larvae benefited from onshore winds. When the wind blows onshore, the water is warmer and the zooplankton is two to three fold that in the cooler water upwelled by offshore winds. Further, they showed that the number of predators was three to twenty fold greater in the upwelled cooler water than in the warmer water driven by the onshore winds. Frank and Leggett (1983) tackled the question of

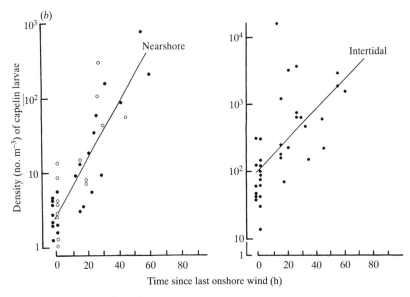

Caption for fig. 6.5 (*cont.*)
(*b*) the density of larvae as function of hours since the last onshore wind (after Frank and Leggett, 1981*a*);

whether the emergence of the larvae was active or passive. Larvae were hatched in tanks flooded with sea water by the tide. They emerged at the same time as those on the beach, during periods of onshore winds, presumably a direct effect of temperature (Figure 6.6).

Leggett *et al.* (1984) noted that the dates of spawning varied from year to year, with consequent differences in dates of hatching. Larval condition varied inversely with beach residence time. They summarized their findings:

YC = 16.1 − 0.19 (Onshore wind) +0.19 (Tempsum),

where YC is year class strength and Tempsum is the sum of degree days (temperature in degrees × days) in the top 20 m between July and December. The result is shown in Figure 6.7.

Thus, a good year class depends on onshore north-easterly winds, which release larvae in good condition into warm water with high zooplankton and few predators. With the luck of good science, the larval niche was well described with a mixture of statistical and experimental procedures.

Frank and Leggett (1984) continued their work by finding that capelin

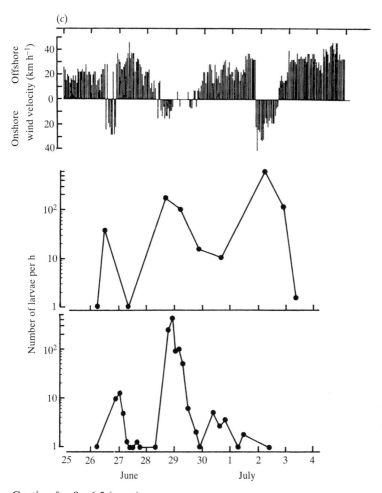

Caption for fig. 6.5 (*cont.*)
(*c*) the strength and duration of the onshore wind in 1978 and 1979 (after Frank and Leggett, 1983).

eggs were a dominant food item in the diet of adult winter flounders (*Pseudopleuronectes americanus*), 1.9% to 5.0%, which contributes 23% to the annual growth of the winter flounder. Subsequently, they found an inverse correlation between the ichthyoplankton and their invertebrate predators over a five-year period (Frank and Leggett, 1985). It was based on the onset of onshore winds and so the correlation does not derive from a predator–prey oscillation, but is probably of evolutionary origin.

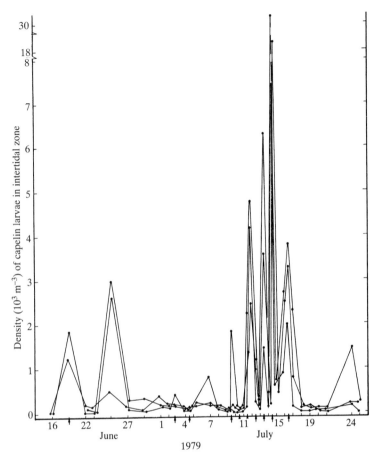

Figure 6.6 The emergence of capelin larvae (after Frank and Leggett, 1981*a*).

The lives of capelin larvae were also observed in the St Lawrence. Fortier and Leggett (1985) conducted a drift study of a patch of capelin larvae, as they followed a drogue through seven distinct water masses; an estimate of mortality was made, corrected for water mass mixture. As distinct from those in Bryant's cove, the capelin larvae emerged at dusk and at dawn, both in the laboratory and in the estuary (Fortier *et al.*, 1987).

Taggart and Leggett (1987*a,b*) studied the loss rates of post-emergent capelin larvae in Bryant's cove with an array of environmental estimates and food and predator density. But the most interesting point was that the mortality of the early larvae was not density dependent, denying

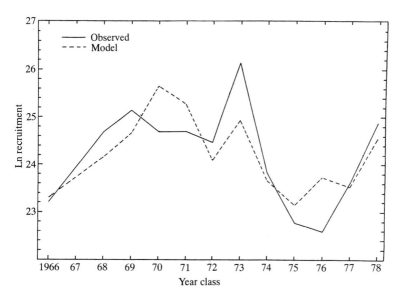

Figure 6.7 The dependence of the recruitment of capelin on the onshore wind and on temperature (after Leggett *et al.*, 1984).

Hjort's first hypothesis that recruitment is generated by the mortality of first feeding larvae, until they learn to feed. Taggart and Leggett (1987*c*), with spectral analyses of wind stress and the density of capelin larvae, found a predominant period of onshore wind stress of 5.3 d. De La Fontaine and Leggett (1988), with mesocosms, investigated the predatory mortality of capelin larvae by jellyfish; they found that it ranged from 5% to 60% and Figure 6.8 shows the seasonal trend in mortality from predation by *Staurophora* and *Aurelia*.

Chambers and Leggett (1987), as noted above, made a study of the growth of larval winter flounders to metamorphosis; in the pleuronectids, metamorphosis can be determined fairly precisely. Eighteen populations were reared in tanks and fed *ad libitum*. Length and age at metamorphosis were correlated, but ages ranged from 44 to 71 days. Growth rates and development rates were correlated, indicating that larvae that grew slowly should metamorphose later; it is possible, but not shown, that the age at metamorphosis was delayed by food lack.

The work of Leggett and his colleagues has advanced our knowledge of larval life considerably. The larvae emerge in different ways in Bryant's Cove (under the onshore winds) and in the St Lawrence (at dusk and dawn) and both mechanisms are active. The eggs are eaten by

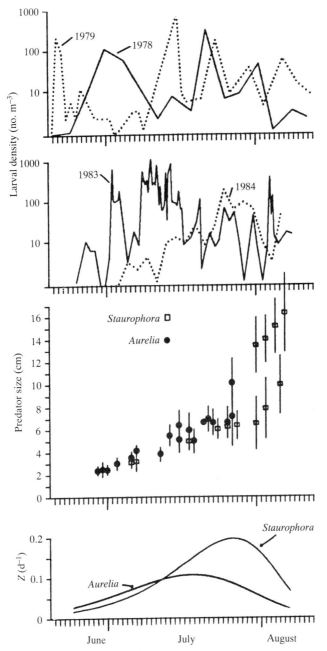

Figure 6.8 The seasonal trend in the mortality of capelin larvae from predation by *Staurophora* and *Aurelia* (de La Fontaine and Leggett, 1988).

winter flounders and the larvae by jellyfish. They showed that the early mortality was not density dependent and that the age at metamorphosis was very variable. Their major achievement was to construct a model of the generation of capelin recruitment based on the evolutionary choice of the onshore winds by the larvae.

The work of Lasker and his colleagues

Much (but not all) of Lasker's work at sea took place in Lasker's Lake a few miles off Point Lomas in San Diego, Southern California. Beers and Stewart (1967, 1969) had described the distribution of microzooplankton in the euphotic zone in the California Current. Lasker *et al.* (1970) reared the larvae of the northern anchovy, *Engraulis mordax* Girard, in the laboratory and they found that the density of larval food needed to obtain moderate larval growth was higher than that usually found at sea by Beers and Stewart. Lasker (1975) proposed four criteria for the successful survival of the anchovy larvae:

1. According to Berner (1959), 70% of the food taken should comprise particles of 60 to 80 μm in diameter.
2. The initial capture efficiency of the anchovy larvae was as low as 10% and they take about 30 days to fully learn to feed (Hunter, 1972*b*) and so high densities of particles at sea are needed for survival.
3. Lasker *et al.* (1970) showed that the anchovy larvae survived well on a diet of the dinoflagellate *Gymnodinium splendens*.
4. Hunter and Thomas (1974) described the feeding and searching behaviour of the larval anchovy and noted that food organisms of the right size and density were needed.

Lasker took the larvae to sea and fed them in various waters. Nauplii were not abundant enough, but for about 100 km between Malibu and San Onofre (in California), the chlorophyll maximum layer comprised mainly *G. splendens*, organisms of 40 μm in diameter at a density of 20 to 40 ml^{-1}. Figure 6.9 shows the layer off San Onofre before and after it was broken by a storm. Hence it appeared that the larval anchovies needed a stable chlorophyll layer for survival; yet that layer should not be composed of *Chaetoceros*. This is the basis of Lasker's stability hypothesis.

Lasker (1981) ranked the anchovy year classes between 1962 and 1977. The best year class, 1976, was hatched in stable seas and the worst, 1975, was hatched during a period of strong upwelling when *Gonyaulax*

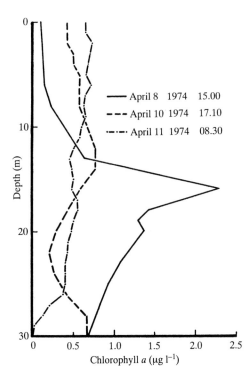

Figure 6.9 The breakup of the chlorophyll layer off San Onofre by a storm (after Lasker, 1975).

polyedra predominated. But Scura and Jerde (1977) had shown that the larval anchovy did not eat this dinoflagellate. Lasker wrote 'stability of the ocean's upper layers during the anchovy spawning season is essential for larval fish aggregation, a prime requisite for larval anchovy survival and thus a successful year class'.

Turbulent mixing occurs at wind speeds greater than 10 m s^{-1}, and an index of calmness comprises four consecutive days of wind speed less than 10 m s^{-2}, now called a Lasker event (Peterman and Bradford, 1987). Peterman and Bradford used larval mortality rates between the ages of 5 to 19 days (4 to 10 mm in length) for the period 1954 to 1984 to relate to that index. Such calm periods allow patches to gather for a short time. In the period of examination, there were 13 years of samples with egg abundances in all months of spawning. Figure 6.10 shows the reduction in larval mortality as function of the number of calm periods per month. The thesis is a very simple one, that the early larvae of the

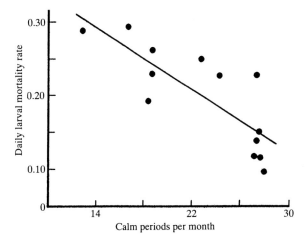

Figure 6.10 Reduction in larval mortality of the northern anchovy as function of Lasker events (after Peterman and Bradford, 1987).

northern anchovy survive better on *Gymnodinium splendens* in the stable waters of Lasker's Lake. Indeed, mortality doubles as the number of Lasker events is halved (Figure 6.10). This is really an effect of the presence or absence of desirable food. Peterman *et al.* (1988) showed no correlation between numbers of 19-day-old larvae and the number of subsequent recruits; but, as shown in the last chapter, Cury and Roy (1989) demonstrated an inverse relationship between turbulence (in this case, wind stress greater than 5 m s^{-1}) and Lasker events.

Butler (1991) has examined the larval mortality of the Pacific sardine, *Sardinops sagax caerulea*, and the northern anchovy, *Engraulis mordax*, in more detail. Larval ages were back-calculated from growth rates estimated from the increments on the otoliths of juveniles. Mortality was estimated from the numbers of larvae produced at successive intervals of length, with a Pareto hazard function:

$$y_t = y_u \, (t/u)^{-\beta},$$

where u is the time of hatching and y_u, the larval production at that time, t is the time of observation and y_t is the larval production at that time, and β is the instantaneous mortality.

Figure 6.11(a) shows the mortality rate of both species from 1951 to 1967; Figure 6.11(b) illustrates the dependence of sardine mortality rate on anchovy biomass. During this period, in general one of stronger upwelling, the anchovy stock was increasing after the collapse of the

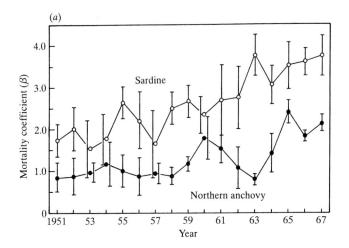

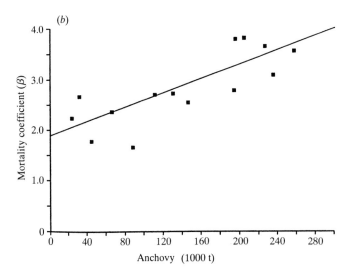

Figure 6.11 (*a*) mortality rates of sardines and anchovies from 1951 to 1967; (*b*) dependence of sardine mortality on anchovy biomass (after Butler, 1991).

sardines. Lasker's stable ocean hypothesis was not confirmed for the sardine larvae by Butler (Peterman and Bradford worked on the northern anchovy), nor was Hjort's transport hypothesis. Sardine recruitment at age two was not correlated with the larval mortality rate, but it was correlated inversely with anchovy biomass and with the combined biomasses of sardine, anchovy and mackerel, that is total pelagic predation. In the previous chapter, the survival of larval sardine and larval anchovy was linked to the nature and the strength of upwelling, anchovy preferring the stronger and intermittent upwelling and the sardine, the weaker and more persistent form. The sardine recruitment is possibly controlled by the total pelagic biomass, but the recruitment of the anchovy may still be partly controlled by Lasker events.

The work of Solemdal and his colleagues on the Arcto–Norwegian cod

From 1979 to 1985 extensive surveys of the larvae and early juveniles (about two months old) were made between the Vestfjord (in northern Norway) and the Barents Sea. Ellertsen *et al.* (1987) showed that the adults spawned on average on 31 March; they spawn in the subsurface thermocline where the temperature does not vary very much from year to year. The larvae were sampled with vertically hauled nets from a depth of 50 m and a large pump; the early juveniles were sampled with midwater trawls. The cod eggs develop in 15 to 40 days, according to temperature. In 1981 and 1983, the distribution of the first feeding larvae was centred on a date about a month later than that of the spawning distribution.

Figure 6.12 shows the dependence of the date of peak abundance of

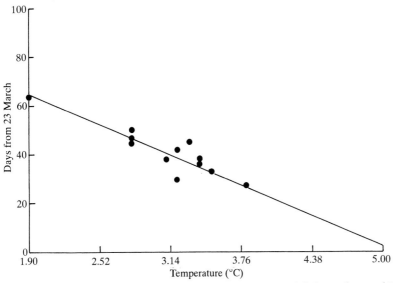

Figure 6.12 Dependence of the date of peak abundance of *Calanus finmarchicus* on temperature between 1 April and 24 May (after Ellertsen *et al.*, 1987).

copepodite stage 1 of *Calanus finmarchicus* on temperature between 1
April and 24 May; this range of nearly two months is much more than
that expected from the dependence of development rate on temperature. Temperature may be a proxy for a set of processes, including
the rise of *Calanus* from 600 m in spring and for the time of onset of the
spring outburst. Figure 6.13 illustrates the dependence of the numbers of
prey organisms per larval gut on the number of nauplii l^{-1}, in the form of
an Ivlev-like curve. At a density of 5 to 10 nauplii l^{-1} we may assume
that the larvae are adequately fed.

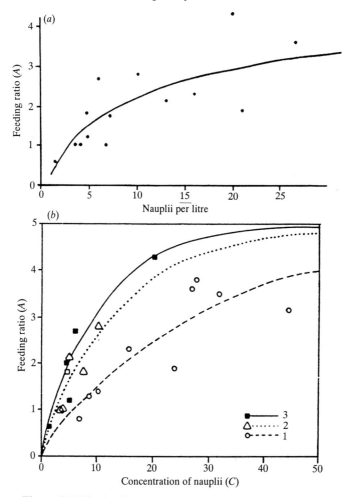

Figure 6.13 The feeding of cod larvae: (*a*) dependence of numbers of prey/gut, or
feeding ratio, on the number of *Calanus* nauplii l^{-1} (Ellertsen *et al.*, 1989); (*b*)
feeding ratios at three levels of turbulence, 1 low, 2 medium, and 3 high.

Table 6.1. *Dry weight at 6 mm in length, temperature and the Julian date of peak abundance of* Calanus *stage 1 copepodites*

	Dry weight at 6 mm in length	Temperature (°C)	Peak date of abundance of *Calanus* stage 1 copepodites
1982	98.75	2.10	149
1984	107.35	3.05	123
1985	120.00	3.50	114
1983	146.33	3.60	109

After fixation, the guts and liver were removed and then the larvae were dried. The dry weights differed from year to year as a function of length. Table 6.1 gives the dry weights (μg) at 6 mm in length, the temperature and the date (in Julian days) of peak abundance (Cushing, 1990*b*).

The higher the temperature, the earlier the peak date of abundance of *Calanus* stage 1 copepodites and the greater the dry weight at a length of 6 mm. This suggests that the relationship shown in Figure 6.13 is based on the generation of larval food during the spring outburst. As noted below, the 1983 year class was a strong one.

Figure 6.14 (after Ellertsen *et al.*, 1989) shows the match or mismatch of larval production to that of their food, with data, for six years. It shows the percentage distribution in time of the cod eggs, the percentage distribution in time of first feeding larvae, and the distribution in time of the density of nauplii l^{-1}. The figure also shows the density of 5 nauplii l^{-1} as an index of adequate feeding. The two mismatched year classes, 1960 and 1981, turned out to be poor ones. That of 1980 was also poor, perhaps because of the passage of the great salinity anomaly of the seventies, the Great Slug. The other three year classes (1983, 1984 and 1985) appear to be matched to the same degree, but the 1983 class was twice as strong as the other two.

Bjørke and Sundby (1987) sampled the early juveniles and found that many of them were retained over the Tromsøflaket in the southern Barents Sea. They also established that the numbers of early juveniles were correlated with the number of 0-groups, which are themselves correlated with the number of three-year-old recruits (Figure 6.15). Thus, many of the processes by which recruitment is generated were completed by the age of two months, the early juveniles.

Sundby *et al.* (1989) showed that between the early juveniles and the

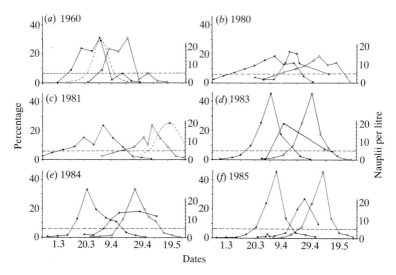

Figure 6.14 Match or mismatch of larval production to that of their food in the Arcto-Norwegian cod; percentage distribution of newly spawned eggs (●), of first feeding larvae (○) and of the production of copepod nauplii (■). The dotted horizontal line represents a density of 5 nauplii l^{-1}, at which the cod larvae are quite well fed (after Ellertsen *et al.*, 1989).

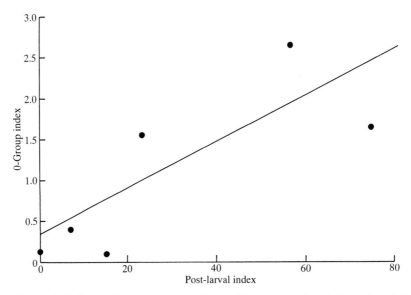

Figure 6.15 Correlation between numbers of 0-groups and early juveniles of the Arcto-Norwegian cod (after Sundby *et al.*, 1987).

0-groups, the mortality was density dependent (Figure 7.3). The conclusion from this remarkable work is that recruitment is in the main determined by the age of two months, that there is a little evidence that the production of cod larvae should be matched to the abundance of their food and that the most successful year class, out of four, grew fastest when the production of the stage 1 copepodites of *Calanus* was early (with relatively high temperature). Again the most important point is the presence or absence of food in the system.

The match/mismatch hypothesis

The original thesis was based on two observations: (a) spring and autumn spawning herring released their larvae in spring or autumn outbursts as shown by the 'greenness' of the plankton recorder material (a method validated by Gieskes and Kraay, 1977) (Cushing, 1967), (b) Cushing (1969) showed that the herring, plaice, sockeye salmon and cod spawned at the same time each year and that the peak date of spawning had a standard deviation of about a week. The period of spawning may last for two or three months; Rothschild (1986) noted that a spread of two standard deviations may describe the average of evolutionary processes, but that the standard error of the mean indicated the most important time of spawning. Recently, Page and Frank (1989) have shown that the peak spawning date of haddock on George's Bank is inversely related to temperature and that the differences from year to year amount to two or three months. On Brown's Bank, however, the fish spawn at the same time each year and there is no relation with temperature. Hutchings and Myers (1993) show that the timing of cod spawning on the northern and southern Grand Bank varied in response to temperature.

Colebrook (1965) showed that the time of onset of the spring outburst (from the continuous plankton recorder material) can vary in the North East Atlantic by up to four to six weeks from year to year. Dickson *et al.* (1988*a*) found that there were trends of that order from decade to decade in the North Sea and the North East Atlantic. Then, if fish spawn at the same time each year, the chances of their larvae finding food must vary from year to year. It is obvious from the last chapter that fish stocks respond to climatic change, sometimes dramatically. The climatic factors can act directly as for example, the survival of cod of all ages off west Greenland is directly affected by temperature (Hermann, 1967; Hovgård and Buch, 1990). But more generally, if climatic change is to affect the recruitment of fish stocks, it would work most efficiently during larval life, through differences in the times of onset of spring or autumn bloom or an upwelling plume. The Sverdrup mechanism, cited by Dickson *et al.* (1988*a*) to explain the delay in production in the western North Sea between the fifties and the seventies and its reversal in the eighties, suffices to explain the changes in time of onset of the spring bloom, either from year to year or in a trend.

The original thesis was restricted to the spring and autumn outburst of temperate seas, but Cushing (1990*b*) extended it to tropical and subtropical seas, including the upwelling areas. Spring outburst and upwelling are analogous processes, but the nature of reproductive effort in the waters equatorward of 40° lat. is quite different from that in waters towards the pole. Many subtropical fishes produce many batches of eggs in a year. In the stickleback (*Gasterosteus aculeatus* L.), Wootton (1977) showed that the interval between batches varied inversely with ration. So in upwelling areas more eggs would be produced from well-fed parents.

When recruitment is related to indices of upwelling, positive correlations are assumed to originate in the availability of food and negative ones are assumed to stem from offshore drift. This assumption is too simple for two reasons. The first is that production is greater when the upwelling velocity is low. Cushing (1971) developed an equation to express the production in the rising water, based on the Steele and Menzel equation for production at depth. It was shown that this production was inversely related to the ascending velocity, that is, the slower the upwelling velocity, the greater the production. So the decoupling between production and consumption is maximized.

The second reason for finding complex relationships between recruitment and indices of upwelling is that with increasing turbulence, recruitment was reduced from a peak (Figure 5.21, Cury and Roy, 1989). Thus the positive correlations between recruitment and upwelling may well depend upon the availability of food, but the negative ones may be due to offshore drift, to turbulence or to a high upwelling velocity. As noted in the last chapter, Bakun and Parrish (1980) and Husby and Nelson (1982) both refer to the slow upwelling habitat preferred by sardines.

A number of correlations have been published between recruitment and indices of upwelling. The stocks include the sardine (Bakun and Parrish, 1980), Pacific mackerel (Parrish and MacCall, 1978), Dungeness crab (Peterson, 1973; Botsford and Wickham, 1975) and hake (Bailey,

1981). The correlations are simple, multiple, positive or negative. Because the upwelling system is complex, the array of correlations imply that the subtropical species in upwelling areas match their production to that of their food.

The hypothesis of Cury and Roy has been extended by Ware and Thomson (1991). They linked the Marr (1960) and MacCall (1979) series of recruitments to the Pacific sardine population off California between 1932 and 1960. With distance-weighted least squares, three complex regressions were established: (a) diatoms sampled off Scripps' Pier on mean wind speed in m s^{-1} (April, May and June), which peaked at 7 m s^{-1} ($p = 0.03$), (b) recruitment/stock on diatoms ($p = 0.05$), and (c) recruitment/stock on mean wind speed in April, May and June ($p = 0.01$). The optimal window in the sense of Cury and Roy peaks at 7 m s^{-1}. The survival of the Pacific sardine appears to depend on an optimal wind speed for upwelling, the consequent production of diatoms and presumably the production of larval food.

Ware and Thomson also examined the long-term time series of fish scales in the anoxic sediments off California (Soutar and Isaacs, 1974, as revised by Smith, 1978). Dunbar and Wefer (1984) recorded the quantities of $^{18}O_2$ in the shells of *Globigerina bulloides*: cool water or strong upwelling is indicated by positive values. The time series from 1750 to 1970 shows the variation in upwelling, which is correlated with hake/sardine biomasses with a five-year lag. As the biomasses of the two stocks are themselves positively correlated, both populations must depend upon the upwelling as long as it is not too strong.

Cushing (1989) suggested that there were two primary ecosystems in the ocean, one based on the very small grazers of the microbial food loop and the other based on larger food particles ($<5\,\mu$m in diameter). The microbial food loop is characteristic of the oligotrophic ocean and the well-stratified waters of seas and lakes in summer. The second system is the traditional food chain, diatom/copepod/fish, which is found where the water is weakly stratified in the early stages of a spring bloom, in the later stages of an autumn bloom, in an upwelling plume or in an oceanic divergence. Because fish larvae take organisms greater than 5 μm in diameter, the regions of the microbial food loop are, in general, not part of our concern. This includes the oligotrophic ocean or the well-stratified waters of the temperate summer; it might also apply to the poorer regions of the Antarctic and the apparently stable system of the subarctic North Pacific.

Brander and Hurley (1991) make a distinction between (a) the timing

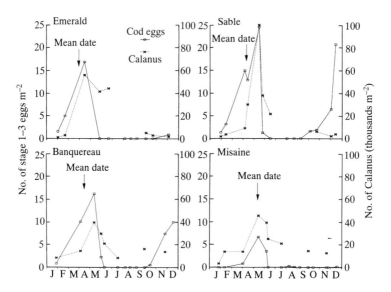

Figure 6.16 The spawning of cod and the production of *Calanus* off Nova Scotia (after Brander and Hurley, 1991).

of spawning, which is coupled to the timing of plankton production as in the original study of the spring and autumn spawning herring, and (b) the variability in the timing of plankton production which may be linked to variation in recruitment. Figure 4.4 shows the spawning distributions (from egg surveys) of seven species in the Southern Bight of the North Sea, plaice (*Pleuronectes platessa* L.), cod (*Gadus morhua* L.), sandeel (*Ammodytes* spp.), whiting (*Merlangius merlangus* L.), bib (*Trisopterus luscus* L.), flounder (*Pleuronectes flesus* L.) and dab (*Limanda limanda* L.). The figure also shows the development of 'greenness' from the continuous plankton recorder material, 1948–62 and 1963–83. All seven species spawn when their larvae will on average find food from the spring outburst. In Figure 6.16 are shown the distributions of cod eggs and *Calanus finmarchicus* (averaged for three years) on four banks on the shelf off Nova Scotia in spring and autumn; again, the cod laid their eggs to give their larvae the best chance of finding food (Brander and Hurley, 1992). Thus, the spawning of some fish appears to be linked to the timing of plankton production. Myers *et al.* (1993) found that cod in the North West Atlantic spawned at the peak time of *Calanus* production.

Earlier in this chapter, the work of Leggett, Lasker and Solemdal and their colleagues was described in some detail. Common to all three considerable studies is the conclusion that larval mortality is modified by the presence or absence of food, in the onshore winds in Bryant's Cove, in Lasker's Lake or in the *Calanus*-rich waters of the southern Barents Sea. In the last chapter the gadoid outburst and the effects of the great salinity anomaly of the seventies were described. The increment in gadoid stocks during the sixties may well have been associated with the delay in the production of the stock of *Calanus finmarchicus* in the North Sea; it is possible that the gadoid stocks moved from a mismatched state to a matched one. It will be recalled that Dickson *et al.* (1988*a*) showed that the zooplankton in the North Sea and the North East Atlantic declined between the fifties and the seventies under an increase in the northerly wind which delayed the spring outburst. The gadoid outburst was generated by this delay, and after 1961 the stock of zooplankton peaked about a month later. The relationship decayed between 1970 and 1975 and the cause is uncertain. The most important point here is that a change in the pattern of recruitment to the North Sea stock was the result of a change in the pattern of the production of *Calanus* in the North Sea, from mismatch to match. During the passage of the Great Slug the recruitment to a number of fish stocks was significantly reduced, possibly because the production cycle was delayed by the reduced temperature, possibly delaying the onset of stratification. In the Arcto-Norwegian cod stock, some evidence was produced to show that matches and mismatches occur, the latter being related to poor year classes. Delay in the production of zooplankton may have increased gadoid recruitment from 1962 onwards for a short period and delayed production during the passage of the Great Slug may have reduced recruitment to many stocks in the North Atlantic. In the Arcto-Norwegian cod stock, mismatch generated two poor year classes and match produced one high one.

The most important point of all is whether recruitment is indeed generated by processes of this form. First, recruitment has been forecast at the larval stage (Burd and Parnell, 1971, on the Downs stock of herring; van der Veer, 1986, on the numbers of larval plaice in the flood waters of the Wadden Sea, although this series does not include the 1969 year class which as late larvae was very strong, but average as 0-groups and recruits). However, Parmanne and Sjøblom (1984, 1987) sampled herring larvae with a Gulf V net at seven stations in four areas off the coast of Finland, each week from May to August between 1974 and 1986. Table 6.2 shows the correlations between numbers of larvae and

Table 6.2. *Correlation of numbers of herring larvae in June off the coast of Finland and subsequent recruitment*

	Statistical areas		
Length	29	30	32
<10 mm	0.714++	0.492++	0.209
10–15 mm	0.644+	0.480	0.798+
>10 mm	0.649+	0.464+	0.828++
>15 mm	0.632+	0.240+	0.792++

Note:
$+ \ p < 0.05, \ ++ \ p < 0.01$.

subsequent recruitment (estimated by virtual population analysis).

Thus, there is a link between recruitment and the numbers of herring larvae at the age of one or two months. The work of Leggett and his colleagues was described earlier: under the onshore winds, the capelin larvae emerge in the relatively warm water where there is more zooplankton for the larvae to eat. In a simple way, recruitment depended on the frequency of onshore winds and temperature, but as shown above, Leggett and his colleagues analysed the processes in great detail.

The recruitment to the Arcto-Norwegian cod appears to be determined at the age of about two months; the mechanisms have not been fully established, but they appear to be linked to the time of onset of the spring outburst and/or the ascent of *Calanus* from their overwintering depth, either of which may appear to depend on temperature.

Bollens *et al.* (1992) tested the match/mismatch hypothesis in a study of ichthyoplankton, potential larval food and invertebrate predators at two stations in Dabob Bay, a temperate fjord in Washington State in the US. The spring spawning icthyoplankton peaked a little before and during the spring outburst (Figure 6.17). Invertebrate predators became abundant later in June, July and August, which led to the attractive suggestion that the fishes spawned in spring to avoid the summer predation on the larvae; indeed, as evolution achieves its multiple ends, this must be true. Figure 6.17 shows the distribution in time (in 1985–1987) of the ichthyoplankton as abundance and as taxa richness and of the larval food ($>73 \, \mu m$ to $216 \, \mu m$). The larval food distributions peaked after those of the ichthyoplankton and taxa richness and indeed,

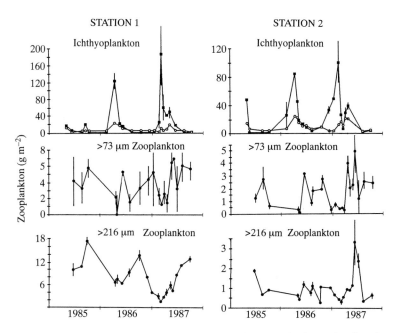

Figure 6.17 The occurrence of ichthyoplankton and the reduction in zoo-plankton in Dabob Bay, Washington. ■, Abundance (nos./100 m³); ○, taxa richness (Bollens *et al.*, 1992).

the latter was inversely correlated with the larval food. A glance at Figure 6.17 suggests that the ichthyoplankton might have eaten the larval food.

Figure 6.18 shows the contrast between strong year classes in cold water and poor ones in warm water for four tuna groups in the subtropical ocean. The tuna tend to live in the peripheral regions of the subtropical gyres where the wind stress is greatest and where the relatively cool water at the surface indicates divergences and the presence of food. Matsumoto (1966) and Nishikawa *et al.* (1985) have shown that the tuna larvae are distributed across the subtropical Pacific for nine months of the year; the adults may well spawn and feed at the same time.

The spring outburst off Maria Is., in western Tasmania, varies in time of onset from year to year by as much as three-and-a-half months as the atmospheric high over Western Australia shifts its position, often linked to ENSO events. Figure 6.19 shows the zonal westerly winds over Tasmania from 1945 to 1985, the time of onset of the spring outburst off

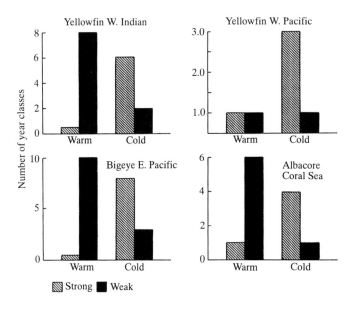

Figure 6.18 The occurrence of poor tuna year classes in warm water and rich ones in cool water in the subtropical ocean (after Yamanaka, 1978).

Maria Is. (between September and December) and the catches of spiny lobsters lagged by seven years (Harris *et al.*, 1988). The catches (in differenced observations) off Tasmania were correlated with late spring blooms, but those off New South Wales were linked to early blooms. The significant cross-correlations were as follows:

	r	lag yrs
Zonal westerly winds/timing of spring outburst	0.50	
Timing of spring outburst/maximal temperature Maria Is	0.70	
Maximal temperature Maria Is/Tasmanian lobster catches	0.37	7
Maximal temperature Maria Is/N.S. Wales lobster catches	0.35	5

Thus, recruitment of spiny lobsters in Tasmania and New South Wales depends lightly upon the timing of the spring outburst at Maria Is. The range of timing is high, the time series is long and only a small proportion of the variance is accounted for. The conclusion is that the process probably exists, but less variable material would be needed to elicit it.

Mertz and Myers (1994) have made a model of match/mismatch which

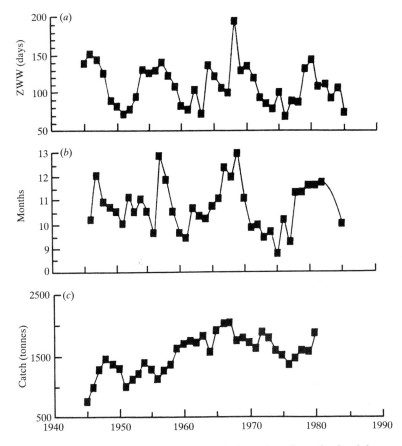

Figure 6.19 The link between westerly winds and catches of spiny lobsters: (*a*) changes in zonal westerly winds between 1945 and 1985; (*b*) dates of onset of the spring bloom at Maria Is. off Tasmania; (*c*) time series of catches of spiny lobsters of Tasmania lagged by seven years to indicate the timing of the spring bloom (after Harris *et al.*, 1988).

describes the window of spawning and that of food production. They used the Calanus V–VI material from the continuous plankton recorder for a period of ten years to show a weak link between the spawning window and subsequent recruitment.

There have been five steps in the development of the hypothesis: (a) the fixed season of spawning and the variation in time of onset of the spring bloom; it is likely that the timing of cod spawning is not as precise in the North West Atlantic as at Lofoten, (b) the link between spawning distributions and those of food for the larvae, (c) the demonstration that the presence or absence of food is essential to the generation of recruitment, (d) the fact that delays actually have occurred, and (e) one or two demonstrations that the processes governing recruitment could be described.

The variation in reproductive capacity

Rijnsdorp *et al.* (1991) studied the changes in reproductive capacity of plaice, sole and cod in the North Sea. They made use of time series of growth maturation and fecundity between the late fifties or early sixties and the present day. Beverton and Holt (1957) found that the plaice of 1945–6, aged between 10 and 13 years, weighed up to 20% less than those in the thirties. In recent decades the range of stock sizes has been low, ×2 for plaice, ×3 for cod and ×5 for sole. The soles increased in length between the late fifties and early seventies by nearly 20%, perhaps because the use of the beam trawl may have increased the numbers of polychaetes, their preferred food (de Groot, 1971; de Veen, 1976; Reise, 1982). Houghton and Flatman (1984) published evidence of density-dependent growth in cod. The explanations for this differ, that the juveniles compete between year classes and that juveniles compete with adults. Rijnsdorp *et al.* found little evidence of density-dependent growth in the three species; in plaice, however, the increment in length in the second year of life was inversely linked to recruitment.

The age of maturation might be expected to depend on recruitment, but it did not. In plaice it depended to some degree on somatic growth and the percentage mature was inversely linked with the length at four years of age. So what evidence there is from the 30-year time series is that the age of maturation may change in response to food. The evidence on fecundity is not very clear, but as Horwood *et al.* (1986) suggested, size-specific fecundity increases as conditions for growth improve. But the three stocks have decreased by factors of 2 to 3 during the period. In cod, the reduction was compensated by an increment of smaller eggs of 16%, in plaice by an advance in maturation of 19% and in sole by a small proportion due to growth and to maturation. But the sole stock changes most by a factor of 5. The most important conclusion of Rijnsdorp *et al.* is that the variation in population fecundity spreads the risk in an adverse environment and that density-dependent effects are of less importance.

Food and fecundity

Recruitment might well be affected by the initial number of eggs spawned. The possibility of density-dependent fecundity and changes in the length or age of maturation will be discussed in the next chapter. Here more direct effects are examined. Many studies have been made on this subject, experimental (Alm, 1946), in lakes and in the sea, but we shall restrict our view to some fish species in the North Sea.

Rijnsdorp *et al.* (1983) found that the fecundity of plaice in the late seventies and early eighties was much greater than during the fifties and in the first decade of the century. Horwood *et al.* (1986) investigated the fecundities of plaice on five spawning grounds: eastern English Channel, Southern Bight, central North Sea, German Bight and Flamborough in 1977 to 1980. With ANCOVAs, significant differences in fecundity were established between 1979 and 1980, ×1.44 in the German Bight and ×1.15 in the Southern Bight, more than might be expected from stock changes from year to year. Fecundities at all lengths and ages were compared between 1947 to 1949 and 1977 to 1980 and an increment of ×1.60 was found in the Southern Bight and of ×1.36 on the Flamborough ground. During this period, the total North Sea stock *increased*, mainly in the eastern North Sea; it should be recalled that there is much migration between grounds in each season.

Rijnsdorp (1990) used the results from tagging experiments to analyse the allocation of energy to somatic growth, reproductive investment and surplus production. If somatic growth is doubled, condition rises by less than 10%, which suggests that surplus production in the female does not affect size-specific fecundity or reproductive investment. So it is growth at the juvenile ages which is important.

Rijnsdorp *et al.* (1991) examined the vital parameters of three stocks in the North Sea, plaice, sole and cod, for the period 1958 to 1988. In the plaice there was a systematic increase in the length of four-year-olds and a slight decrease in that of eight-year-olds. The length increment of one-year-old fish (as at 10 cm in length) depended inversely on recruitment, but this might have been due to the shorter growing season with strong year classes (Zijlstra *et al.*, 1982). The proportion maturing at four years of age depended positively on the somatic growth of the juveniles, but not on year class strength. Further, the relative fecundity (fecundity per g wt) differed between areas and years, possibly linked to the pre-spawning condition factors.

The lengths of female soles (aged 4, 6 and 8 years) increased from

1958 to the seventies and subsequently they remained steady. de Veen (1976) had attributed the enhanced growth to more food organisms, in particular polychaetes, as a result of beam trawling, which stirs the sea bed. The lengths of the juveniles did not depend on year class strength, but the adult condition factors were inversely correlated with adult biomass. In the cod, there were no trends in somatic growth of adults or juveniles or in the age at first maturity. However, relative fecundity increased as the spawning stock declined by a factor of 3. The annual feeding level did not appear to change much during the period of stock decline.

The object of this work on three major fish species in the North Sea was to investigate density dependence. Changes in the growth rate of juvenile plaice and adult soles were more important than changes in density. The relative fecundity in cod increased with decreased stock with no changes in growth or maturation. Across the broad North Sea there must be many differences between areas in the food distributions which provide opportunities for changes in growth or fecundity independent of density.

Conclusion

Three technical problems have been discussed: the effect of temperature on growth, the study of daily rings on the larval otoliths and the sampling of fish larvae. Five approaches to the study of recruitment have been presented: the work of Leggett, Lasker and Solemdal and their colleagues, the match/mismatch hypothesis and the variation in reproductive capacity.

Crucial to the work of Leggett, Lasker and Solemdal is the proposition that the presence or absence of food is essential to the generation of recruitment, but physical processes play essential parts: the frequency of onshore winds in Bryant's Cove, the occurrence of calm periods in Lasker's Lake and the physical events that determine the production of *Calanus finmarchicus* in the sea of Solemdal.

Differences in reproductive capacity from year to year revealed unexpected results, that considerable differences in fecundity occur from year to year, that large changes in fecundity occurred between the fifties and the late seventies with an increase in stock, and that changes in juvenile growth rate and in adult growth rate played predominant parts in determining the fecundities of plaice and sole. Differences in fecundity might well be generated by food differences across broad areas.

7

Density-dependent processes

Introduction

In the next chapter the stock recruitment relationship is discussed. Here one point only will be made, that most such relationships based on long data sets in exploited stocks imply strong density dependence (Shepherd and Cushing, 1980). Indeed, the existence of density dependence follows from the fact that stock numbers often remain steady with relatively low coefficients of variation (although, of course, collapses and dramatic rises do occur as described in Chapter 5). The necessity of density dependence was stated by Haldane (1953) and by Moran (1962).

Let $\delta N = B - D + I - E$,

where N is the number in the population, B is the number of births y^{-1}, D is the number of deaths y^{-1}, I is the number of immigrants y^{-1}, E is the number of emigrants y^{-1}. Then $\delta N/N = b - d + i - e$ in relative rates of change (b is the relative rate of change in the number of births y^{-1}, d in the number of deaths, etc.). If $\delta N/N$ is to approach zero over a number of generations, some or all of the variables must be functions of N unless $(b - d + i - e)$ is always identically zero, which is unlikely. Thus, a population persisting for many generations must be regulated by density dependence. It would be desirable to identify the sources of density dependence independently of the averaged estimates from a stock/recruitment relationship. Traditionally, in fisheries research, five forms of density dependence have been proposed. The first two are associated with predation on the eggs, larvae and juveniles. Predators might aggregate on groups of fishes of any age, but such processes have not been shown very often; the aggregation of trout on the fry of the Pacific salmon as they emerge from river to lake was proposed by Ricker (1958) as a cause of density-dependent mortality and was part of his derivation of a stock recruitment relationship. Cushing (1955) showed that herring aggregated on to patches of *Calanus*; on a broad scale of about 100 km, aggregation took about three weeks after which the fish dispersed, a slow process. More usually, the predator is considered to take its prey adventitiously. The Ricker–Foerster thesis (Ricker and Foerster, 1948) states that larvae or juveniles which grow at less than the maximal rate suffer predation for longer. The growth of fishes is potentially density-dependent, possibly leading in the adults to density-dependent fecundity and in juveniles to delay in the age of maturation. The five forms of density dependence, aggregative predation, adventitious predation, density dependent fecundity, age at first maturation and the Ricker–Foerster thesis have all been a little difficult to establish, despite the strong density dependence revealed in the stock recruitment relationship. Fish can die from causes such as disease, cold water or old age. In very old fish the myofibrils in the white muscle become thinner, reducing the attack capacity; 'watery plaice' were found off Newfoundland (Templeman and Andrews, 1956), thin and bony plaice were caught on the Dogger Bank in the North Sea (Garstang, 1900–3) and in the Skagerak 'praeste flynderne' (priestly flounders) were found (Petersen, 1894). So senescence exists and in an unexploited stock would serve to end the life of a cohort.

Reddingius (1971) pointed out that if a population proceeded by a random walk with no density dependence, it must eventually become extinct. Hence, persistence implies density dependence and the maintenance of some form of equilibrium. Reddingius also pointed out that only a weak density dependence was needed to establish equilibrium. Sinclair (1988) gave a brief history of the development of concepts from Howard and Fiske (1911) to Smith (1935), Solomon (1949) and Lack (1954). May (1977) wrote that in terms of the logistic equation, $(N = rN (1 - N/K)$, where N is number in the population, r is the intrinsic rate of increase and K is the carrying capacity), K is the equilibrium to which numbers, N, return after a perturbation; if $N < K$, dN/dt is positive and if $N > K$, dN/dt is negative.

In the last decade or so, there has been much discussion on the detection of density dependence (Bulmer, 1975; Royama, 1977; Dempster, 1983; Hassell, 1985, 1987; Gaston and Lawton, 1987; Pollard and Lakhani, 1987; Mountford, 1988). Bulmer's method is autoregressive and stationarity is demanded of the population. In many studies, much use was made of the equation $N_{t+1} = \alpha N_t^\beta$, where N_t and N_{t+1} are numbers in successive generations, where α provides an estimate of density independent survival and β is a coefficient of density dependence (survival, $N_{t+1}/N_t = \alpha N_t^{\beta-1}$). It is a somewhat unsatisfactory form of stock recruitment relationship (see Chapter 8). If density dependence is weak there are statistical problems in estimating β. Varley and Gradwell (1960, 1968) had suggested that two regressions were needed, of ln survival against ln density (where the slope should be significantly less than one) and of ln density against ln survival (where the slope should be significantly greater than one), so that the sampling errors in both components were minimized. Bartlett (1949) and Ricker (1975) devised regressions in which errors were distributed about both axes so the Varley–Gradwell procedure was not needed. A second equation used is $M = a + b N_t$, where M is the instantaneous mortality rate, a estimates density independent mortality and b is a coefficient of density dependence; $M = \ln (N_{t+1}/N_t)$, so there is an element of bias in the procedure. Further, it is usually assumed that the number of predators remain constant during the period of examination.

Recently, Hassell *et al.* (1989) relaxed the Varley–Gradwell conditions and have shown that of the mortalities from 63 data sets of 58 insect species, more than half were density dependent. But the most interesting point was that density dependence was more prominent in studies of longer duration, more than ten generations. Then, Solow and Steele (1990), with Bulmer's method, in a simulated and stationary population, showed that up to 30 generations were needed to establish density dependence. Fish live much longer than insects, so we should not be surprised if density dependence, of whatever form, is somewhat difficult to detect with the very short time series available. The fish populations that we study are nearly always well exploited and so the stock numbers are considerably less than in the natural state (by a factor of 10 or more). This alone reduces the chance of detecting density dependence. Now that the difficulties of estimating density dependence have been pointed out, the evidence so far published will be reviewed.

The mortality of eggs and larvae

The production of fish eggs and larvae (and, incidentally, of copepodite stages) is derived by estimates of abundance in time divided by the development rate at the temperature observed. From the ratio of the productions at different stages, estimates of mortality can be derived (see for example Harding *et al.*, 1978 on the mortality of larval plaice, based on the method of Buchanan–Wollaston, 1926). However, this procedure assumes that predatory mortality is constant during the period in which abundance is estimated. The MacKendrick–von Foerster equation (MacKendrick, 1926; von Foerster, 1959) describes the condition in which the number of predators may vary with time during the period of sampling. Wood and Nisbet (1991) have developed a method of estimating mortality from successive measures of abundance with a cubic spline supported by a cross-validation spline. One of the interesting points raised is that the expected very large number of samples is not needed. The estimates of the mortality of eggs, larvae and juveniles made so far are potentially biased, but the extent of this bias is not yet known.

Given that the eggs, larvae and juveniles of plaice are probably well sampled in space and time, we may examine their apparent mortality rates in the Southern Bight of the North Sea and the eastern English Channel and on the Balgzand in the Wadden Sea in northern Holland. The original material was taken from Harding *et al.* (1978), Rauck and Zijlstra (1978), Zijlstra *et al.* (1982), van der Veer (1986) and van der Veer and Bergman (1987). Eleven series of cruises to sample eggs and larvae were made between 1947 and 1971; from 1947 to 1957, Hensen nets or Helgoland larva nets were used and in 1962, 1963, 1968, 1969 and 1971 the Lowestoft high speed sampler was used (Beverton and Tungate, 1967). Eggs were well sampled in all cruises, but the larvae was only fully sampled between 1962 and 1971. On the Balgzand, the 0-groups were sampled with a 1.9 m beam trawl, during immature life, after settlement, from 1973 to 1982. Figure 7.1 shows the mortality rates averaged in time for the first two years of life. An arbitrary curve is fitted to the data:

$$M_t = M_{max}/(1 + bt),$$

where M_{max} is the initial and maximal mortality of eggs ($= 0.09$), M_t is the instantaneous mortality rate at time t, and b is the rate constant at which mortality declines with time ($= 0.013$). The average mortality between egg stages I and V (for 11 years of observation) was 0.09; there was no decline in mortality during the egg stages, which implies that the decline in mortality observed in the larval stages occurred as they grew through successive predatory fields of larger and fewer predators. There is a break at about 120 to 130 days, when the 0-groups were eaten by shrimps at settlement causing a density-dependent mortality (van der Veer, 1986), of about 8% of the total; with enclosure experiments, van

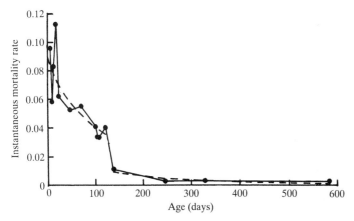

Figure 7.1 Mortality of eggs, larvae and juveniles of the plaice in the southern North Sea. The curve fitted to the observation (solid line is observations, broken line fitted curve), is $M_t = M_{max}/(1 + bt)$, where M_t is the mortality at time t, M_{max} is the maximal mortality rate, that of the eggs, and b is a constant.

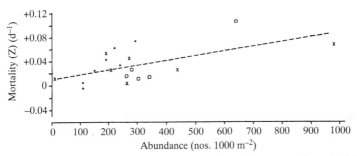

Figure 7.2 The density-dependent mortality of metamorphosed plaice as they settle to the seabed where they are eaten by shrimps (after van der Veer, 1986).

der Veer and Bergman (1987) showed that the newly settled plaice were eaten by shrimps. Indeed, the high year classes of the cold winters (see Cushing, 1982) may be associated with the migration offshore of the predatory shrimps. Figure 7.2 shows this mortality estimated in the years 1980, 1981 and 1982; the density-independent component amounts to 0.104 and the density-dependent one, at the mean, to 0.024. As shown below, Beverton and Iles (1992) have found that the mortality of the juveniles on the beaches is also density dependent.

It might appear that density dependence appears only when the larvae settle from the plankton to the seabed, from the open water to a surface.

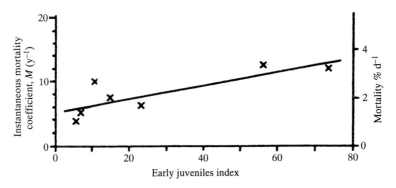

Figure 7.3 The dependence of the mortality of the early juvenile cod (of two or three months of age) on their density, in the Barents Sea (Sundby *et al.*, 1989).

However, Sundby *et al.* (1989) have shown that the mortality of the early juveniles of cod is also density dependent. The animals were sampled with a midwater trawl at an age of about two months, when they were still feeding on *Calanus finmarchicus*. Figure 7.3 shows the relationship. So density-dependent mortality does not demand settlement to the seabed, although the predators may conveniently aggregate there. Sundby *et al.* (1989) found that the numbers of 0-group cod in the Barents Sea were positively correlated with numbers of early juveniles. The densities of 0-groups are themselves correlated with recruitment ($r^2 = 0.67$, 1966 to 1981). Consequently, density dependence may occur at the age of two or three months, during the late larval stages and early juvenile stages (Russell, 1976, used the phrase post larvae for the late larval stages, but the early juveniles here have just metamorphosed; see Blaxter, 1988, for a summary of the present terminology).

It will be recalled that capelin recruitment can be predicted from the frequency of onshore winds and the temperature, which implies that recruitment depends upon the zooplankton in the warmer water blown onshore. The major regulative process occurred during the larval stages. van der Veer (1986) sampled the plaice larvae in the plankton in stage 4 with a plankton net as they entered the Wadden Sea, i.e. before they had settled. For six years between 1974 and 1982, the abundances of these larvae were correlated with an index of year class strength on the Balgzand ($r_s = 0.83$; $p < 0.05$, where r_s is the Spearman rank correlation coefficient). As noted above, plaice larvae are well sampled. The index of year class strength is that established by Rauck and Zijlstra (1978), who related the densities of settled 0-group plaice to recruitment as estimated by virtual population analysis; this index was also shown to be correlated with an index from the International Wadden Sea Program (van der Veer, 1986).

van der Veer *et al.* (1991) extended this series to the flounder *Platichthys flesus*. They found that the late larval stages in the plankton over the Balgzand were significantly correlated with the demersal stages in August ($r_s = 0.99$; $p < 0.05$). There are no data on the adult stock and by analogy with the plaice, van der Veer and his colleagues suggested that recruitment was also determined in the larval stage for the flounder.

Jenkins *et al.* (1991) examined the growth rates of larval southern bluefin tuna in a patch near the spawning ground off the north-west coast of Australia, following it for a short period of time. They found that growth rates of individual fishes depended inversely on log abundance ($r^2 = 0.37$). The study illustrates a possible view of the processes that might take place.

As described in Chapter 6, Butler (1991) examined the larval mortality of sardines and of anchovy off Southern California between 1951 and 1967. He found that this mortality was positively correlated with the anchovy biomass, the stock of which predominated in the seas off California during this period (Figure 6.11). Presumably the larvae were victims of the juvenile and adult anchovies. It is an unusual form of density-dependent mortality in that it is generated by two species. Butler also compared the growth rates of the northern anchovy in 1980 with that in 1983, the year of the strong El Niño; in 1983, the larvae grew more slowly, as might be expected, but the greatest differences between the two years was found amongst the later larvae, that is from 25 mm in length and older.

Watanabe and Lo (1989) investigated the larval production of the Pacific saury (*Coliolabis saira*) across an area of 2.5 million square miles to the east of Japan (see Figure 4.14(*e*)). The survey was made with surface nets, at all times of the day, for 15 years. The data were averaged for 10 times of day and for 11 length groups (7.5 to 57.5 mm) and so correction factors were calculated. From growth increments on the otoliths, Watanabe (1988) was able to convert the observed lengths to ages. Overall mortality rates were calculated and they were correlated with the initial larval production ($r_s = 0.704$, $n = 14$; $p < 0.05$). Thus the larval mortality rate is density dependent.

During the eighties considerable work was carried out on the Walleye pollock (*Theragra chalcogramma*) which spawns in the Shelikof Strait,

Table 7.1. *Correlations between abundances of larvae, 0-groups and two-year-olds, Pearson above the diagonal and Spearman below it*

	Larvae	0-groups	Two-year-olds
Larvae		0.82++ (6)	0.36 (8)
0-groups	+		0.69+++ (11)
Two-year-olds	n.s.	+++	

Note:
$+ (p<0.05)$; $++ (p<0.025)$; $+++ (p<0.01)$.

between Kodiak Is. and the Alaska Peninsula, where the larvae are retained on the shelf. Between 1.3 and 6.0 billion fish spawn near Cape Kekurnoi at about the same time each year between late March and mid April; the batches of eggs are laid in a depth of 150 m, but the larvae migrate to the upper 50 m where they grow at about 2 mm d^{-1}. There is a large patch of eggs or larvae off Semidi Is. by the end of May and off the Shumagin Is. by June or July with lengths of 20 to 30 mm (Figure 4.14(*i*)). Later the juveniles are found in the bays along the Alaska Peninsula.

Brodeur *et al.* (1991), expressing gastric evacuation as a linear function of time, showed that cannibalism on the eggs of the Walleye pollock was very low. But Dwyer *et al.* (1989) had found that the larger fish (>40 cm in length) took up to 400 billion 0-group fish. A negative exponential gastric evacuation function was used, so the estimate might be overestimated (see below) but the quantity eaten remains high.

Bailey and Spring (1992) compared abundances of larval, 0-group and two-year-old pollock. Larvae were sampled with bongo nets, 0-groups with shrimp trawls and the abundance of two-year-olds was determined with virtual population analysis (and hydroacoustic surveys). Table 7.1 displays the relationships between them.

Abundance of larvae (of 15 mm in length) is linked to that of the 0-groups, which in their turn are linked to that of the two-year-olds. But the numbers of larvae are not correlated with the numbers of two-year-olds. Bailey and Spring suggest that the relationships were blurred by the cannibal losses of the 0-groups. This result suggests that recruitment is generated from larval and juvenile life.

It will be recalled that the recruitment of the Baltic herring can be forecast from the abundances of larval stages (Parmanne and Sjøblom, 1984, 1987). There is a little evidence that the processes that determine recruitment can occur during the larval stages and that some are density dependent. Obviously, they can continue during the juvenile stages but the larval stages can no longer be excluded merely because they have been badly sampled.

The growth of flatfish larvae to metamorphosis (when fed *ad libitum*) has been described by a number of authors. Metamorphosis is an important event in the life of fishes. Plaice make the most profound morphological change as they pass from pelagic life to one on the seabed and clupeids sprout fins and start to shoal. Policansky (1982) found that in the starry flounder (*Platichthys stellatus*), the ages at metamorphosis varied from 30 to 64 d, with a weak inverse relationship between length and age at metamorphosis. Seikai *et al.* (1986), with the Japanese flounder, *Palalichthys olivaceus*, showed that the standard deviation of length increases towards metamorphosis and also that it increases with temperature. As noted in the last chapter, with *Pseudopleuronectes americanus*, the winter flounder, Chambers and Leggett (1987) also found that the coefficient of variation of length was less than that of age at metamorphosis. Figure 7.4 shows their major result, the dependence

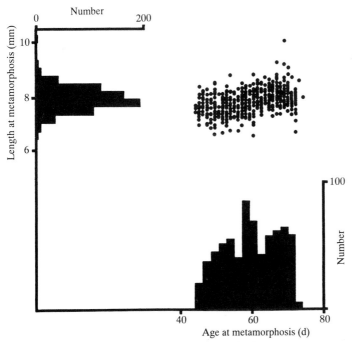

Figure 7.4 The dependence of length at metamorphosis of the winter flounder (*Pseudopleuronectes americanus*) on age (after Chambers and Leggett, 1987).

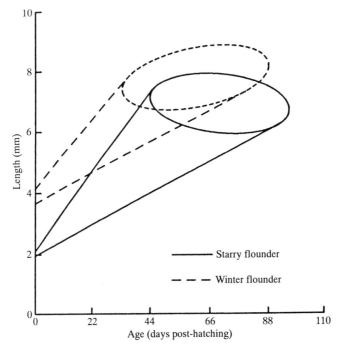

Figure 7.5 Metamorphic envelopes for the starry flounder and winter flounder (after Chambers and Leggett, 1992).

of length at metamorphosis upon age; the variation in length implies differences in growth rate. In an extensive review of the problem, Chambers and Leggett (1992) expressed results with metamorphic envelopes based on 95% isoclines for the starry flounder and the winter flounder and Figure 7.5 shows such envelopes for both species. An interesting point is that variation increases from hatching to metamorphosis. Although the larvae were fed *ad libitum*, faster-growing larvae would metamorphose earlier. With less food such differences in growth must become more pronounced.

Processes at three trophic levels

The work on plaice larvae in the Southern Bight of the North Sea revealed no relationship between the instantaneous mortality rate and the numbers of plaice larvae (Harding *et al.*, 1978). Following Ricker and Foerster (1948), Beverton and Holt (1957) proposed that density-dependent processes occurred during a critical period. Cushing and

Horwood (1977) suggested that it was the cumulative mortality during the critical period that was important. Shepherd and Cushing (1980) extended the argument in the following way. Critical periods are essential components of stock/recruitment relationships. Here the critical period is based on density-dependent growth during larval life, or indeed juvenile life if appropriate. It is a representation of the Ricker–Foerster thesis that animals that grow more slowly suffer more predation.

When fish are fed *ad libitum* at constant and optimal temperatures, growth rate is maximal, G_{max}. Then density-dependent growth may be expressed (Shepherd and Cushing, 1980):

$$G(N) = \frac{G_{max}}{(1 + N/K)},$$

where G_{max} is the maximal instantaneous growth rate, $G(N)$ is growth rate at the population density, N, and K is the density of fish larvae at which G_{max} is reduced to half by competition for food.

Let

$$\frac{dW}{dt} = GW$$

where W is the weight of a fish larva.

Let

$$\frac{dN}{dt} = -\mu N$$

where μ is the instantaneous rate of mortality due to predation; the predator is assumed to be an adventitious one which does not aggregate.

Eliminating time,

$$\frac{dW}{GW} = -\frac{dN}{\mu N}$$

and

$$\frac{dW}{W} = -\left(\frac{G_{max}}{\mu}\right) \frac{dN}{N(1 + N/K)}$$

Expanding the right-hand side in partial fractions,

$$\frac{dW}{W} = -\frac{G_{max}}{\mu}\left[\frac{1}{N} - \frac{1}{(N+K)}\right]dN$$

so

$$(\ln W)\frac{W_1}{W_0} = -\frac{G_{max}}{\mu}[\ln N - \ln(N+K)]\frac{N_1}{N_0}$$

and

$$(\ln W_1/W_0) = -\frac{G_{max}}{\mu}\ln\left[\frac{N_1}{(N_1+K)}\Big/\frac{N_0}{(N_0+K)}\right]$$

The constant (W_1/W_0) defines the size range through which the larvae grow during a critical period.

$$A = (W_1/W_0)^{-\mu/G_{max}} = \exp\left(-\frac{\mu}{G_{max}}\ln\frac{W_1}{W_0}\right) = \exp - \mu T_0$$

So

$$T_0 = \frac{\ln(W_1/W_0)}{G_{max}}$$

T_0 is the time to grow through this critical period if food were superabundant and A represents the fraction surviving at the end of that period. It is a cumulative density-independent mortality. If $G < G_{max}$ the critical period lasts longer and the cumulative mortality becomes density dependent. Indeed, $T_1(>T_0) = \ln(W_1/W_0)/G$. Thus, both density-dependent and density-independent mortalities are cumulative.

Proceeding with the argument,

$$\frac{N_1}{N_1+K} = A\frac{N_0}{N_0+K}$$

and

$$\frac{1}{N_1} = \frac{1}{A}\left(\frac{1}{N_0} + \frac{1}{K}\right) - \frac{1}{K}$$

or

$$N_1 = \frac{A\,N_0}{[1 + (1 - A)N_0/K]}$$

The relationship is shown in Figure 7.6, relating N_1 to N_0. Figure 7.6(a)

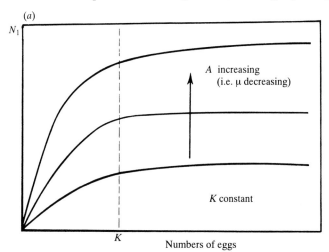

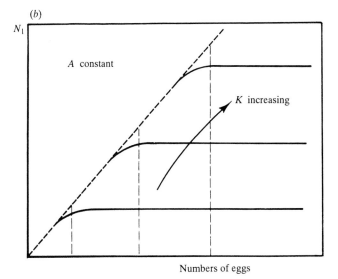

Figure 7.6 Dependence of numbers at the end of the critical period, N_1, on those at the beginning, N_0, using the model of Shepherd and Cushing (1980): (a) with K constant and A increasing, or how density-independent factors affect survival; (b) with A constant and K increasing or the effect of food on survival.

shows how the density-independent factors increase the general survival and Figure 7.6(b) illustrates the effect of food on that survival. Figure 7.6(a) gives the relationship with K constant, but A increasing (as mortality decreases) and Figure 7.6(b) shows it with A constant and K increasing. It is a model in three trophic levels: predators, fish larvae and their food. It has obvious implications in the study of stock and recruitment, but here one is concerned with the biology of the fish larvae.

One needs to know the maximal growth rates at a range of observed temperatures (as discussed in Chapter 6) and also the ages of the larvae so that the growth rates can be measured properly. Further, one needs to estimate the mortality rates of the larvae. So far, the latter has been a very expensive estimate and methods are needed to devise cheaper ones. The suggestion of Wood and Nisbet (1991) that good estimates of mortality do not require large numbers of samples is of some relevance.

The argument has been extended by Shepherd (1994). Consider the satiation of the food of predators. There are N predators per unit volume, X, and one predator searches a volume, V, per unit time. The total number of predators is NX, the total volume searched is NXV and the fraction of the volume searched per unit time is $NXV/X = NV$.

If f_0 is the maximal fraction of kills to encounters, then the maximal mortality rate of the prey, $M_0 = f_0NV$. But as predators become sated, that mortality rate will be reduced. If $f < f_0$, kills of larvae per unit time per predator is fVn, which corresponds to a feeding rate, $fVnw$, where n is the number of larvae per unit volume and w is the weight of a larva. If $f_0 > R$, the predator ration, the encounter rate is reduced, so $f = R/nVw$ and

$$\frac{1}{f} = \frac{1}{f_0} + \frac{nVw}{R}$$

Then,

$$f = \frac{f_0}{1 + f_0 nVw/R}$$

and

$$M = \frac{NVf_0}{(1 + f_0 nVw/R)}$$

Table 7.2. *The larval mortality rates of plaice in the Southern Bight of the North Sea*

	Larvae 1	Larvae 2	Larvae 3	Larvae 4
	Egg V	Larvae 1	Larvae 2	Larvae 3
1962	0.083	0.033	0.099	0.037
1968	0.028	0.095	0.048	0.039
1969	0.025	0.024	**0.001**	**0.009**
1971 (Southern Bight)	0.056	0.062	0.048	0.054
1971 (Eastern Channel)	0.118	0.058	0.078	0.065
Mean	0.062	0.054	0.055	0.041

Source: Harding *et al.*, 1978.

Let $\mu = f_0 V$, the maximal mortality exerted by a single predator and $r = R/w$, the predator ration in numbers of prey per unit time, then

$$M = \frac{\mu N}{(1 + \mu n/r)},$$

The numbers of larvae decrease with time and with a constant number of predators, mortality would increase with time. As mortality decreases with time the number of predators must do so also. Table 7.2 summarizes our information on plaice larval mortality rates in the Southern Bight of the North Sea.

The mean mortality rate declines with time. An important point is that the larval mortalities of the 1969 year class from Larvae 1 to Larvae 4 (in bold in Table 7.2) are distinctly lower than the mean values. The mortality, presumably predatory, declined to very low levels indeed. This is either because the predators disappeared or they were sated and the alternatives cannot be distinguished. Let $K' = r/\mu \; (= R/(f_0 w V))$, the density at which satiation beings) and recall that $G = G_{max}/(1 + N/K)$. Then, proceeding as Shepherd and Cushing (1980):

$$\frac{dw}{w} = -\frac{G \, dn}{M \, n}$$

$$= -\frac{G_{max}}{(1+n/K)}\frac{1+\mu n/r}{\mu N}\frac{dn}{n}$$

$$= -\frac{G_{max}}{\mu N}\left[\frac{1+n/K'}{1+N/K}\right]\frac{dn}{n}.$$

Taking partial fractions,

$$\frac{dw}{w} = -\frac{G_{max}}{\mu N}\left[\frac{1}{n}-\frac{\beta}{(n+K)}\right]dn$$

where

$$\beta = 1 - K/K'.$$

Then, integrating,

$$\ln[w]\Big|_{w_0}^{w_1} = \frac{G_{max}}{\mu N}\left[\ln(n) - \beta\ln(n+K)\right]_{n_0}^{n_1},$$

and

$$A = (w_1/w_0)^{-\mu N/G_{max}}.$$

Then

$$\frac{n_1}{(n_1+K)^\beta} = A\frac{n_0}{(n_0+K)^\beta}.$$

If $n_1 << n_0$ and if $n_0 \approx K$ (that is, the stock is roughly matched to its carrying capacity), $n_1 << K$,

$$n_1 = \frac{A n_0}{(1+n_0/K)^\beta}.$$

This is very close to the versatile stock/recruitment relationship proposed by Shepherd (1982). The important point is that the stock/recruitment relationship can be based on the three trophic level model, predator, fish larva and larval food and an important component is the satiation of the predator. Shepherd gives differential equations for the cases when the predator dies in time or grows in time, either of which could provide a metaphor for the possibility that larvae grow through a succession of predatory fields.

The two models also reveal that the structure of larval life can be described quite fully and they illuminate the form of data that should be collected. The major result is that Shepherd's stock/recruitment equation can be virtually derived from processes in larval life. The real point is that this is a useful metaphor for the probable decline in predation during larval life.

Are fish larvae too dilute to affect their food densities?

Some evidence has been given that recruitment can be forecast from larval numbers and indeed that larval mortality can be density-dependent. Cushing (1983) tried to answer the question of whether fish larvae are too dilute to affect their food densities. Laurence (1982) summarized his experimental work on haddock larvae: their growth rate at temperature, weight/length relationship and rate of digestion. Further, he developed a model of searching and also a model of the daily ration in numbers of food organisms.

Blaxter and Staines (1971) observed that the forward vision of a fish larva extends in the vertical plane to 90° upwards and 45° downwards; hence, about two-thirds of the potential area forward is available for looking ahead. Laurence's expression for the volume searched by a fish larva per unit time is:

$$v = (2/3)\pi\delta^2 al \text{ ml s}^{-1},$$

where l is length in cm, δ is a search constant, a is a constant of swimming speed. The larvae feed visually for 12 h d^{-1} and $A = (2/3)\pi\delta^2 a\, 36000 \times 12 \text{ cm}^2 \text{ d}^{-1}$, so $V = Al \text{ ml d}^{-1}$. Laurence gives a weight/length relationship, $W = 0.044\, l^{4.476}$ (where W is dry weight in µg and l is length in mm). Then $l = 0.2009\, W^{0.2234}$ (with l in cm) and $V = 0.0154\, W^{0.2234} \text{ m}^3 \text{ d}^{-1}$, the volume searched by one larva d^{-1} (in the original publication a mistake was made in the estimate of V which is corrected in Cushing and Horwood, 1994).

The larva grows exponentially, G being the instantaneous growth rate. The larvae die:

$$N_{t+1} = N_t \exp - Mt$$

where N_t is the number of larvae m^{-3} at time t, N_{t+1} is the number of larvae m^{-3} at time t, and M is the instantaneous mortality rate. The volume searched by N larvae m^{-3}, $P_t (= N_t V_t)$ at time t, after a period of growth and death, is given by

$$P_t = (N_t)\ 0.0154\ (W_0 \exp G_t)^{0.2234}.$$

The larvae grow and die in the model and then the volume searched by N larvae m^{-3}, P_t, is calculated for three growth rates at three temperatures ($G = 0.04$, 4 °C; $G = 0.06$, 7 °C; $G = 0.12$, 9 °C) and two to four mortality rates ($M = 0.01$, 0.02 at 4 °C; $M = 0.01$, 0.02 and 0.03 at 7 °C; $M = 0.03, 0.05, 0.07$ and 0.09 at 9 °C). After 40 days and 75 days the estimates of P_t were:

| | $G = 0.12$ | | | $G = 0.06$ | | | $G = 0.04$ | |
| | M | | | M | | | M | |
Days	0.09	0.05	0.03	0.03	0.02	0.01	0.02	0.01
1	0.032	0.033	0.034	0.033	0.033	0.034	0.033	0.034
40	0.001	0.006	0.013	0.017	0.026	0.039	0.022	0.032
75				0.004	0.009	0.019	0.007	0.014

As mortality must decrease as the animals grow into the warmer water, the most likely values of P_t are about 0.03, decreasing to above 0.01. Thus by the time of metamorphosis, the volume searched has been maintained at 1% to 3% d^{-1}. This should affect the density of the food of the haddock larvae, the nauplii and younger copepodite stages of *Calanus*.

This thesis is based on the idea that the larvae do not grow at their maximal rate and that their mortality rate is affected by their food intake, the Ricker–Foerster thesis: larvae that grow more slowly due to food lack suffer predation for a longer time. Pepin (1991) studied the effect of temperature and size on mortality rates in a broad range of published material on the eggs and larvae of marine fish. A positive relationship was established between mortality rate and temperature and the residuals were inversely dependent on length; $M = 0.25 \exp 0.67\ T.\ L^{-0.68}$, where M is the instantaneous mortality rate, T is temperature in °C and L is length in mm. So predation is more active at higher temperatures and it declines as length increases. Again, M was positively related to G (growth rate) and the residuals were inversely dependent on ln length, so $M = 5.17 G^{0.74} L^{-1.17}$. Thus, mortality is a power function of growth rate and an inverse one of length, so mortality and growth are linked and mortality declines with length. An interesting point was that the mean growth rate was greater than the mean mortality rate, so biomass increases during larval life as might be expected.

Lough and Laurence (1981) estimated the number of food organisms needed for each larva, R, from

$$R_t = (\Delta W_t + KW^n)/(1 - \alpha)\beta w,$$

where K is a metabolic constant, n is a coefficient, α is the fraction lost in metabolism, β is a digestive coefficient, w is the dry weight of the food organism.

One may proceed (Cushing and Horwood, 1994). First the mortality rate is taken from that given above, declining with time as the larvae grow through successive predatory fields (Figure 7.1).

The number of food organisms decline as they are eaten by the larvae:

$$F_{t+1} = F_t(1 - P_t),$$

F_t is the number of food organisms at time t, and F_{t+1} is the number of food organisms at time $t+1$. The food organisms grow, from nauplii to copepodite of *Calanus finmarchicus*.

$$w_{t+1} = w_t \exp(gt)$$

where w_t is the weight of the food organism at time t, w_{t+1} is the weight of the food organism at time $t+1$, and g is the instantaneous growth rate of the food organisms. The ration taken by the fish larva, R_t, is given by:

$$\Delta R_t = P_t \cdot F_t/N_t,$$

and then the increment in growth:

$$W_t = R_t(1 - \alpha)\beta w_t - K W_t n.$$

The model is run from an initial estimate of P_t, the volume searched, to estimate the numbers of food organisms, F_t, the daily ration, R_t, and then the daily weight increment, growth rate, mortality rate and then the volume searched again; each day the food organisms grow at a standard rate as if the fish larvae were growing with a cohort of their food as suggested by Jones (1973). Constants are given in Table 7.3.

A maximal dry weight of about 3000 µg is reached in about 40 days at

Table 7.3 *Constants used in the larval growth program*

a	1.5	m	0.625
δ	0.75	j	0.002
k	0.2009	W_0	33 µg dry weight
c	0.2234	w_0	0.6 µg dry weight
n	0.67	g	0.0875
α	0.40	M_{max}	0.09
β_{max}	0.80		

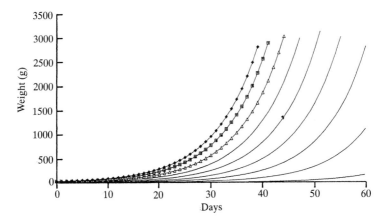

Figure 7.7 The growth of haddock larvae at different levels of food (1000, 2000, 3000, 4000, 5000, 6000, 7000, 8000 and 10 000 m^{-3}) for periods of up to 60 days. The initial number of larvae was 1 m^{-3} and the coefficient of larval mortality, $b = 0.005$ (after Cushing and Horwood, 1994).

the maximal growth rate. This is assumed to be the weight at metamorphosis (see earlier discussion, Figure 6.8). The program is run from 40 to 60 days which yields a 20-day spread in the date at which metamorphosis takes place. If the growth rate is high, metamorphosis is early and vice versa. The main consequence is that as growth rate is reduced, mortality endures for longer, essentially the Ricker–Foerster thesis. It might be considered arbitrary that the model is based on the time to reach metamorphosis but the same would be true of any critical period.

Figure 7.7 shows the growth of haddock larvae for a range of food organisms; the very slow growth rates at the lowest food densities are probably not observed at sea. Figure 7.8 displays the dependence of average growth rate on numbers of food organisms. Figure 7.9 illustrates

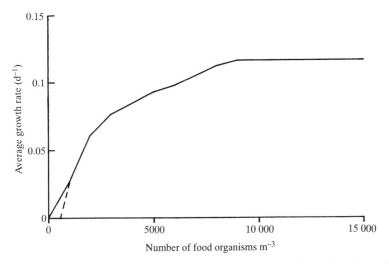

Figure 7.8 The dependence of growth rate, averaged to the time of met-amorphosis, on the numbers of food organisms (after Cushing and Horwood, 1994).

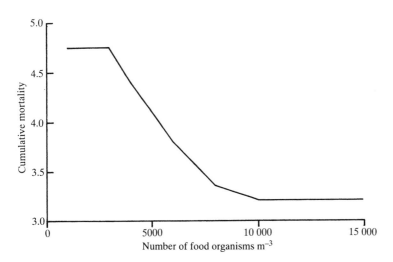

Figure 7.9 The dependence of cumulative instantaneous mortality on number of food organisms; at low food, mortality is constrained by the limit of the age of metamorphosis to 60 days (after Cushing and Horwood, 1994).

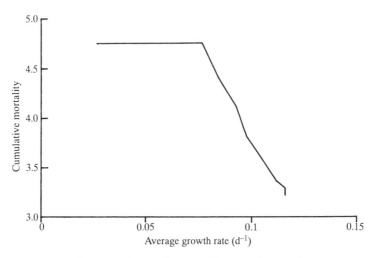

Figure 7.10 The dependence of cumulative mortality on the average growth rate (after Cushing and Horwood, 1994).

the link between cumulative mortality and number of food organisms m^{-3} and Figure 7.10 shows the relationship between cumulative mortality and average growth rate. The structure of Figures 7.9 and 7.10 is constrained by the maximal growth rate and by the upper limit to the age of metamorphosis of 60 days.

Figure 7.11 shows the dependence of the proportion surviving at metamorphosis on the number of food organisms. This proportion is an estimate of recruitment at metamorphosis. In Figure 7.12 is displayed the dependence of recruitment at metamorphosis on initial numbers of larvae at different levels of food. In Figure 7.13 is given the dependence of cumulative mortality on initial numbers of larvae (0.1 to 7.5 m^{-3}) at different levels of food (2000 to 10 000 m^{-3}); density-dependent mortality is high. Figure 7.14 shows the dependence of recruitment at metamorphosis on cumulative mortality at different levels of food and different initial numbers of larvae. In Figure 7.15 is displayed the dependence of recruitment at metamorphosis at different levels of food and different initial numbers of larvae, with the period of examination extended from 60 to 100 days. The estimate of stock, as the higher initial numbers of larvae, is much larger than recorded estimates of the George's Bank haddock; however, Herrington (1948) showed low recruitment at high stock.

In these examples, density-dependent mortality accounted for about

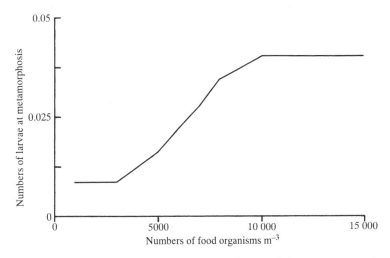

Figure 7.11 The dependence of the proportion surviving at metamorphosis on the number of food organisms, an estimate of recruitment as at metamorphosis (after Cushing and Horwood, 1994).

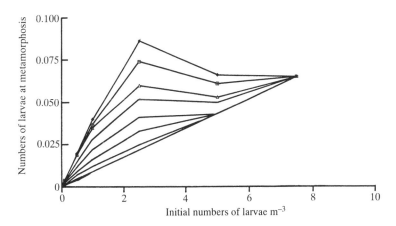

Figure 7.12 Dependence of recruitment at metamorphosis on initial numbers of larvae at different levels of food (Cushing and Horwood, 1994).

one-third of the total cumulative mortality, as the spread of time to metamorphosis from 40 to 60 days is one-third of the total life. Lough and Laurence (1981) show that gadid larvae are found on George's Bank at a density of as much as 3 m^{-3}. The initial density would have been higher and that in the unexploited stock, very much higher.

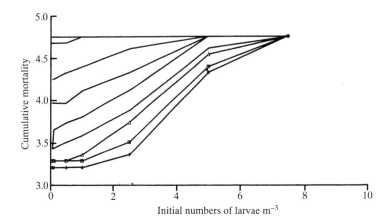

Figure 7.13 Dependence of cumulative mortality upon initial numbers of larvae (0.1 to 7.5 larvae m^{-3}) and different levels of food (2000 to 10 000 m^{-3}); the density-dependent mortality is prominent (after Cushing and Horwood, 1994).

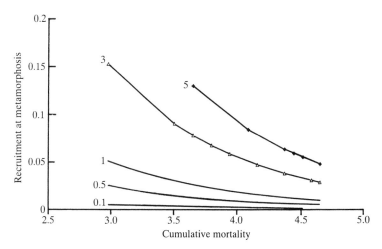

Figure 7.14 Dependence of recruitment at metamorphosis on cumulative mortality at different levels of food and different initial numbers of larvae. Plots are labelled with the number of food organisms, in thousands/m^3 (after Cushing and Horwood, 1994).

The time to reach metamorphosis depends on growth rate and there are two constraints, the upper limit in days to reach metamorphosis (60 d) and the maximal growth rate. Between the two constraints, the cumulative mortality is density dependent and recruitment (at metamorphosis) is a function of food.

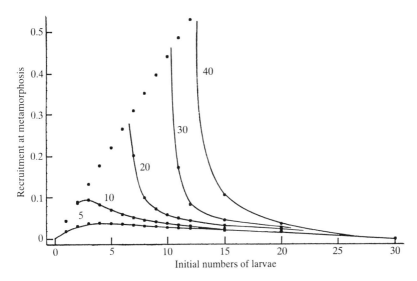

Figure 7.15 Dependence of recruitment at metamorphosis on initial numbers of larvae at different levels of food when the period up to metamorphosis is extended from 60 to 100 days. Number of food organisms/m³ are indicated on the plots (after Cushing and Horwood, 1994).

The decline of mortality with time derives from Figure 7.1, which is based on a large amount of data on the plaice in the Southern Bight of the North Sea. Lough (1984) used a Macness net to sample haddock larvae on George's Bank and he estimated that the daily mortality of the larvae was about 6%: this corresponds to a value of $b = 0.01$, not very different from that used in the model. If the mortality were greater, the density dependence would be sharper and vice versa. Recruitment (at metamorphosis) obviously depends on mortality, higher at low mortality and vice versa.

Lough and Laurence (1981) showed a peak density of gadid larvae on George's Bank of $2.5\ \mathrm{m}^{-3}$. This was confirmed by Lough (1984) who found that the haddock larvae were 6 mm in length, 18 to 24 d post hatch. At the mortality rate of 6% d^{-1}, the initial number of larvae would have been $4.3\ \mathrm{m}^{-3}$, with an average of about half that value. Figure 7.15 shows that at $2.0\ \mathrm{m}^{-3}$ there is maximal recruitment at 10 000 food organisms m^{-3}, which implies some density-dependent mortality (see Figure 7.13).

The model of larval growth presents a quite imaginary picture of larval growth and death and its justification lies in the high cost of collecting good data. The premises are not unrealistic and the conclusions reveal that density-dependent mortality can occur during larval life to a possibly significant degree.

The life of fish in their first summer on the nursery ground

Zijlstra *et al.* (1982) made a very thorough study of the growth of 0-group plaice on the Balgzand in the western Wadden Sea in northern Holland. A beam trawl (1.9 m in width) was worked at 36 to 40 stations for between 7 and 17 times a year for a period of seven years. Fonds (1979) had reared immature plaice fed *ad libitum* at a range of temperatures; he found that $l = 0.12\ T + 0.05\ l - 0.4$, where l is length in cm and T is temperature in degrees Celsius. Figure 7.16 shows the observed and simulated mean lengths of 0-group plaice on the Balgzand in the summers of 1973 to 1979. Thus in June, July and August the little fish were fully fed and there is no evidence of density-dependent growth. Zijlstra *et al.* applied the same method to earlier studies in Loch Ewe in Scotland (Steele and Edwards, 1970) and in Filey Bay on the English North Sea coast (Lockwood, 1972) and established that the 0-group plaice there were also fully fed.

However, in March, April and May, the mean length of the newly settled plaice clearly depended on density in numbers. This is the period of settlement and the little fish suffer a density-dependent mortality on settlement as noted above (van der Veer, 1986). Presumably, the shrimps eat the larger fish that settle first. A multiple regression of mean length against temperature and density in numbers showed that 83% of the variance was due to abundance. Settlement takes place over a period of two to three months (which is roughly the period of spawning), but after it is complete by the end of May, the 0-group fish feed at maximal rate and grow rapidly. On the nursery ground the little fish may well obtain maximal growth at least risk of predation.

Iles and Beverton (1991), Beverton and Iles (1992,a,b) have studied the mortality of three species of 0-group flatfish (plaice, dab and turbot) on the beaches in the North Sea, Irish Sea, west coast of Scotland and in the Kattegat. The populations recruited to the beach were sampled from settlement for a period of some months before emigration. Gear efficiencies were estimated and densities in numbers were expressed in logarithms to stabilize the variances. Figure 7.17 shows a good example of the decline in sample density with time. Regressions were fitted from the time of maximal numbers on the beach to January in the following

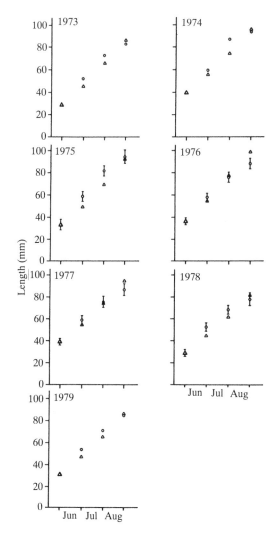

Figure 7.16 Observed (○) and simulated (△) mean lengths of 0-group plaice on the Balgzand (Zijlstra *et al.*, 1982).

year. Figure 7.18 illustrates a life table to the end of the first year of life.

In density, D,

$$(1/D)(dD/dt) = -(\mu_1 + \mu_2 \ln D).$$

Integrating,

$$D_t = \exp(\mu_1/\mu_2)[(\exp -\mu_2 t) - 1](D_0 \exp -\mu_2 t).$$

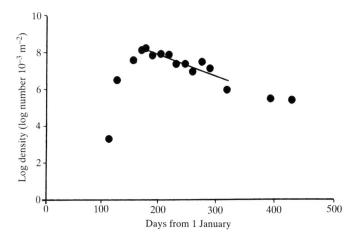

Figure 7.17 Decline in sample density of immature plaice (after Beverton and Iles, 1992*a*).

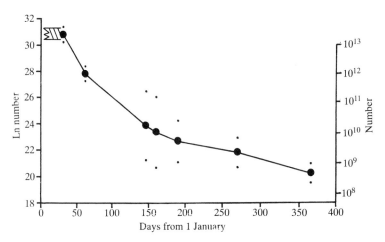

Figure 7.18 Life table of plaice from spawning to the end of the first year of life (after Beverton and Iles, 1992*b*).

This expression is curvilinear. Figure 7.19 shows mortality, $\overset{\circ}{M}\,d^{-1}$, against ln maximal density (ln numbers $10^{-3}\,m^{-2}$); the regression is corrected for the light curvilinearity and for gear efficiency. As might be expected, the extensive data set is a little variable and includes information from perhaps three or more stocks, as usually defined, but, equally, the broader population might represent the density-dependent

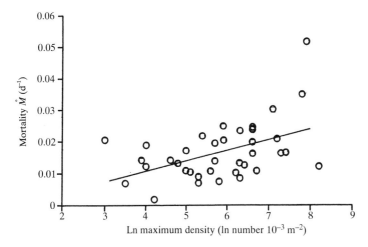

Figure 7.19 Dependence of the mortality of immature plaice on density (after Beverton and Iles, 1992*b*); M is the seasonal mortality rate.

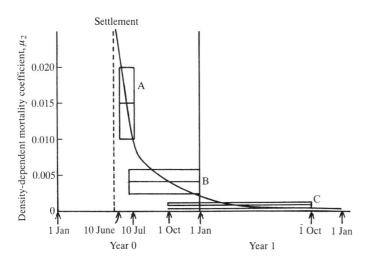

Figure 7.20 The decline of density-dependent mortality of the immature plaice. A, June to July; B, July to October; C, October to October (after Beverton and Iles, 1992*b*).

processes a little more fully. Figure 7.20 shows the decline of mortality with time.

Density dependence was shown in the following way. With the logarithmic form of the equation given above, ln D_0 (as 0-groups from surveys made by the International Council for the Exploration of the

Sea) was plotted against ln D_1, a constant time of one year apart (which avoids some of the errors referred to above). The slopes were less than unity, suggesting weak density dependence, but the confidence limits were high. A similar analysis of the Wadden Sea material revealed lower confidence limits, cited as evidence of strong density dependence.

A very interesting analysis of damping was made. Let V_t be the ratio of the largest year class to that of the lowest when they enter the density dependent phase. Then $V_t = V_o \exp - \mu t$. The variation in the life table is maximal at the end of the larval phase. Figure 7.20 shows the decline of density-dependent mortality, the cumulative mortality amounting to 1.211 between settling and January in the following year.

The earlier work of Zijlstra and his colleagues showed that the 0-group on the beaches of the Wadden Sea were fully fed and that growth played no part in the modification of mortality which must have been predatory. Beverton and Iles have shown that this mortality may well be density dependent. Myers and Cadigan (1993) have examined survey material for 17 populations of demersal fishes on both sides of the North Atlantic. They found that the juvenile mortality of cod, plaice, sole and whiting, of up to 0.4, was significantly density dependent.

Density-dependent growth in juvenile and adult fishes

As food becomes short with increased stock, growth might become density dependent; for example, with the weight of plaice at infinite age, W_∞ (based on samples of 10.7 to 13.7 years of age), Beverton and Holt (1957) found a difference between 3177 g in 1945 and 1946 and 2195 g in the years before the Second World War. This is one of the few observations made on an exploited stock compared with those on a more or less unexploited one.

Amongst exploited stocks, density-dependent growth might be expected in the younger fishes merely because they grow more quickly. The growth of I-group Norway pout (Raitt, 1968) and 0-group herring (Burd, 1985; Hubold, 1978) were both shown to be density dependent. Deriso (1980) found that the increment of weight of juvenile eight-year-old halibut increased by a factor of 2.5 for a reduction in numbers by a factor of 4. Southward (1967) wrote of the Pacific halibut: 'from the first summer onwards the degree of density dependence appears to decline, consequently from generation to generation numbers cannot be controlled by density dependent fecundity'. But much more recently, Hagen and Quinn (1991) examined the first five growth zones of 745 halibut from 26 year classes. They were able to separate year effects from those of year classes and linked annual growth to sea surface temperature. They found no evidence of density-dependent growth but there was a link between the growth in ages 1 and 2 and year class strength, but none for the 0-groups. In such a large fish, density-dependent growth should be detectable, but the evidence is so far a little thin.

Amongst mature animals, the detection of density-dependent growth is more difficult. Perhaps the most detailed study was made by Houghton and Flatman (1984). Their estimate of the growth rate of cod in the west central North Sea was based on ln (W_{t+1}/W_t) against W_t, simple exponential growth for two increments in each year class between 1963 and 1977 (but there is a danger of bias with $1/x$ against x). With an analysis of covariance a common slope was estimated and differences in the intercepts were plotted against stock. The method extracts the maximal amount of information from somewhat variable material. As numbers of stock were halved, the intercept of growth rate increased by about 20%. However, with the same material, Daan *et al.* (1990) showed that in 1, 2 and 3-group cod, there was no dependence of annual weight gain on total stock density. However, the weight gain in 1-groups was negatively correlated ($p = 0.05$) with the density of the 0-groups indicating the possibility of density-dependent growth. Bromley (1989) has shown that the apparent differences in growth with density in North Sea gadoids were generated by differences in sampling area.

Daan *et al.* (1990) tabulated the evidence for density-dependent growth in the North Sea for eight species for periods of 8 to 45 years (Table 7.4). The estimate for haddock (Jones and Hislop, 1978) was based on a time series of 45 years in the ages one to five. Numbers were estimated by virtual population analysis and the density-dependent growth was described for mainly juvenile fish. Burd (1984, 1985) published an analogous relationship for herring based on a time series of 34 years on three-year-old fish. Each of these long time series is effectively a study of juvenile growth when the growth rate is high.

In general, the good correlations occur in samples of juvenile fish. That of Houghton and Flatman on cod from 1963 to 1977 has been discussed above. Daan *et al.* point out that feeding rates or food availability were not recorded in any of the studies in Table 7.4. The North Sea is heavily exploited, each of the eight species being fished hard. A relatively low fishing mortality reduces stock by an order of magnitude. Hence the food should perhaps be superabundant. This

Table 7.4. *Some estimates of density dependent-growth in North Sea fishes*

Species	Ages	Period	Density	Correlation
Cod	1–5	1958–70	CPUE	$p < 0.05$
	1–13	1963–77	VPA	$p < 0.05$
	3	1962–76	VPA/CPUE	$p < 0.05$
	1–5	1968–82	CPUE	$p < 0.05$
Whiting	1–5	1958–74	CPUE	$p < 0.05$
Haddock	2	1928–74	CPUE	n.s.
	1–3	1925–36	CPUE	$p < 0.05$
	1–5	1926–71	VPA	$p < 0.05$
Saithe	4	1958–73	VPA	$p < 0.05$
Norway pout	1	1930–66	CPUE	$p < 0.05$
	1–3	1960–83	CPUE	$p < 0.05$
Sole	3–6	1960–70	VPA	n.s.
	3–15	1957–73	VPA	n.s.
	0	1955–72	VPA	n.s.
	1–15	1957–73	VPA, CPUE	$p < 0.05$
Plaice	10–13	1935–46	CPUE	$p < 0.05$
	3–5	1931–50	CPUE	n.s.
	2–4	1934–42	CPUE	n.s.
	0	1955–73	VPA	$p < 0.05$
	4–7	1965–82	VPA	n.s.
Herring	3	1947–81	VPA	$p < 0.05$

Note:
VPA, an estimate of stock by virtual population analysis;
CPUE, catch per unit of effort.
References are given in Daan *et al.* (1990).

simple view masks the fact that heavy fishing may lead to a food surplus being eaten by other animals. There are many competitors for the food of planktivorous fish, both vertebrate and invertebrate. For some demersal fish the competition may not be quite so pronounced. However, to establish the density dependence of growth we really need to show that food was fully available.

Another possible form of regulation is a change in the age of maturation. The fullest study of this problem is given by Jørgensen (1990) on long-term changes in maturation in the north-east Arctic cod for the year classes 1923 and 1976. The median age of maturation was calculated from maturity ogives weighted for the fishing mortalities of adults and juveniles each year. Figure 7.21 shows the median age of maturation for the year classes 1938 to 1976 for males and females; the

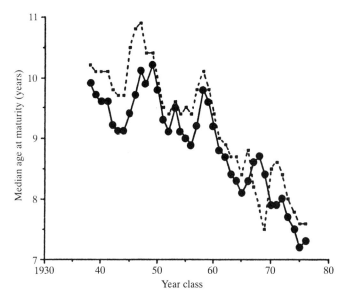

Figure 7.21 Median age of maturation of the year classes 1938 to 1976 in the Arcto-Norwegian cod (after Jørgensen, 1990).

full line shows the median age adjusted for mature and immature fishing mortalities and the dotted line, that unadjusted. During this period stock biomass was reduced by a factor of about 5 (Figure 7.22). The change in age of maturation may well be a compensatory response to stock decline but the expected increase in length is not as great as might have been expected.

The increased rate of exploitation will not of itself account for the decline in median age of maturation, particularly as the changes in adult length for age do not appear to be density dependent. Jørgensen believed that there was a compensatory change based on circumstantial evidence. The age of maturation is really an effect of immature growth. With reduced stock, the juveniles may grow faster and then mature earlier in their lives. The growth rates of the juveniles are more likely to be density dependent than those of the adults. It would be interesting to show this for the Arcto-Norwegian cod stock.

The most important point in this discussion of density-dependent growth is that in exploited stocks, density-dependent growth can be found amongst juvenile fishes but rarely, if ever, amongst the mature ones. This means that density-dependent fecundity is an unlikely form of regulatory mechanism in exploited stocks; even in a large fish like the

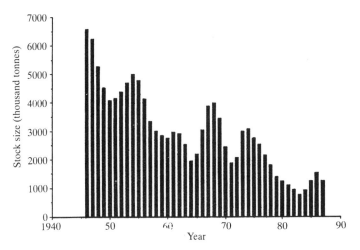

Figure 7.22 Decline in stock of the Arcto-Norwegian cod (Jørgensen, 1990).

Pacific halibut it must be very hard to detect. In unexploited stocks it may play a part, perhaps in the perch in Windermere in north-west England.

Cannibalism

Harris (1975) distinguished stock-dependent mortality from the density-dependent form; the former derives from the abundance of the adults and not on the contemporaneous abundance of young fish, which may not be related at all to that of the parents. Because adults, eggs and larvae drift away from the spawning ground in a common current there is an opportunity for cannibalism by parents on their offspring. It cannot last very long because the two components of the population do not migrate in the same way. For example, the plaice eggs and larvae in the Southern Bight of the North Sea drift slowly in the midwater towards the north-west, but the adults travel away quickly under selective tidal transport. Spent herring are carried away from the spawning ground, leaving their eggs on the seabed. However, in the slow eastern boundary currents the opportunity for cannibalizing eggs and larvae by sardine and anchovy is perhaps greatest.

The study of cannibalism is today based on the experimental analysis of digestion. Gastric evacuation is often displayed as a concave curve in time. Bromley (1988) drew attention to the censoring of the data with the inclusion of zeros, from fish with empty guts. He showed, for the whiting (*Merlangius merlangus* (L.)), that digestion could be a linear function of time. There are important consequences for the estimation of cannibalism from gut contents; for example, the results of Hunter and Kimbrell (1980) tend to overestimate the degree of cannibalism. Bromley and Last (1990) showed that sometimes small fish are eaten in the trawl and that a cover should be used to separate the small fish from their predators, another source of overestimation. Daan *et al.* (1984) examined the predation of cod and plaice eggs by herring. They were taken mostly by the more numerous younger fish. The stomach was assumed to empty in 12 hours. The fractions eaten of the number produced were low (0.0011 for cod and 0.0135 for plaice).

Hunter and Kimbrell (1980) and Alheit (1986) reported the number of eggs/g wet weight eaten (and corrected for the rate of digestion) of the northern anchovy, *Engraulis mordax*, as proportion of the ovary-free wet weight, about 2% to 3%. This fraction was raised by the proportion of females spawning each night and by the sex ratio to reach 17.2%. The rate of evacuation was obtained by plotting ln (number of eggs) against time in hours. If plotted on a linear scale, the censoring by zero values becomes very clear. Further, the numbers observed in the stomachs at sea are biased upward by one or two high values. The raising factor used may be overestimated. Valdes *et al.* (1987) proceeded in a slightly different way. The cannibal mortality was expressed as the ratio of number of eggs eaten per fish per day (with the Hunter and Kimbrell figure) to the egg production per day. Mortality was estimated from a regression of mortality against age, 45% of the total mortality of eggs estimated from numbers m^{-2} against age in hours. (I am grateful to Dr Peter Bromley for his advice on this problem.)

The most likely form of cannibalism is that exerted by adult fish on the juveniles when they live in the same area. Cook and Armstrong (1986) have shown that the recruitment of North Sea haddock and whiting is reduced by previous high year classes. The older fish probably eat the younger as they settle to the seabed. Dwyer *et al.* (1989), with gut evacuation experiments, found that, in the south-east Bering Sea (but not elsewhere), 4.10^{11} 0-group Walleye pollock were eaten by older pollock as they settled to the seabed (but the authors did not record whether a cover on the trawl was used). The proportion of the total stock cannot be estimated until this stock question has been elucidated.

Santander (1987) estimated the ratio of egg production at sea to population fecundity of the Peruvian anchoveta (*Engraulis ringens*) and plotted it against parent stock (which is part of the estimate of population fecundity). The inverse regression suggested density dependence at the egg stage. A second figure showed egg production as function of parent stock with no obvious relationship, which suggests that the first regression may be biased by ($1/x$ on x).

Cannibalism is a most attractive mechanism to regulate a population, particularly if the cannibals are the only predators. The problem is to distinguish the two forms of predation, cannibal and adventitious. In the extreme, one would almost need evidence of motive to establish the difference. To proceed, the evidence for the forms of predation should be analysed in detail. Fish are often cannibals and large numbers are taken in this way, as for example, the Alaska pollock. The point is of some importance to the problem of stock and recruitment because a domed curve can arise from cannibalism and this mode of regulation cannot yet be ruled out.

Conclusion

Five density-dependent processes have been identified: aggregation by predators, the effects of the adventitious predator, the Ricker–Foerster thesis that slow growers suffer greater mortality, density-dependent fecundity (as consequence of the density-dependent growth of adults), and age at first maturation (as consequence of the density-dependent growth of immatures). Cannibalism has also been discussed. A short survey showed that density-dependent mortality could be detected in the larvae of the Japanese saury, the Walleye pollock and in the early juveniles of the Arcto-Norwegian cod.

Two models of processes at three trophic levels are that of Shepherd and Cushing (1980) and that of Shepherd (1994). The first is an extension of the Ricker–Foerster thesis, in which the critical period is defined as that within which a specified increment of growth takes place, slowly or quickly. The second, more extensive model, is taken further in the analysis of predation to the point of satiation. The final result is close to the versatile stock recruitment relationship of Shepherd (1982).

The question of whether fish larvae were too dilute to affect the numbers of their food organisms was raised and it appeared that they indeed were able to reduce the density of their food by as much as 3% d^{-1}. This work was extended (Cushing and Horwood, 1994) following the observation of Chambers and Leggett (1987) that the time at metamorphosis could endure for as long as 20 days in the population. If we assume that metamorphosis occurs at a relatively fixed weight (or fixed length, as shown by Chambers and Leggett), and if larvae are permitted to grow at less than their maximal growth rate, then larval growth becomes density dependent and so does the cumulative mortality. Further, recruitment (as at metamorphosis) is generated as product of growth and mortality. This is no more than an elaboration of the Ricker–Foerster thesis. But the density-dependent mortality was unexpectedly high.

Processes on the nursery ground have been summarized. When the baby plaice settled they are eaten by shrimps and their mortality is density dependent. It had been shown that the growth of the little plaice was at the maximal rate and so their mortality does not follow the Ricker–Foerster course, but the mortality is probably density dependent and presumably the predators must aggregate upon them.

Density-dependent growth of adults has not been established unequivocally, but it appears to take place among some immature fishes although there are exceptions to this rough rule. So density-dependent fecundity in the stocks of exploited fishes is unlikely; it remains possible, of course, in unexploited stocks. The age at first maturation may advance as the result of density-dependent growth of immature fish, therefore exploitation may lead to an earlier age of maturation as the immature stock is reduced; the effects of adult and immature exploitation cannot yet be separated. Cannibalism has been cited as a possibly strong source of stock-dependent mortality, but the evidence has not yet proved decisive. Large cod eat little cod and South African anchovies eat their eggs but the difficulty is to distinguish this mortality from that from other predators.

8

Stock and recruitment

Introduction

There are two principal models used in fisheries science. The first is that which describes production, developed from Schaefer (1954, 1957) by Pella and Tomlinson (1969), Deriso (1980) and Schnute (1977). The earlier versions were calculated in catch per effort and expressed in yield and so the variation in recruitment was concealed. In the later age-structured models the variation in recruitment was made explicit, but the stock recruitment relationship was subsumed in general density dependence. The second model comprises a small suite of sequential population analyses derived by Gulland (1965) and Pope (1972), themselves a development from work by Beverton and Holt (1957). Much of the routine work today is based on yield per recruit assessments and the stock recruitment relationship is expressed separately; in some multispecies models it is an explicit component (for example, see Shepherd, 1988). The production models tend to be used on the Pacific coast of North America, whereas in the Atlantic the sequential population analyses tend to predominate.

Hilborn and Walters (1992) summarize methods used today. Anon (1992) gives a substantial account of methods used by management scientists throughout the world.

The variability about the stock recruitment relationship has generated dismay of two kinds. The first, more common some years ago, was that the relationship did not seem to exist because the best fit appeared to have an intercept with no slope and the necessary passage of the slope through the origin was ignored. Yet, as noted in the previous chapter, such a distribution is evidence of strong density dependence (Shepherd and Cushing, 1980). The second source of dismay, of more recent origin, states that there is no stock recruitment relationship unless there are enough observations to describe the slope at the origin, which may mean that the fishery should collapse before the relationship is recognized.

The usual stock recruitment relationship may be expressed as a curve with a convex rising slope, an asymptote or as a dome. All express density dependence. As noted in the last chapter, persistence of the population implies density dependence and the stock recruitment relationship does no more than try to separate density-dependent mortality from the density-independent survival. There are considerable statistical difficulties; the number of observations is low, 30 to 60 at best and usually much less, but the real handicap is that there are rarely enough observations to describe the slope at the origin, which is needed to prevent recruitment overfishing. The description of the stock recruitment relationship at high stock exists in the data but it is not very interesting, primarily because the trend of natural mortality with age at high or unexploited stock is unknown. In the previous chapter, the first indications of independent estimates of density-dependent processes were discussed but they are, as yet, far from the parameters of the stock recruitment relationship. In any population study, its regulation in the face of environmental change is of prime interest, but the stock recruitment relationship merely averages the two components of density-independent survival and density-dependent mortality for the period of observation. As the period of observation lengthens, environmental changes of shorter duration become embedded in the material.

The four models of stock and recruitment

Four stock recruitment relationships have survived: in order of their appearance, those of Ricker (1954, 1958), Beverton and Holt (1957), the Power Law (Cushing, 1971) and Shepherd (1982). Shepherd wrote the general form:

$$R = aPf(P/K),$$

where R is recruitment in numbers, a is the slope at the origin, (in recruits/biomass) an estimate of density-dependent survival, P is spawning stock biomass in tonnes, a proxy for the number of eggs produced, K is the threshold biomass, at which recruitment is half that under density-independent processes only; above this value density-dependent processes predominate. In this form the four equations are (with the original form in brackets):

$$R = aP \exp - (P/K)$$

(Ricker: $R = aP \exp - bP$; where b is a coefficient of density dependence);

$R = aP/(1 + P/K)$

(Beverton and Holt: $R = 1/(a + bP)$);

$R = aP(P/K)^{-b}$

(Cushing: $R = aP^b$);

$R = aP/(1 + (P/K)^b)$

(Shepherd).

The coefficient b expresses compensation for an excessive recruitment. Figure 8.1 displays the characteristics of all four curves. That of Beverton and Holt approaches an asymptote in recruitment at high stock, that of Ricker is domed and reaches low recruitment at high stock, and the power law rises in recruitment. The Shepherd equation can express the other three forms as the coefficient of density dependence increases (power law, $b < 1$; Beverton and Holt, $b \approx 1$; Ricker $b > 1$).

Dome-shaped curves have been described from time to time (for example, George's Bank haddock (Herrington, 1948) and the Skeena sockeye salmon (Shephard and Withler, 1958), but as more data become available and as the observations were fitted logarithmically, the domes

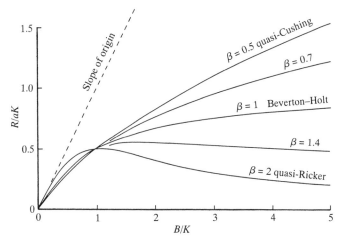

Figure 8.1 The characteristics of all four stock recruitment relationships as expressed by the Shepherd curve (after Shepherd, 1982) $\beta = b$.

in these two examples disappeared (Myers *et al.*, 1994*a*). In the last chapter, the possible role of cannibalism was inconclusively reviewed: the most likely cannibals are probably gadoids, such as Walleye pollock, cod or whiting which do not distinguish their food from their children. Because the dome can arise from cannibalism it cannot yet be rejected. Density-dependent fecundity has not been generally detected, but it might well appear if ever stocks are exploited at a lower rate. So the Shepherd equation is to be preferred because of its versatility, despite the fact that there are three constants to be fitted.

Hennemuth *et al.* (1980) and Garrod (1983) showed that recruitment was probably distributed lognormally because the number of recruits is the sum of a long sequence of deaths. Rothschild (1986) suggested that each egg has an independent probability of survival and that the distributions might be contagious; Shelton (1992) proposed that the recruitment of bet-hedging fishes might not be lognormally distributed. However, Myers *et al.* (1990) published the time series of the recruitments of one hundred stocks; each was expressed in three forms: $\log_{10}$, in detrended $\log_{10}$ and in first differenced $\log_{10}$. Autocorrelation and the persistence of anomalous recruitments were expressed graphically. Myers and his colleagues standardized the observations and arranged them in ascending sequence; they tested normal quantiles of $\log_{10}$ recruitment on sample quantiles (together with the detrended and first differenced values) and showed that of their hundred stocks, nearly all the recruitments were lognormally distributed.

Further, Myers *et al.* (1994*a*) have summarized data on the stock and recruitment for 170 stocks, world-wide. Time series are displayed of recruitment in numbers and of stock in tonnes and of catches, together with fishing mortality. Of 36 stocks with 20 or more observations, the distribution of significant autocorrelations with age at different lags were as follows:

Lags	1	2	3	4	5	6
Autocorrelations	24	11	–	–	1	–

That with autocorrelation extending to five years of mature age is the Pacific halibut. A stock is no more than the sum of successive recruitments and autocorrelation is to be expected particularly if the stock is heavily exploited. The general conclusion is that observations of recruitment are not independent of stock; in unexploited stocks the autocorrelation might be much higher. Four models were used: power law, Ricker, Beverton and Holt, and Shepherd; the stock recruitment relationships were fitted logarithmically by maximum likelihood; gamma distributions were also used. For each stock, a table is given of the constants of each model derived with logarithmic and with gamma distributions.

Myers *et al.* (1994*a*) investigated the possibility of depensation with sigmoid stock recruitment relationships (for example, $R = aP^\delta \exp - bP$, where δ expresses the degree of depensation). Of 105 stocks for which there were 15 or more observations, only one, the Iceland spring spawners, displayed any apparent depensation and the authors suggested that there was an environmental effect. As described in Chapter 5, Cushing (1988*b*) suggested that the collapse of the Iceland spring spawners may have been due to the passage of the great salinity anomaly of the seventies through the waters north of Iceland. However, a much more important point is that evidence for depensation does not really exist; the stock recruitment relationship itself is sufficient to describe the world-wide collapses of fish stocks under exploitation.

Nine well-known stocks

The stock recruit relationships of nine well-known stocks, for which data are available for more than 30 years, are now examined. The data are fitted with the Shepherd curve. As noted above, the Shepherd curve is used because it is open to all possible trends and is not constrained by its structure. With one exception, the data have fitted the constants given by Myers *et al.* (1994*b*) (solid lines), which has the advantage of common treatment. The exception is the North Sea herring for which I have included the very large 1991 year class and the data were fitted logarithmically with least squares using Solver in Microsoft Excel 4.

Some are fitted with the Shepherd restrained method and are shown in dashed lines. The reason for using this procedure is to weight the fitting towards the slope at the origin, which is the interesting part of the stock recruitment relationship. Shepherd (personal communication) has approached the problem in the following way. Logarithmic residuals from the Shepherd curve are estimated. Two weights are calculated: (a) the difference between the slope at the origin and the mean slope ($\ln a - $ mean $\ln R/\mathrm{SSB}$), and (b) the difference between the compensatory coefficient and zero ($\ln b - \ln 1$). The first is raised by the ratio of the variance of the fitted residuals to the observed and the second by the ratio of the fitted residuals to the compensatory coefficient. These are added to the sums of squares of the residuals and new constants are calculated non-linearly by least squares with Solver in Microsoft's Excel 4.

The nine stocks are the Skeena sockeye salmon, the Downs stock of herring in the North Sea, North Sea herring, North Sea sole, North Sea plaice, Pacific halibut, Iceland cod, Arcto-Norwegian cod and the Northern cod (Figure 8.2). For each stock there are three diagrams: the time series of recruitment (with geometric means and standard deviations), the trend of survival in time, and the stock recruitment relationship itself. For most stocks, it was assumed that eggs weighed 1 mg wet weight (except for the Skeena sockeye salmon and the Pacific halibut; see below). The data for the nine stocks came from The International Council for the Exploration of the Sea (courtesy of Dr Roger Baily) or from Dr R.A. Myers.

Skeena sockeye salmon (Shephard and Withler, 1958; Figure 8.2(*a*))

Between 1908 and 1952, strong year classes appear roughly every five years or so. The survival was estimated from (numbers in the stock raised by the fecundity -4000) and reduced by the wet weight of the eggs, 86.3 mg (Foerster, 1968). The stock recruitment relationship is very slightly domed and is quite different from that originally proposed by Shephard and Withler, which displayed a very sharp dome indeed (when fitted by moving means).

Downs stock of herring (Cushing, 1992; Figure 8.2(*b*))

The Downs herring was sampled between 1923 and 1987 and it suffered from recruitment overfishing in the late thirties and from 1955 onwards when the 1952 year class entered the fishery; it recovered during the eighties after exploitation had been banned from 1977 to 1982. During the sixties and seventies the year classes were very low indeed. The time series of survival is very different, most remarkably in higher survivals at very low stock after 1965, by up to a factor of 20. The stock recruitment relationship is a nearly linear one, which explains why the stock was so sensitive to recruitment overfishing. All three figures show different aspects of the response of the stock to heavy fishing. The question arises why the full recovery has not taken place to the abundance of the thirties, late forties and early fifties. The partial recovery shows that the

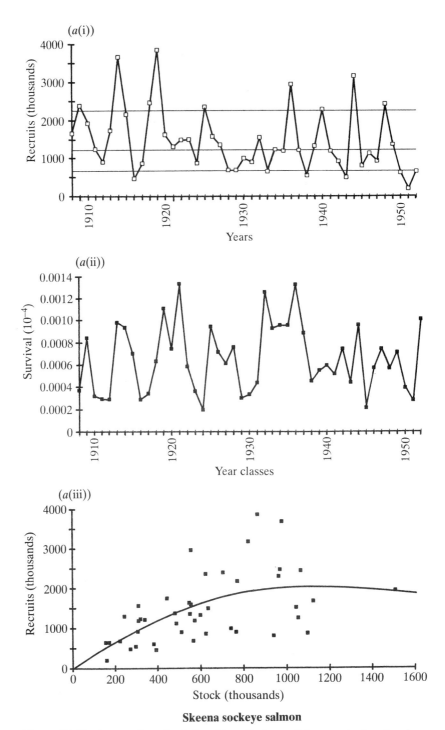

Skeena sockeye salmon

Figure 8.2 Stock recruitment relationships of nine well-known stocks, showing the time series of recruitments with geometric means and logarithmic standard deviations, survival and the stock recruitment relationships. The nine stocks are:

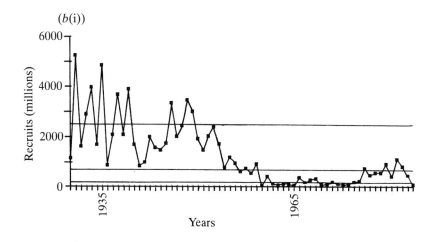

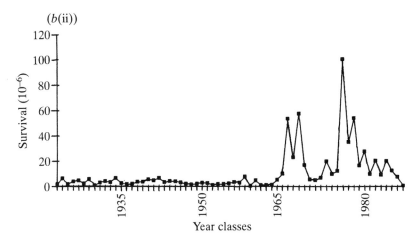

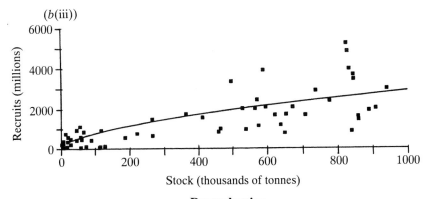

Downs herring

(*a*) the Skeena sockeye salmon; (*b*) the Downs herring; (*c*) the North Sea herring; (*d*) the North Sea sole; (*e*) the North Sea plaice; (*f*) the Pacific halibut; (*g*) the Iceland cod; (*h*) the Arcto-Norwegian cod; and (*i*) the Northern cod.

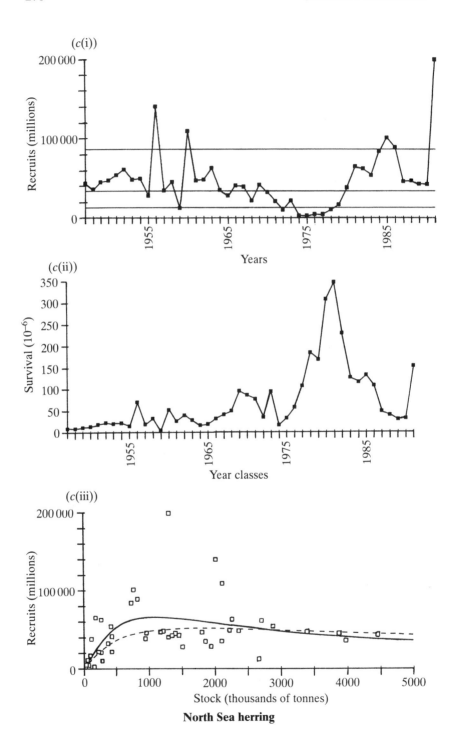

North Sea herring

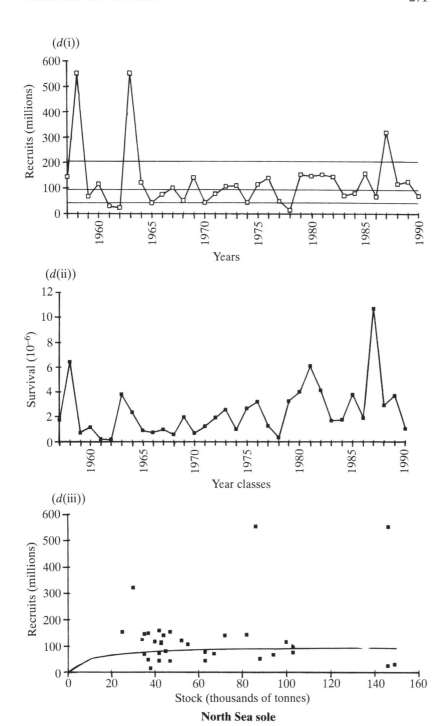

North Sea sole

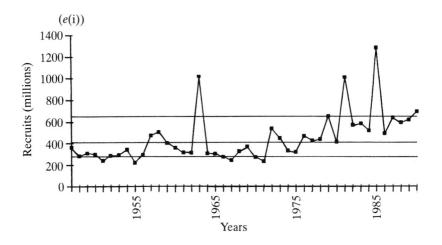

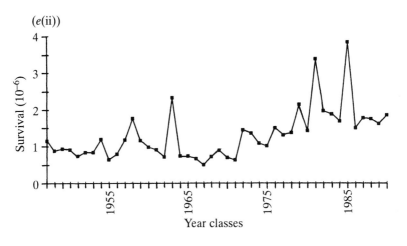

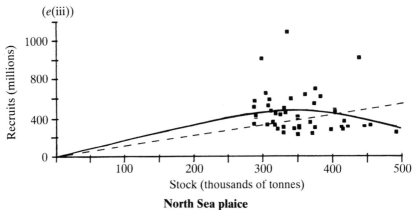

North Sea plaice

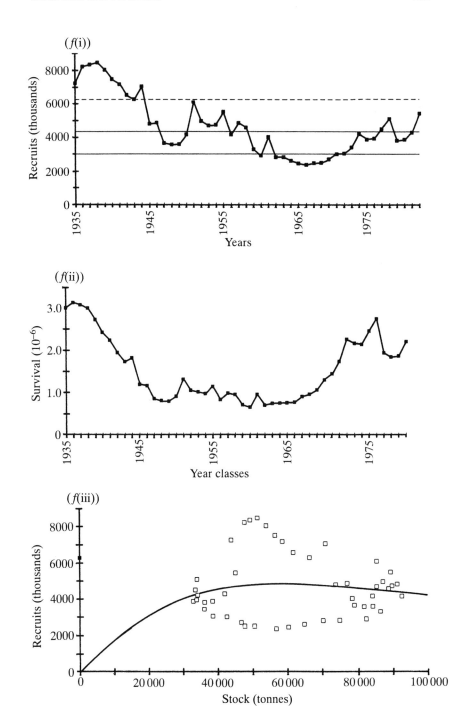

Pacific halibut

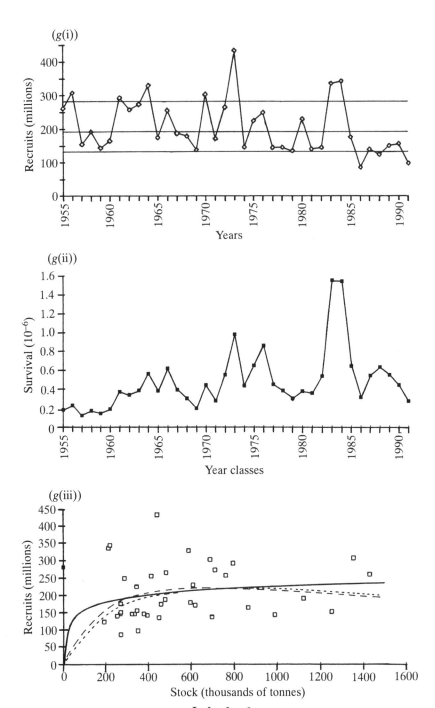

Iceland cod

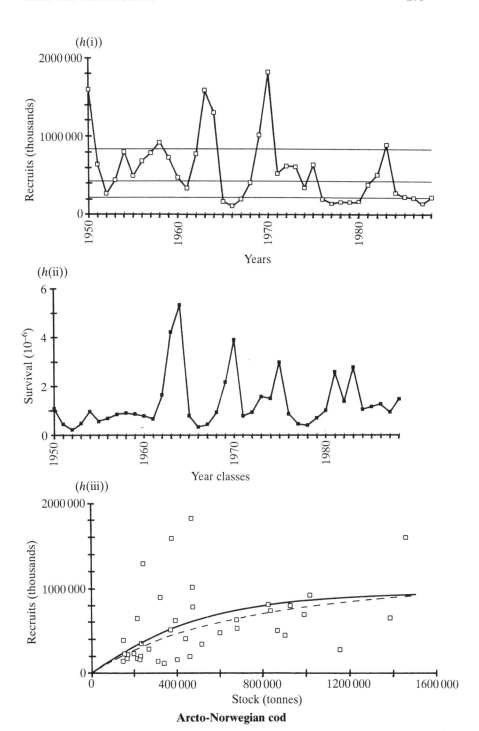

Arcto-Norwegian cod

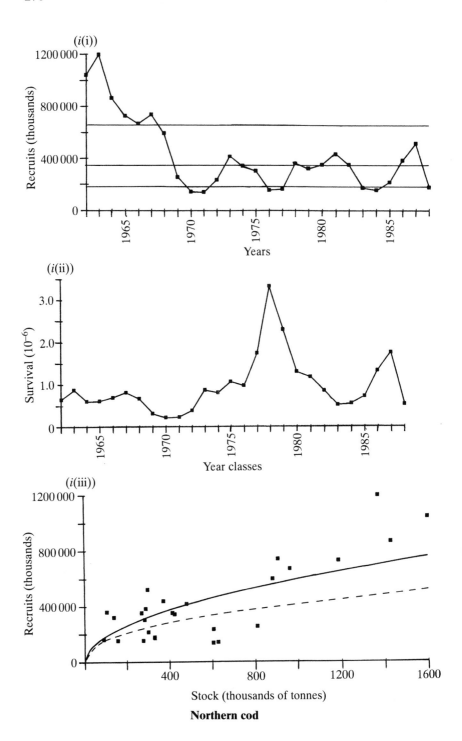

Northern cod

stock suffered from recruitment overfishing and that possibly a source of exploitation remains.

North Sea herring (I.C.E.S., 1993, personal communication from Dr R. Bailey; Figure 8.2(*c*))

(The North Sea herring estimates include catches from the Downs stock, which is probably distinct from that further north – Cushing and Bridger, 1966; Cushing, 1992 – but since the mid-fifties the catches in numbers are dominated by the Bank stock in the central and northern North Sea.) From 1947 to 1991, the time series reveals two events, recruitment overfishing in the seventies and the three outstanding year classes, 1956, 1960 and 1991. The survival increased a little in the late sixties and by much more in the late seventies and early eighties when the stock was very low; recovery followed a ban on catches between 1977 and 1982. The stock recruitment relationship is well described because the stock was collapsed by recruitment overfishing. In the figure the observations are fitted by least squares and by the Shepherd restrained method. Because of stock collapse the relationship is well described.

The North Sea sole (I.C.E.S., 1993; Figure 8.42(*d*))

The time series of year classes shows the strong ones that originated in the cold winters of 1958, 1963 and 1987 in the southern North Sea when the predatory shrimps may migrate into deeper water (van der Veer and Bergman, 1987). Survival tends to be a little lower after the strong year classes and it increased slowly from the late sixties onwards. The stock recruitment relationship comprises the three strong year classes and the rest. The Shepherd restrained method fits the observations well and eliminates the little peak at low stock (outside the range of observations) in the logarithmic fit by maximum likelihood.

The North Sea plaice (I.C.E.S., 1993; Figure 8.2(*e*))

The recruitment time series shows three strong year classes and an increase from 1972 onwards. One of the strong year classes occurred in a cold winter (1963), but the other two did not. In survival, an increase started in about 1967 by a factor of about two. As noted in the last chapter, trawling with heavy beam trawls may have raised the productivity of the seabed. The stock recruitment relationship is slightly domed when fitted logarithmically by maximum likelihood, as a consequence of increased recruitment at middle stock, an artificial dome. With the Shepherd restrained fit it disappears.

The Pacific halibut (Myers *et al.*, 1994*a*; Figure 8.2(*f*))

All three diagrams show the same event, a decline in survival and recruitment to a low in the sixties followed by some recovery in the seventies, which has not been fully explained.

Iceland cod (I.C.E.S., 1993; Figure 8.2(*g*))

There were three or four good year classes, but a decline from 1986 onwards. Survival increased slowly to 1983–4, after which it declined quite sharply, by a factor of three. The stock recruitment relationship until recently has always been stable with low variation, but of course, high density-dependent mortality. The recent fall in recruitment has given cause for concern. There are three curves fitted: (a) fitted with arithmetic residuals by the Marquardt algorithm, (b) fitted logarithmi-cally by maximum likelihood, and (c) fitted with Shepherd's restrained method. From this limited treatment the logarithmic fit is to be preferred for the purpose of approaching the slope at the origin.

The Arcto-Norwegian cod (I.C.E.S., 1993; Figure 8.2(*h*))

A number of strong year classes appeared at fairly regular intervals. The passage of the Great Slug into the Barents Sea reduced recruitment in the late seventies and early eighties. The stock recruitment relationship shows a very slight dome, but nothing like that displayed earlier with fewer data points (see for example, Cushing and Harris, 1973).

The Northern cod (Myers *et al.*, 1993; Figure 8.2(*i*))

Recruitment declined sharply during the sixties, after which it remained rather low. Survival decreased to 1969 to 1972 and it increased in the late seventies. The stock recruitment relationship is near linear which must mean that environmental factors dominate the life history of the cod in this severe region.

The three sets of figures show changes in recruitment, survival and the stock recruitment relationship. In recruitment, there are long-term declines followed by recoveries, as recruitment overfishing has been ameliorated. Such a decline in recruitment can occur at constant fishing mortality, as for example in the Icelandic spring-spawning herring (see Rothschild, 1986). If a stock is heavily exploited, three poor year classes in succession can generate recruitment overfishing. It does not matter whether they were reduced by an environmental effect or by reduced stock, for the consequence is the same, recruitment overfishing. The sporadic or periodical high year classes sustain the stock and might even generate recovery. Those of the North Sea sole, Arcto-Norwegian cod and the North Sea herring are much larger than the upper standard deviation and because they sustain the stock for a number of years, it would be desirable to forecast them.

The stock recruitment relationship displays three distinct processes: the variation in recruitment, the restraint of recruitment by density-dependent mortality and the decrease in recruitment at low stock (if shown). Survival is affected by all three processes, but reveals them in a different way: the sharp increase in the survival of the Downs stock from 1965 onward, the increase in survival of the North Sea plaice from the late sixties onward and higher survival in the Iceland cod until 1985. That of the Downs stock demonstrates density dependence but that of the North Sea plaice does not because the stock did not change. The better survival at constant stock generated better recruitment.

The stock recruitment relationships are now convincing, as in the Skeena sockeye salmon, the North Sea herring and in the Downs herring, the last two of which suffered recruitment overfishing. The three flatfish relationships (North Sea sole, North Sea plaice and Pacific halibut) are less attractive because distinct environmental events are embedded within them: the large cold winter year classes of the North Sea sole, the rise in recruitment and hence, stock, of the North Sea plaice and the decline and slow recovery in the Pacific halibut. Those of the three cod stocks are profoundly different: the near linear one of the Northern cod, the low variation of recruitment to the Iceland cod, with possible decline at low stock, and the high and the quasi-periodic variation of the recruitment to the Arcto-Norwegian cod stock.

Survival was estimated as the ratio of numbers of recruits to numbers of eggs of their parents and so the estimates can be compared between stocks. Table 8.1 displays the average survivals for the nine stocks.

The survivals arrange themselves in three groups: salmon, herring,

Table 8.1 *Average survival of each of the nine stocks*

Skeena	6.54 ($\pm$3.1) 10^{-4}
Downs	10.4 ($\pm$16.7) 10^{-6}
N. Sea herring	70.5 ($\pm$76.5) 10^{-6}
N. Sea sole	2.4 ($\pm$2.2) 10^{-6}
N. Sea plaice	1.2 ($\pm$0.7) 10^{-6}
Pacific halibut	1.4 ($\pm$0.8) 10^{-6}
Iceland cod	0.5 ($\pm$0.3) 10^{-6}
Arcto-Norwegian cod	1.3 ($\pm$1.1) 10^{-6}
Northern cod	1.0 ($\pm$0.7) 10^{-6}

and together, the flatfish and cod. The salmon spawn on secluded reeds high in the river systems. The herring lay their eggs on the seabed. The flatfish and cod have pelagic eggs which apparently may suffer greater mortality.

A remarkable stock recruitment relationship has been described by Elliott (1985); he caught sea trout by electrofishing and with block nets in Black Brows Beck in the Lake District in north-west England. They were sampled as eggs, 0-group parr in May to June and August to September and as 1-groups in the same months. Numbers at each age of parr were plotted against the numbers of eggs in the parent stock and dome-shaped Ricker curves were fitted to the data (Figure 8.3); the confidence limits for the early stages are remarkably low.

In eight of the years of observations, the little fishes were sampled much more frequently and a critical time was established at 30 to 70 days from first feeding. During this period density-dependent loss rate diminished sharply. At the critical time, dead parr were seen drifting downstream. Territories were established both before and after the critical time. For four year classes, direct counts were made in two small areas at very short time intervals. By the critical time, the numbers of fish per group and the number of groups fell to fairly constant values.

If these processes were merely a change in survival associated with territorial behaviour, an asymptotic stock recruitment relationship would be expected. The dome-shaped curve can arise by cannibalism or by the aggregation of predators, but the parr were not found in the stomachs of their parents or other predators. A slight dome can also appear if mortality and growth are linked (Cushing and Horwood, 1977), but there is no relation between the growth of parr and parent stock up to the critical time.

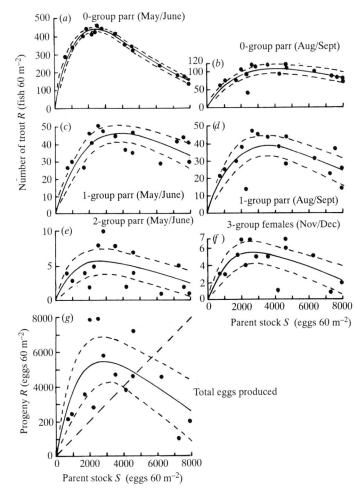

Figure 8.3 The stock recruitment relationship of the migrant trout of Black Brows Beck in Cumbria in north-west England. The parr were sampled in May to June and in August to September as 0 groups and as 1-groups and as older fish. The standard stock recruitment relationship is shown in (g).

The variation in growth was inversely related to the density-dependent mortality. So the variation in growth changed with the duration of the critical period. Perhaps there is a threshold size at which territory can be established and those fish that fail to reach that size die and are drifted downstream. Hence, an effect of food on the growth of the 0-group parr might have been concealed if the death rate of the poorly fed parr were stock dependent (analogously to cannibalism). Then a dome might not be unexpected.

The stock recruitment relationship of the Downs herring also displays a point of interest. Many years ago, Cushing and Bridger (1966) suggested that the Downs stock was a distinct one, based on the contrast in vital parameters with the Bank group to the north. With a long time series, 1923 to 1989, no correlation was found between recruitment to the Downs stock and that of the Bank stock (Cushing, 1992).

Dickson *et al.* (1988*b*) found that under the increased stress of northerly winds in spring between the fifties and the seventies, the production of zooplankton (including *Pseudocalanus*) was delayed in spring in the western North Sea and reduced in abundance in the whole North Sea and North East Atlantic. In the eighties there was a general recovery but not in the Southern Bight of the North Sea. Hardy (1924) had shown that the larval herring in the southern North Sea depended almost entirely on *Para/Pseudocalanus*. Recruitment to the Downs stock was positively correlated with *Para/Pseudocalanus* ($r^2 = 0.19$; $p < 0.01$). An ANOVA on ln R (recruitment) on ln SSB (spawning stock biomass) showed that both constant and slope were significant. When the effect of *Para/Pseudocalanus* was included (ln R on ln SSB and ln *Para/Pseudocalanus*), the constants were about the same but the effect of ln *Para/Pseudocalanus* was not significant. So the environmental effect was suppressed by that of exploitation, by recruitment overfishing.

Beddington and Basson (1993) constructed stock recruitment relationships for elephant, red deer and reindeer. That of the elephant, as might be expected, showed weak density dependence, deriving from juvenile mortality, interbirth period and age at first maturation. For the red deer and the reindeer, the stock recruitment relationships show pronounced domes; juvenile mortality and fecundity were the sources of density dependence. The interest for fisheries biologists lies in the appearance of a dome when adult parameters are invoked. Beddington and Basson compared the yield/mortality relationships for red deer and herring; the red deer suffered dramatic decline in yield at high mortality rate as the breeding stock was reduced. In herring, the yield declined much more slowly with mortality. The herring is more resilient in the sense that numbers of eggs are greater. The stock recruitment relationship of the red deer resembles that of a fish stock, with about the same degree of variability, in contrast to that of the elephant.

The stock recruitment relationships described above are diverse. They display a number of different conditions and it is now clear that they express real biological differences. Indeed, a comparative study of these relationships may well yield information of great value.

Recruitment overfishing

The main problem in the management of the stock recruitment relationship is to forecast or prevent the decline in recruitment before it happens. Since the early fifties many pelagic stocks have collapsed by recruitment overfishing; the invention of the power block allowed the purse seiners to work offshore, to catch the shoals detected by echo sounder or sonar and to increase catchability. Today, it is possible that some gadoid stocks suffer in an analogous way.

A distinction is often made between 'growth overfishing' and 'recruitment overfishing' (Petersen, 1894; Cushing, 1971, 1975*b*). The first occurs when fish are caught before they have had time to grow. The second takes place when recruitment is reduced as spawning stock biomass diminishes under the pressure of fishing. It is an obvious definition, but it is also imprecise.

The scatter plot of recruitment on parent stock can be dissected into a group of cells, to each of which is assigned a probability from the number of observations in each cell. Then, from the allocation of probabilities, an expected recruitment for each cell is calculated. Each observation of recruitment is assumed to be independent of stock so the conclusions are not biased by the use of a particular model. Getz and Swartzman (1981) used this method on the stocks of George's Bank haddock, Namibian anchovy and the yellowtail flounder of southern New England. Swartzman *et al.* (1983) used it on the stock of Pacific whiting and Overholtz *et al.* (1984) on that of the George's Bank haddock. Rothschild and Mullen (1985) extended the method to estimate the return times of five stocks.

The purely statistical approach to the stock recruitment relationship requires an estimate of the errors of measurement. Smith and Walters (1981) and Ludwig and Walters (1981) created highly variable stock recruitment relationships with simulated recruitments with large errors in measurement. However, in the better examples, estimated by sequential population analysis, the errors are probably fairly low. Perhaps the problem was overstated; certainly there is no suggestion that recruitment overfishing can be masked by errors in measurement.

If the stock recruitment relationship is described well enough, as in those in Figure 8.2, the decline in recruitment is shown, as for example in

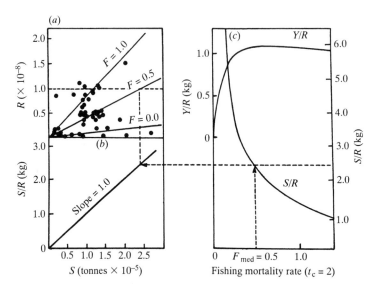

Figure 8.4 The use of biological reference points to secure a stock from recruitment overfishing. $F_{0.5}$ or F_{med} is defined as that at which half the observations lie above the line and half below, in the stock recruitment relationship (*a*). The fishing mortality is defined on the yield per recruit and spawning stock biomass per recruit shown in (*c*). The dotted line traces the relationship between the fishing mortality (against yield per recruit) and the number of recruits expected. Recruitment overfishing occurs when recruitment falls as fishing mortality increases.

the Downs herring (Figure 8.2(*b*)) and in the North Sea herring (Figure 8.2(*c*)). A method was needed to prevent this decline happening. In Figure 8.4 is shown a structure which links fishing mortality to yield per recruit and to the stock recruitment relationship (in this case, the George's Bank haddock; Sissenwine and Marchessault, 1985). The method was developed by Sissenwine and Shepherd (1987). On the left is shown the stock recruitment relationships (excluding one very high year class). On the right is displayed the yield per recruit diagram together with that of biomass per recruit. F_{med} (or $F_{0.5}$ on the stock recruitment relationship) is defined as the replacement line with half the observations above it and half below. With the yield per recruit and biomass per recruit, that fishing mortality is traced graphically along the dotted line. In the stock recruitment relationship this line cuts the replacement line (of F_{med}) and the equilibrium number of recruits is defined. Given an adequate number of observations, the method provides a system by which recruitment overfishing might be prevented.

If fishing mortality were increased, the distribution of recruits about the horizontal dotted line in Figure 8.4 remains the same, and the stock survives. If however, recruitments tend to fall below that line, recruitment overfishing has supervened, whatever the proximate cause of the reduced magnitudes of recruitment.

In recent years, thresholds in stock have been devised above which recruitment overfishing is unlikely. Thresholds are: (a) the threshold stock size, S_b, at which the 90% recruit per stock line intersects the 90% level of recruitment (Serebryakov, 1991) and (b) the 20% proportion of the virgin stock (Beddington and Cooke, 1983). Myers *et al.* (1994*a*) have used their extensive array of stock recruitment relationships to examine such thresholds. They conclude that the first, half the maximal recruitment, is probably the safest, with S_b being frequently useful. Obviously any threshold must be used with care.

Such are the methods that might prevent recruitment overfishing. It would be desirable to estimate the decline in recruitment at the earliest possible stage or even to estimate the slope at the origin before it is established in the run of observations.

A stochastic model

Shepherd and Cushing (1990) stressed the importance of the slope at the origin of the stock recruitment relationship. The departure of the stock recruitment relationship from the slope at the origin displays the strong density dependence. Figure 8.5 shows a stock recruitment relationship with lines of (R/SSB) increasing with increasing fishing mortality. Possible equilibria exist where the (R/SSB) intersect the stock recruitment relationship. But at high fishing mortality, the line of (R/SSB) can exceed the slope at the origin of the stock recruitment relationship. Because R/SSB = 1/BPR, (where BPR is biomass per recruit, an index of survival), survival fails with disastrous consequences.

Reddingius (1971) had shown that only weak regulation is needed to prevent extinction. When $F < M$, fish stocks do not expand rapidly, nor do they collapse when $F > 2M$. Indeed, fish stocks can remain stable even when $F \approx 5M$. Weak regulation is all that is needed at high stock, but very strong regulation is needed to sustain equilibrium at high fishing mortality and to prevent collapse.

If we look at the stock recruitment relationships of the plaice, North Sea herring and Arcto-Norwegian cod (Figure 8.2), recruitment may be relatively low at high stock, itself a small proportion of the unexploited

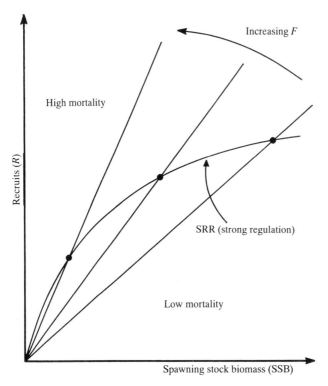

Figure 8.5 The departure of the stock recruitment relationship from the slope at the origin is evidence of strong density dependence. The line (R/SSB) increases with rising fishing mortality and the population is stable where it intersects the stock and recruitment relationship. (R/SSB) = (1/BPR), where BPR is biomass per recruit, an index of survival. Then, as (R/SSB) increases, survival declines, leading to collapse. (After Shepherd and Cushing, 1990.)

stock. Shepherd, in the article by Shepherd and Cushing (1990), noted that the arithmetic mean of a lognormal distribution is greater than the geometric mean by about $(\sigma^2/2)$ and the excess would increase as stock decreased.

A stock comprises a succession of recruitments and becomes their arithmetic mean. If the geometric mean is determined by a weak regulator, then larger average ratios of (R/SSB) will appear at low stock. As stock declines, rare extreme recruitments should occur. A simulation model was constructed on the basis of the Shepherd stock recruitment relationship. Let $a = 2\,R_0/P_0$; $K = P_0$ and $\beta = 0.5$ and let R_0 and P_0 be the mean observed recruitment and spawning stock biomass of the North Sea plaice. An expression was devised to make the variance inversely

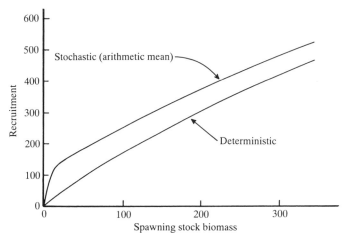

Figure 8.6 Two forms of stock recruitment relationship. One is deterministic based on the median and the second is stochastic based on the arithmetic mean (by raising the Shepherd equation by (exp $\sigma^2/2$)).

proportional to biomass and the median recruitment is raised by exp $(\sigma^2/2)$; Figure 8.6 shows the two stock recruitment relationships, first from the median which is deterministic and, second, from the arithmetic mean which is stochastic.

Simulations were made for one hundred years at different levels of fishing mortality, with constant and stock-dependent variability, which showed the range of variation of recruitment expected. However, the most remarkable simulation was that at high fishing mortality which endured for a thousand years: the stock appears to become extinct, but recovers with a strong year class after some centuries, entirely due to chance. This does not deny the results described in Chapter 5 on the effects of climate, but may amplify them. Figure 8.7 shows a stock recruitment relationship (with $F = 0.3$, 0.5 and 0.8) and the trend in survival (as ln (R/SSB) at $F = 0.8$) for a period of one hundred years: it does not differ much from the real data shown in Figure 8.2.

Animals and plants live in patches in the sea and the vital parameters of the populations must be distributed in a patchy manner. This is enough to provide a stochastic element in any case. Indeed, one result of a lognormal distribution of predators is that the distribution of the prey will be described by the median, or the geometric mean. The important point is that within the changing climate system a chance event may generate sharp increases in recruitment at very low stock which are subsequently amplified.

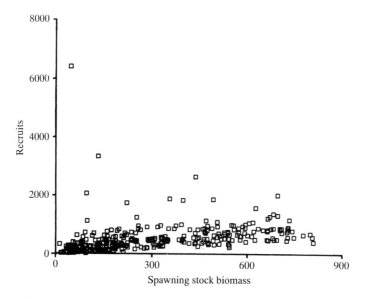

Figure 8.7 A stock recruitment relationship generated in a simulation model (based on the North Sea plaice) based on a lognormal distribution of recruitment. Under high fishing mortality for a thousand years, the stock was driven to very low levels and after centuries, recovered.

The biology of stock recruitment relationships

As fitted to the array of observations, the stock recruitment relationship represents an average condition for the period under examination and the deviations show differences in recruitment from year to year. It provides information that can be used to analyse environmental effects which are also governed to some degree by this average. As the time series lengthen, some secular changes will become shorter than the period averaged. For example, the west Greenland cod stock was very abundant for about 15 years or so, when it supplied migrants to the Iceland cod stock. The equilibrium implied by the models may in practice shift as more data become available.

Cushing (1979) studied the effect of a pollutant impact and assumed that it reduces density-independent survival, for example, of the number of eggs. Then, with a fixed (or relatively fixed) fecundity, the density-dependent mortality is also reduced, compensating for the impact and even yielding higher recruitment at high stock under some circumstances. Any environmental effect would act in the same way and so the density dependence will depend upon it. Such factors can operate at any point in the life cycle between spawning and recruitment, but as mortality diminishes with age their power will diminish.

In Chapter 7 a model of the growth and mortality of haddock larvae was discussed. The essence of the model is that the larvae may grow at less than the maximal rate and that they would metamorphose later (assuming that metamorphosis occurs at a fixed weight, which is approximately true). It was shown that cumulative mortality depended inversely on growth rate and that it was density dependent. Because the cumulative mortality depends inversely on growth rate, it also depends inversely on food. Hence the density-dependent mortality is greater with less food.

Perhaps the most interesting result of the model is to show that recruitment (as at metamorphosis) is determined by food and by numbers of larvae. To put it another way, as has been long expected, recruitment is determined partly by the density-dependent mortality; but, of course, there is also a component of density-independent mortality, of which we are almost completely ignorant. The death of the pelagic eggs is probably not density dependent and should perhaps be treated as a straight loss, perhaps one-third of the total density-independent mortality; the loss of eggs on the seabed may be density dependent, but this has not yet been shown. The most important consequence is that there may be a single process at the heart of the generation of recruitment, the growth and death of the larvae.

In the model the density-dependent mortality is a function of the relative lack of food. The remainder of the total mortality is density independent if the fecundity is more or less fixed. Hence, the ratio of the two forms of mortality in the model also depends on food if the growth rate is less than maximal and food is superabundant. As shown above, variation in impact will generate differences in the proportions of the two forms of mortality. So both components affect each other and a constant density-dependent mortality does not exist in this argument. An average density-dependent mortality does exist and the differences in recruitment from year to year reflect the differences in both forms of mortality and their mutual interactions.

Another output of the model is an expression of the match/mismatch hypothesis in that it was run in a matched condition, as the haddock larvae were allowed to grow with a cohort of their food. A mismatch would take place if the food cohort started earlier or later. Then density dependence is the partial result of mismatch. However, to test the model, work is needed on the larval stages and this is expensive.

In Chapter 5, in the discussion of climate and fisheries, it was shown that the great changes in the fisheries, changes in recruitment, are linked to climatic factors. They affect the production cycle in the sea directly, advancing or delaying it in time, and it is this that provides the basis of the match/mismatch hypothesis. Hence the model of the stock recruitment relationship is peculiarly fitted to describe the effect of climatic factors on recruitment, to lay the bases of the match/mismatch hypothesis and perhaps to separate density-dependent mortality from the density independent.

Conclusion

The study of stock and recruitment remains the central problem of fisheries science. If the stock is collapsed by recruitment overfishing, the stock recruitment relationship can be described, but the biological information needed to elucidate it is largely absent. Beverton and Iles (1992*b*) have described the density-dependent mortality of some flatfish in European coastal waters. The model of Cushing and Horwood (1994) raises the possibility that the mortality of the later larvae is density dependent. It is an imaginary construct, but one conclusion is that the density-dependent mortality of the late larval stages might be greater than that on the beaches. Further, the estimates of density-independent survival do not exist despite the fact that they are needed to establish the slope at the origin of the stock recruitment relationship.

The problem is a biological one in that more information is needed on the growth rate and feeding rates of fish larvae, and more observations on their mortality rates as function of food and predation. With the use of traditional methods of ship surveys, the cost might be prohibitive with perhaps only a low chance of success. New methods are needed by which the sampling power is increased by orders of magnitude and which can be sustained at sea for long periods, perhaps using acoustic methods and data assimilation procedures.

9

Conclusion: Fisheries and marine ecology

Introduction

In this last chapter, the ideas put forward in earlier ones are reviewed, summarized and to some degree, amplified. The book has two objects. The first is to broaden fisheries science to include more marine biology or biological oceanography. The second object is to explore the nature of the regulation of fish populations at all stages of their life history. My interests have included both the details of stock assessment and the open field of marine biology and so my approach should not be unexpected.

The effects of climate

Fisheries have appeared for a few decades; for example, halibut were abundant on George's Bank between 1830 and 1850. Day (1880–94) noted that herring appeared off Cromarty in northern Scotland between 1690 and 1709. From time to time, traditional herring fisheries have disappeared, such as that in Loch Fyne on the west coast of Scotland (Day, 1884) and that in the Firth of Forth on the east coast of Scotland (Mitchell, 1864). Cushing (1961) described the collapse of the Plymouth herring stock during the thirties of the present century. Johansen (1926) gave an account of the appearance of large numbers of haddock in the Baltic in the early twenties; they disappeared in a few years because they cannot spawn there in the low salinity water. Goode (1884–95) described the appearance of the tilefish off the Eastern Seaboard of the United States, large numbers of which were killed in 1882. Such events are part of fishermen's lore and although catches depend on the patterns of migration, radical and secular changes are part of their tradition.

There have been great changes in the fisheries through the centuries, particularly in Scandinavia and off Japan. Ljungman (1880) recorded the alternation between the Norwegian and Swedish herring periods for nearly a thousand years; we now know that during the Swedish periods, North Sea herring penetrated the Skagerak, to the Bohuslan coast of Sweden. During the Norwegian periods, the Norwegian herring were abundant in the Norwegian Sea (Høglund, 1972). Devold (1963) described the recent alternation in some detail from 1760 to 1960. The Swedish periods last about 40 years which suggests a link with the North Atlantic Oscillation. The mechanisms that govern these events remain obscure.

Off Japan, catches of the Hokkaido herring and the Japanese sardine have alternated for centuries (Uda, 1957) and the latter correspond roughly to the Norwegian herring periods. In Chapter 5 was described the alternation between anchovy and sardine in various parts of the world ocean, shown most dramatically in the Kawasaki diagram (Figure 5.23). The Japanese sardine returned in the early seventies, nearly 40 years after its collapse in the late thirties. Zupanovitch (1968) linked the high and low catches of the Adriatic sardine since the sixteenth century with Uda's changes in the catches of the Japanese sardine. The long-term changes in the pelagic stocks have been recorded for a long time.

Ottestad (1960) analysed cod catches from 1885 to 1939 in the Vestfjord in northern Norway. Four periods were detected by Fourier analysis from the widths of the annual rings on the pine trees in the area and their components were added each year. The resulting curve was fitted by least squares to the time series of the cod catches, lagged by seven years to allow for the age of recruitment. If this analysis were right, the cod year classes originated in the same conditions experienced by the pine trees each year. Factors common to the two systems are solar radiation, wind strength and direction, which dominate events in larval life.

In Chapter 5 the effects of climatic factors upon differences in recruitment from year to year were established in the work of Shepherd *et al.* (1984), Hollowed *et al.* (1987, 1992), Koslow *et al.* (1987) and Myers (1991). Such are at the roots of the long-term changes described above. The most dramatic changes are those that determined the rise and fall of the west Greenland fishery. Because the stock was maintained and sustained from Iceland it depended on the winds from the east across the Denmark Strait and in their turn upon the presence of the Greenland high. It built up slowly for a period of 40 years, since the beginning of the century, and collapsed in the late sixties. Since then the west Greenland cod stock has only been maintained sporadically. The proximate causes were the transport of larvae and juveniles from Iceland in the Irminger Current under the easterly wind and the existence of relatively warm water at west Greenland.

As the Greenland high collapsed, it first contracted to a band over east Greenland with very cold northerly winds over the east Greenland Current, making it temporarily a polar current. Here was created the great salinity anomaly of the seventies, a large mass of cool water. It retained its identity for 14 years, as it passed from Iceland to west Greenland, Labrador, the edge of the Grand Bank and across the Atlantic to the Faroe–Shetland Channel, the Barents Sea and back to Iceland. During its passage the recruitments to a number of fish stocks were reduced and I suggested that this occurred because the production cycle was delayed in the cooler water (Cushing, 1988*b*).

At about the same time, a ridge of pressure difference established itself between Iceland and Morocco, so that northerly winds blew in spring across the North Sea and the North East Atlantic. In the central North Sea, under the influence of gales, production was reduced and delayed and in the western North Sea it was merely reduced. As a consequence, the phytoplankton and zooplankton in the North Sea were reduced between the fifties and the seventies and both recovered in the eighties. The production of *Calanus* in the North Sea was delayed and the gadoid stocks may have profited with higher recruitment between the early sixties and the early eighties.

The most remarkable events were the concomitant rises and falls of three sardine stocks across the Pacific (and there may have been links with such stocks in the other subtropical oceans). Anchovy and sardine tend to exchange prominence on a decadal scale and this may well be a consequence of the adaptation of the two fishes to different forms of upwelling systems, one stronger but intermittent and the other weaker but persistent. Bakun and his colleagues have described the systematic differences in upwelling between the various habitats in the subtropical fishes (see Chapter 5).

The great changes in fish stocks appear to be linked to climatic changes, sometimes quite clearly as with the west Greenland cod stock and sometimes by implication as with the alternation between sardine and anchovy. Because climatic factors affect events in the mixed layer of the ocean directly where the larvae live, recruitment may be mainly determined during that period of the life history. Hence, the study of production in the sea is of great importance to fisheries biologists if they wish to understand how the recruitment of fishes is generated.

Production of life in the sea

In the last two decades, great advances have been made in marine biology and in the study of productive processes in the sea. A fair proportion of marine production is that due to the cyanobacteria, organisms discovered as recently as the late seventies. A new part of the oceanic food chain has been revealed, the microbial food loop, found in the oligotrophic ocean and in the temperate seas of summer. It is a long chain from the free-living bacteria (also discovered quite recently), cyanobacteria and eukaryotes, to the protozoa that feed on them and the ciliates that eat the protozoa in their turn. Finally, the ciliates are taken by copepods. The organisms are all quite small and the transfer of energy is quick. They are dispersed, nearly all are eaten and the transfer of energy is efficient; the euphotic layer is deep and the ocean is deep blue. However, the organisms are small, many of them too small to attract the attention of copepods and fish larvae (the exception being the ciliates). This might well explain why fisheries are in the main found where bigger algal cells are produced, as in the spring outburst in the North Atlantic, in upwelling areas and at fronts.

A second advance lay in the use of continuously recording instruments to discover the position of tidal fronts around the British Isles and elsewhere. The same methods revealed that production in the summer temperate seas was found mainly in the region of the thermocline. However it was continuous there whereas it had been believed earlier that such production was sporadic, merely because the sampling bottles were lowered to fixed depths, which might find the thermocline or might miss it. Perhaps the most important consequence was the collaborative work that started between physicists and biologists, which is now commonplace.

The fronts separate waters that are stratified in summer from those that are mixed. In the former, there is a spring outburst as the stratification develops but in the mixed region production continues throughout the summer, if slowly. It is interesting to note that the herring larvae, which grow slowly, do so in the mixed waters within the tidal fronts. We might expect to find slow-growing larvae in the mixed waters and fast-growing ones in the stratified sea on the other side of the front. One of the most interesting developments at the present time is the very detailed model studies under way on George's Bank on the survival of gadid larvae with respect to the position and shifting of the tidal front there. Another is the use of transport models within the North Sea to track the movement of larvae or tracers from spawning ground to nursery ground. Indeed, detailed studies are starting to show that fish do spawn in productive areas of the sea, to give their larvae optimal chances of survival.

Models of productive processes in the sea have been made since the time of Lohmann (1908). The intermediate models in development, those of Riley and his followers (see Chapter 3), described the productive processes in the sea, but in the more recent ones two important steps were made. Evans and Parslow (1985) devised a model which runs successfully from year to year and with which they could experiment to establish the possible ways in which processes dominate production. A model developed at the same time by Fasham (1985) was based not on the quantities of biomass at the different trophic levels, but on the fluxes between them. Later developments of this model have been used with the general circulation models to describe the distribution of production across the Atlantic Ocean. In another direction, Platt and his colleagues (see Chapter 3) have shown how satellite estimates of colour can be used to estimate the oceanic production, which will be of considerable importance when the new satellites start to fly. Lastly, Woods and his colleagues (see Chapter 3) have developed Lagrangian models of production within the mixed layer by which the detrainment of diatoms below the thermocline was described. Their more recent work has shown that the compensation depth may well track the thermocline in early summer and that the Sverdrup critical depth may not be quite right, because some production must continue during winter.

If fish larvae do depend upon the productive processes to survive and if the generation of recruitment depends upon them, then the models described above should be part and parcel of the trade of fisheries science, as indeed it was during the thirties. The major problems of fisheries science today are the generation of recruitment and the stock recruitment relationship. Within both lie the processes that govern production in the sea.

The population in the sea

The plaice in the southern North Sea maintain themselves in the same position from year to year with a variety of devices. They spawn at the same location at the same time each year, although the season of reproduction is spread over a period of two or three months. The maturing fish migrate south in late autumn by selective tidal transport until they reach their spawning ground. After spawning, the adults return to the north by the same mechanism in reverse. The larvae are drifted away in midwater towards the nursery ground on the flats on the coasts of northern Holland and Germany. They sink to the seabed and migrate into the Wadden Sea again by selective tidal transport. Later, from there, they diffuse seawards and in depth until they start to join the adult stock and the cycle of migration starts again.

There is a low leak from spawning group to spawning group within the North Sea. There is, however, a distance from the spawning ground, which bounds the area within which all larvae drifted from it have reached the seabed. Hence the unity of the stock is defined within the circuit of migration, by the position of the fixed spawning ground and the range of greatest larval dispersal.

It has long been hoped that genetic studies would resolve the problem of the unity of a stock. For about 20 years, the results have established differences which are perhaps obvious, but, in the last five years or so, when the sampling has been extended considerably, three successful studies have been made. (1) Stocks of the barramundi over a range of more than a thousand miles in northern Australia have been separated by rivers, each of which is a little more than a hundred miles from the next. (2) Similarly, stocks of chinook salmon in the Columbia River and its tributaries have been separated successfully. (3) The same methods have led to the discovery that two stocks of haddock in the North Sea are separated by the Greenwich meridian; the *raison d'être* of this observation is that the western group is supported from the west coast of Scotland by the Fair Is. current and the Shetland current. The eastern group is supported from the Faroe Is. by the Tampen Bank current. Thus, advances have at last been made. One remarkable discovery was that the return of the George's Bank stock of herring was not a colonization from elsewhere, but was a resurgence of the original stock; it was established by the presence of a rare gene.

An examination of extensive tagging experiments on cod across the Atlantic has shown that there is a low level of distant recaptures that could be regarded as a low-level leak or stray which is roughly comparable to a genetic loss. A study of Thompson's results on the Grand Bank and off Nova Scotia (see Chapter 5) suggested that the fishes returned to their ground of first spawning and that after two or three years there were a few distant recaptures. In general, however, a spawning group was restricted to an area of about 300 miles across.

The generation of recruitment

One of the restraints to the study of recruitment is the earlier inadequate sampling of the larval fish. Certain larvae are well sampled and some gears in present use are in the main adequate. The real trouble is that the older gears sampled the smaller larvae properly, but growth and mortality may only become effective on the older animals, which were poorly sampled. Again, the most attractive method of ageing the larval fish by the daily rings on their otoliths has taken a little time to reach practical validation. Perhaps the most promising outcome will be the more extensive use of distributions of hatch dates in association with estimates of survival.

The major advances have been made by three groups: Leggett, Lasker, Solemdal and their colleagues. The extensive studies on the emerged capelin larvae by Leggett's group have shown that the warmer water with richer zooplankton under onshore winds provide better survival for the capelin larvae than the upwelled cooler and poorer water under the offshore winds. Further, the larvae emerged under the onshore wind perhaps because the water was warmer. Eggs were eaten by winter flounders and the larvae by jellyfish. The system on the east coast of Newfoundland was compared with that in the St Lawrence estuary where the larvae emerged at dusk and dawn. Mortality of the newly emerged larvae was not density dependent, denying Hjort's first hypothesis (Hjort, 1914). It is likely that the larvae of the winter flounder metamorphose at a relatively fixed weight and that the spread in time of metamorphosis is a consequence of differences in growth rates.

Lasker and his colleagues (see Chapter 6) showed that the larvae of the northern anchovy needed a number of food particles of a particular size and so, at sea, he was able to classify water as bearing enough food or not. His cruise off San Onofre showed that there were enough food organisms within the thermocline, but that after a storm the food organisms were dispersed. Subsequently, an index of calm weather was devised and survival was greater during such periods.

Solemdal and his colleagues (see Chapter 6) showed that the cod larvae depended on the numbers of *Calanus* nauplii in their guts in an Ivlev-like curve. They then established that the peak abundance of

Calanus clearly depended on temperature, the higher the temperature the earlier the date of peak *Calanus* abundance. They compared the distribution of larvae in time in the Vestfjord with that of *Calanus* nauplii for a number of years and showed that in some years the distribution of larvae matched that of their food and in others they did not. They established that the mortality of the early juveniles (at an age of about two months) was density dependent and it can be shown that numbers of recruits were correlated with numbers of the early juveniles.

Common to all three groups is the belief that the presence or absence of food plays a crucial part in the life of the larvae and of the juveniles. Indeed, the generation of recruitment appears to be a consequence of feeding success. This is a most important conclusion and it forms the basis of my match/mismatch hypothesis. As recently developed, this hypothesis (Cushing, 1992) now includes events in regions of upwelling because in subtropical pelagic fishes the interspawning interval appears to depend on food. In temperate waters, fish spawn at relatively fixed (if rather extensive) seasons and in upwelling areas they do so when food is adequate. Hence there is a greater chance of match there. But the differences remain and mismatch of larval production to that of their food may well play a part in the generation of recruitment.

Density-dependent processes

Density-dependent processes play a prominent part in the populations, as shown quite clearly by any stock recruitment relationship. It is difficult to establish density-dependent mortality because information is needed for many generations, often more than have been sampled. Similarly, density-dependent growth in adults has often been cited as a source of population regulation but the evidence dissipates as more becomes available. The study of density dependence has been hedged with statistical problems most of which have been recently relaxed and clarified. In the study of fishes, most have believed that the larvae could not affect the density of their food, despite the proposal that haddock larvae grew with a cohort of their food organisms, *Calanus*. Yet there are sporadic observations that recruitment can sometimes be predicted from numbers of late larvae.

Shepherd and Cushing (1980) had proposed that by the end of a critical period, numbers were determined by the mitigation of predatory mortality by growth, in essence, the Ricker–Foerster thesis. With abundant food, the growth rate is maximal and the critical period is least. When food is short, the growth rate is less and the critical period longer. The equation developed by Shepherd is a model of the stock recruitment relationship based on the three trophic levels within which the animals live. Shepherd (1994) extended the argument in terms of the search capacity of the predators. Thence emerges the satiation of the predator which might imply an increase of mortality with age if the numbers of predators were constant. As it diminishes with age, the numbers of predators must decline. In the field, satiation cannot be distinguished from emigration of the predators. The argument was developed on much the same lines as in the earlier paper in terms of the three trophic levels: food, larvae (or juveniles) and predators. The result was an equation very close to the three-parameter stock recruitment relationship developed by Shepherd (1982).

A review of the biology of fish larvae revealed that numbers of recruits can, here and there, be forecast from the numbers of late larvae (herring, Walleye pollock, plaice, flounder and capelin). There is an element of speculation in the suggestion that recruitment can, in the main, be established during the late larval stages. More surprising, the mortality of larvae of the Japanese saury may well be density dependent. So the idea that population regulation might start during the larval stage should not be abandoned.

In the early eighties the question was raised, whether fish larvae were abundant enough to affect the density of their food, despite the fact that Jones (1973) had suggested that larval haddock grew with cohorts of their food, *Calanus finmarchicus*. With a brief model, Cushing (1983) suggested that the numbers of such a cohort of *Calanus* could be reduced by up to 3% d^{-1}, which could be effective over a period, long enough in temperate waters.

An extension was made by Cushing and Horwood (1994). In their model, haddock larvae were allowed to grow to a constant weight at metamorphosis. At the same time, they were allowed to die at a rate decreasing with time. Because the maximal growth rates and the ration needs were known from the work of Laurence (1982), imaginary cohorts of haddock larvae were grown at different food levels from hatching to metamorphosis. The results were expressed in cumulative mortality and as recruitment at metamorphosis.

The cumulative mortality was partly density dependent and the magnitude of this component was unexpectedly high, higher than that observed on the beaches. The recruitment at metamorphosis depended on food and its distribution as a function of the initial numbers of larvae resembled a stock recruitment relationship. The quantities of larvae and food organisms were of the order observed in the George's Bank haddock. At much higher initial numbers of larvae, the recruitment at metamorphosis was reduced, which superficially resembles the state of affairs reported by Herrington (1948). The model does not represent a stock recruitment relationship because the juvenile mortality has been excluded.

In the model, the haddock larvae were allowed to grow with a cohort of their food, *Calanus finmarchicus*, in such a way that the smallest larvae fed on nauplii and the larger on bigger stages. So the production of the haddock larvae was matched to that of their food. Mismatch would be obtained if and when the nauplii were not available at the right time; if early, the food organisms would be too big and if late the larvae would starve. Thus the model might provide the means of testing the match/mismatch hypothesis.

If the model were true, there is density-dependent mortality during the larval stages, an entirely unexpected result. Up to the present, density-dependent mortality has been detected only amongst the juvenile stages. Sundby *et al.* (1989) demonstrated density-dependent mortality amongst the early juveniles of the Arcto-Norwegian cod. Beverton and Iles (1992,*a*,*b*) examined the mortality of three species of flatfish on the European beaches. They showed that it was density dependent. In both observations the density-dependent mortality amounted to about 0.1 to 0.2, less than posited in the larval model described above.

In the adults, density-dependent growth and its consequence, density-dependent mortality, has often been proposed. It is frequently present among juveniles, but among adults it appears to be absent and so density-dependent fecundity does not in general regulate the fish populations. The density-dependent growth of juveniles is probably of some importance because its consequence is delay in the age of maturation, and so it is a derived form of density-dependent fecundity. It has not been investigated as fully as it might deserve. But the reduction in the age of maturation under exploitation has been frequently observed.

Cannibalism has often been cited as a potential form of population regulation. Currently, there are two forms of candidate, first, the exploitation of the eggs by the adults and, second, the capture of juveniles by larger animals. Both occur but the effect of cannibalism as regulator of the population has not been established, despite the fact that large numbers of victims have been recorded.

There are a number of potential regulators: density-dependent mortality, density-dependent growth as it may affect fecundity in adults, the age of maturation at the end of juvenile life, and cannibalism. Of these, only density-dependent mortality has been established, although the age of maturation remains a candidate. The case for cannibalism has not yet been made because cannibal predation must be separated from the non-cannibal and population regulation has not been shown.

The stock recruitment relationship is the usual way of expressing the nature of population regulation in fishes. It is often misunderstood because the environmental effects are displayed as deviations from a particular model. The implicit question is whether a decline in recruitment at low stock is due to fishing or not. The Iceland spring spawners probably declined as the Great Slug passed through Icelandic waters. If that were true then the few survivors might have been taken anyway, but the origin of decline is never known at the time. So the management problem should be met by use of the biological reference points, such as F_{med} or its analogues. The real danger is that when recruitment declines naturally so does the subsequent stock and usually, in the past, the fishing effort remained the same and then recruitment overfishing supervened. It is no accident that collapses due to recruitment overfishing have often been associated with an environmental decline, if only transient.

The stock recruitment relationship, when well established with a period of recruitment overfishing, displays density-independent survival and density-dependent processes, together with the variability of recruitment during the period of observation. The problem is to separate and to quantify the components. If the density-dependent mortality is of the order expected from the model described above, population regulation starts during the larval phase and is presumably completed during juvenile life.

A number of stock recruitment relationships have been displayed. As the decades of observation lengthen the relationships are no longer mere scattergrams but appear to be more coherent and there are differences between different groups of fishes. The markedly domed relationship has tended to disappear as more information has become available. They remain little more than pictures of biological interest, with management relying on the biological reference points.

One of the early assumptions was that the density-dependent processes were more or less constant. The stock recruitment relationship displays an average condition and the model described above would produce different density-dependent mortalities from year to year and different density-independent survivals. The model relationships imply equilibria which may be illusory. Great changes have been described, originating in changes in recruitment. It is possible that relatively short secular changes are more important than the period needed to establish an average equilibrium.

Recruitment overfishing is destructive because when stocks recover after a ban on catches, the fishermen have gone. Since the forties, stocks throughout the world have been ravaged by recruitment overfishing and methods have been devised to try to prevent it. To do so properly the biological mechanisms need to be understood. Much that is needed is ordinary biological work, but the crucial studies will be those at sea for extended periods with all the sophisticated science that modern oceanographers can provide.

References

Ahlstrom, E.H. (1966) Distribution and abundance of sardine and anchovy larvae in the California Current region off California and Baja California, 1951–64: a summary. *Spec. Sci. Rep. U.S. Fish Wildlife Serv.* **534**: 1–71.

Alcaraz, M., Paffenhöfer, G.A. and Strickler, J.R. (1980) Catching the algae: a first account of visual observations on filter feeding calanoids. In *Evolution and ecology of zooplankton communities* (ed. Kerfoot, W.C.), pp. 241–248. University of New England Press, Hanover, New Hampshire, and London.

Alheit, J. (1986) Egg cannibalism versus egg predation: their significance in anchovies, ICES CM H59 5P.

Alldredge, A.L. and Gotschalk, C.C. (1989) Direct observations of the mass flocculation and diatom blooms: characteristics, settling velocities and formation of diatom aggregates. *Deep Sea Research*, **36**: 159–171.

Alm, G. (1946) Reasons for the occurrence of stunted fish populations. *Medd. fr. Sat. unders. och forsoks. f. sottvat*, **25**: 1–125.

Andersen, P. and Fenchel, T. (1985) Bacterivory by microheterotrophic flagellates in seawater samples. *Limnol. Oceanogr.* **30**: 198–202.

Anderson, G.C., Lam, R.K., Booth, B.C. and Glass, J.M. (1977) A description and numerical analysis of the factors affecting the processes of production in the Gulf of Alaska. *Univ. Washington Dept. Oceanography Spec. Rep*: 1–121.

Anderson, G.C. and Munson, R.E. (1972) Primary productivity studies using merchant vessels in the North Pacific Ocean. In *Biological oceanography of the North Pacific Ocean* (ed. Takenouti, A.Y.), pp. 245–251. Idemitsu Shoten, Tokyo.

Andersson, L., Ryman, N., Rosenberg, R. and Stahl, G. (1981) Genetic variability in Atlantic herring (*Clupea harengus harengus*): description of protein loci and population data. *Hereditas*, **95**: 69–78.

Anon, (1992) Techniques for biological assessment in Fisheries management. *Ber. Okolog. Forsch.* **9**: 1–63.

Arnold, G.P. (1981) Movements of fish in relation to water currents. In *Animal migration* (ed. D.J. Aidley), pp. 55–79. Cambridge University Press, Cambridge.

Arnold, G.P. and Cook, P.H. (1984) Fish migration by Selective Tidal Stream Transport: first results with a computer simulation model for the European continental shelf. In *Mechanisms of migration in fishes* (ed. McCleave, J.D., Arnold, G.P., Dodson, J.J. and Neill, W.H.), pp. 227–261. Plenum Press, New York.

Arnold, G.P., Greer-Walker, M. and Holford, B.H. (1989) Fish behaviour: achievements and potential of high resolution sector scanning sonar. *Rapp. Procès-Verb. Réun. Cons. Int. Explor. Mer*, **189**: 112–122.

Arnold, G.P. and Metcalfe, J.D. (1989) Fish migration: orientation and navigation or environmental transport. In orientation and navigation – birds, humans and other animals. *J. Navigation*, **42**: 367–374.

Aspinwall, N. (1974) Genetic analysis of North American populations of the pink salmon, *Oncorhynchus gorbuscha*, possible evidence for the neutral mutation – random drift hypothesis. *Evolution*, **28**: 295–305.

Astthorsson, O.S., Hallgrimsson, I. and Jonsson, G.S. (1983) Variations in zooplankton densities in Icelandic waters in spring during the years 1961–82. *Rit. Fiskideildar*, **7**: 73—113.

Atkins, W.R.G. (1923) The phosphate content of fresh and salt waters in its relationship to the growth of algal plankton, Part 1. *J. Mar. Biol. Assn UK*, **13**: 119–150.

Atkins, W.R.G. (1925*a*) On the thermal stratification of sea water and its importance for the algal plankton. *J. Mar. Biol. Assn UK*, **13**: 696–699.

Atkins, W.R.G. (1925*b*) Seasonal changes in the phosphate content of sea water in relation to the growth of the algal plankton during 1923 and 1924. *J. Mar. Biol. Assn UK*, **13**: 700–720.

Azam, F., Fenchel, T., Field, J.C., Gray, J.S., Meyer-Reil, L.A. and Thingstad, F. (1983) The ecological role of water column microbes in the sea. *Mar. Ecol. Progr. Ser.* **10**: 257–263.

Backhaus, J., Harms, I. and Krause, M. (1993) A hypothesis concerning the space time succession of *Calanus finmarchicus* in the northern North Sea. *ICES J. Mar. Sci.*

Bailey, K.M. (1981) Larval transport and recruitment of Pacific hake, *Merluccius productus*. *Mar. Ecol. Prog. Ser.* **6**: 1–9.

Bailey, K.M., Francis, R. and Schumacher, J. (1986) Recent information on the causes of variability in recruitment of Alaska pollock in the eastern Bering Sea: physical conditions and biological interactions. *Bull. Int. N. Pac. Fish. Commn*, **47**: 155–165.

Bailey, K.M., Francis, R.C.C. and Stevens, P.R. (1982) The life history and fishery of Pacific whiting, *Merluccius productus*. *CalCOFI Rep.* **23**: 81–98.

Bailey, K.M. and Incze, L.S. (1985) El Niño and the early life history and recruitment of fishes in temperate marine waters. In *El Niño North* (ed. Wooster, W.S. and Fluharty, D.L.), pp. 143–165. Washington Sea Grant.

Bailey, K.M. and Spring, S. (1992) Comparison of larval, age-0 juveniles and age-2 recruit abundance indices of walleye pollock, *Theragra chalcogramma*, in the western Gulf of Alaska. *ICES J. Mar. Sci.* **49**: 297–304.

Bailey, K.M. and Stehr, C.L. (1986) Laboratory studies on the early life history of the walleye pollock, *Theragra chalcogramma* (Pallas). *J. Exp. Mar. Biol. Ecol.* **99**: 236–246.

Bailey, K.M. and Stehr, C.L. (1988) The effects of feeding periodicity and ration on the rate of increment formation in otoliths of larval walleye pollock, *Theragra chalcogramma* (Pallas). *J. Exp. Mar. Biol. Ecol.* **122**: 147–161.

Bakken, E. (1983) Recent history of Atlanto-Scandian herring stocks. *F.A.O. Fish. Rep. 29* **2**: 521–536.

Bakun, A. (1973) Coastal upwelling indices. West coast of North America, 1946–71. NOAA Tech. Rep. NMFS SSRF-671, 103 pp.

Bakun, A. (1985) Comparative studies and the recruitment problem: searching for generalizations *CalCOFI*, **26**: 30–40.

Bakun, A. and Parrish, R.R. (1980) Environmental inputs to fishery population models for Eastern Boundary Current regions. *Intergovernmental Oceanogr. Commn. Workshop Report*, **28**: 67–104.

Bakun, A. and Parrish, R.H. (1982) Turbulence, transport and pelagic fish in the California and Peru current systems. *CalCOFI Rep.* **23**: 99–112.

Bakun, A. and Parrish, R.H. (1990) Comparative studies of coastal pelagic fish reproductive habitats: the Brazilian sardine (*Sardinella aurita*). *J. Cons. Int. Explor. Mer*, **46**: 269–283.

Bakun, A. and Parrish, R.H. (1991) Comparative studies of coastal pelagic fish reproductive habitats: the anchovy (*Engraulis anchoita*) of the Southwestern Atlantic. *ICES J. Mar. Sci.* **48**: 343–361.

Banse, K. (1982) Cell volumes, maximum growth rates of unicellular algae and ciliates and the role of ciliates in the marine pelagial. *Limnol. Oceanogr.* **27**: 1059–1071.

Banse, K. (1991*a*) Rates of phytoplankton cell division in the field and in iron enrichment experiments. *Limnol. Oceanogr.* **36**: 1886, 1898.

Banse, K. (1991*b*) Iron availability, nitrate uptake and exportable new production in the subarctic Pacific. *J Geophys. Res.* **96**: 741–748.

Banse, K. (1991*c*) Iron, nitrate uptake by phytoplankton and mermaids. *J. Geophys. Res.* **96**: 20701.

Banse, K. (1992) Grazing, temporal changes of phytoplankton concentrations and the microbial loop in the open sea. In *Primary productivity and biogeochemical cycles in the sea* (ed. Falkowski, P.G. and Woodhead, A.D.), pp. 409–440. Plenum, New York.

Barlow, R.G., Burkill, P.H. and Mantoura, R.F.C. (1988) Grazing and degradation of algal pigments by marine protozoan *Oxyrrhis marina*. *J. Exp. Mar. Biol. Ecol.* **119**: 119–129.

Barnett, T.P. (1991) The interaction of multiple time scales in the tropical climate system. *J. Clim.* **4**: 269–285.

Barnett, P.R.O., Watson, J. and Connelly, D. (1984) A multiple corer for taking virtually undisturbed samples from shelf, bathyal and abyssal sediments. *Oceanolog. Acta*, **7**: 399–408.

Bartlett, M.S. (1949) Fitting a straight line when both variables are subject to error. *Biometrics*, **5**: 207–213.

Bartsch, J., Brander, K., Heath, M., Munk, P., Richardson, K. and Svendsen, E. (1989) Modelling the advection of herring larvae in the North Sea. *Nature*, **340**: 632–636.

Bartram, W.C. (1981) Experimental development of a model for the feeding of neritic copepods by phytoplankton. *J. Plankt. Res.* **3**: 25–51.

Bartsch, J. (1993) Application of a circulation and transport model system to the dispersal of herring larvae in the North Sea. *Cont. Shelf Res.* **13**: 1335–1361.

Baumgartner, T.R., Soutar, A. and Ferreira-Bartrina, V. (1992) Reconstruction of the history of the Pacific sardine and northern anchovy populations over the past two millenia from sediments of the Santa Barbara basin, California. *CalCOFI Rep.* **33**: 24–40.

Beacham, T.D., Withler, R.E. and Gould, A.P. (1985) Biochemical genetic stock identification of pink salmon (*Oncorhyncus gorbuscha*) in Southern British Columbia and Puget Sound. *Can. J. Fish. Aquat. Sci.* **42**: 1474–1483.

Beardall, J., Foster, P., Voltolina, D. and Savidge, G. (1982) Observations on the surface water characteristics in the western Irish Sea: July, 1977. *Est. Coast. Shelf Sci.* **14**: 589–598.

Beddington, J.R. and Basson, M. (1993) The limits to exploitation on land and sea. *Phil. Trans. Roy. Soc. London* **343**: 87–??.

Beddington, J.R. and Cooke, J.G. (1983) The potential yield of fish stocks. *FAO Fish. Tech. Paper* **242**: 1–47.

Beers, J.R. and Stewart, G.L. (1969) Microzooplankton and its abundance relative to the larger zooplankton and other seston components. *Mar. Biol.* **4**: 182–189.

Beers, J.R. and Stewart, G.L. (1971) Microzooplankton of the plankton communities of the upper waters of the eastern tropical Pacific. *Deep Sea Res.* **18**: 861–883.

Berner, L. (1959) The food of the larvae of the northern anchovy, *Engraulis mordax. Bull. Inter. Amer. Trop. Tuna Commn,* **4**: 1–22.

Betzer, P.R., Showers, W.J., Laws, E.A., DiTullio, G.R., Kroopnick, P.M. and Winn, C.D. (1984) Primary productivity and particle fluxes on a transect of the equator at 153° W. in the Pacific Ocean. *Deep Sea Res.* **31**: 1–11.

Beverton, R.J.H. and Holt, S.J. (1957) On the dynamics of exploited populations. *Fish. Invest. London Ser. 2,* **19**: 533.

Beverton, R.J.H. and Iles, T.C. (1992*a*) Mortality rates of 0 group plaice (*Pleuronectes platessa* L.), dab (*Limanda limanda* L.) and turbot (*Scophthalmus maximus* L.) in European waters. II Comparison of mortality rates and construction of life table for 0 group plaice. *Neth. J. Sea Res.* **29**: 49–59.

Beverton, R.J.H. and Iles, T.C. (1992*b*) Mortality rates of 0 group plaice (*Pleuronectes platessa* L.), dab (*Limanda limanda* L.) and turbot (*Scophthalmus maximus* L.) in European waters. III Density dependence of mortality rates of 0 group place and some demographic implications. *Neth. J. Sea Res.* **29**: 61–79.

Beverton, R.J.H. and Lee, A.J. (1964) The influence of hydrographic and other factors on the distribution of cod on the Spitzbergen shelf. *Int. Commn North West Atl. Fish. Spec. Publ.* **6**: 225–245.

Beverton, R.J.H. and Tungate, D.S. (1967) A multi purpose plankton sampler. *J. Cons. Int. Explor. Mer,* **31**: 145–157.

Beyer, J.E. and Laurence, G.C. (1980) A stochastic model of larval fish growth. *Ecol. Modelling* **8**: 109–132.

Bienfang, P. and Takahishi, M. (1983) Ultraplankton growth rates in a subtropical ecosystem. *Mar. Biol.* **76**: 213–218.

Billet, D.S.M., Lampitt, R.S., Rice, A.L. and Mantoura, R.F.C. (1983) Seasonal sedimentation of phytoplankton to the deep sea benthos. *Nature,* **302**: 520–522.

Bishop, J.K.B., Conte, M.H., Wiebe, P.H., Roman, M.R. and Langdon, C. (1986) Particulate matter production and consumption in deep mixed layers: observations in a warm core ring. *Deep Sea Res.* **33**: 1813–1841.

Bjerknes, J. (1966) Survey of El Niño 1957–8 in its relation to the tropical Pacific meteorology. *Bull. Inter-Amer. Trop. Tuna Commn,* **12**: 3–42.

Bjerknes, J. (1972) Global ocean atmosphere interaction. *Rapp. Procès-Verb. Réun. Cons. Int. Explor. Mer,* **162**: 108–119.

Bjørke, H. and Sundby, S. (1987) Distribution and abundance indices of larval and post larval cod. In *The effect of oceanographic conditions on distribution and population dynamics of commercial fish stocks in the Barents Sea* (ed. Leong, H.), pp. 127–144. Institute of Marine Research, Bergen.

Blacker, R.W. (1982) Recent occurrences of blue whiting *Micromesistius poutassou* and Norway Pout *Trisopterus esmarkii* in the English Channel and Southern North Sea. *J. Mar. Biol. Assn UK*, **61**: 307–313.

Blaxter, J.H.S. (1988) Pattern and variety in development. In *Fish physiology XIa* (ed. Hoar, W.S. and Randall, D.), pp. 1–58. Academic Press, Chichester and New York.

Blaxter, J.H.S. and Staines, M.E. (1971) Food searching potential in marine fish larvae. In *Fourth European Marine Biology Symposium* (ed. D.J. Crisp), pp. 467–485. Cambridge University Press, Cambridge.

Bollens, S., Frost, B.W., Schwaninger, H.R., Davis, C.S., Way, K.J. and Landsteiner, M.C. (1992) Seasonal plankton cycles in a temperate fjord with comments on the match/mismatch hypothesis. *J. Plankt. Res.* **14**: 1279–1305.

Booth, B.C., Lewin, J. and Lorenzen, C.J. (1988) Spring and summer growth rates of subarctic phytoplankton assemblages determined from carbon uptake and cell volumes estimated using epifluorescence microscopy. *Mar. Biol.* **98**: 287–298.

Booth, B.C., Lewin, J. and Norris, R.E. (1982) Nanoplankton species predominant in the subarctic Pacific in May and June, 1978. *Deep Sea Res.* **29**: 185–200.

Botsford, L.W. and Wickham, D.E. (1975) Correlation of upwelling indices and Dungeness crab catch. *Fish. Bull.* **73**: 901–917.

Brander, K. (1992) A re-examination of the relationship between cod recruitment and *Calanus finmarchicus* in the North Sea. *ICES Mar. Sci. Symp.* **195**: 393–401.

Brander, K.J. and Hurley, P.C.F. (1991) Distribution of early stage Atlantic cod (*Gradus morhua*), haddock (*Melanogrammus aeglefinus*) and witch flounder (*Glyptocephalus cynoglossus*) eggs on the Scotian shelf: a reappraisal of evidence on the coupling of cod and plankton production. *Can. J. Fish. Aquat. Sci.* **49**: 238–258.

Brander, K. and Thompson, A.B. (1989) Diel differences in avoidance of three vertical profile sampling gears by herring larvae. *J. Plankt. Res.* **11**: 775–784.

Brett, J.R., Shelbourn, J.E. and Shoop, C.T. (1969) Growth rate and body composition of fingerling sockeye salmon, *Onchorhyncus nerka*, in relation to temperature and body size. *J. Fish. Res. Bd Can.* **26**: 2363–2394.

Brock, J., Sathyendranath, S. and Platt, T. (1993) Modeling the seasonality of subsurface light and primary production in the Arabian Sea. *Mar. Ecol. Progr. Ser.* **101**: 209–221.

Brodeur, R.D., Bailey, K.M. and Kim, S. (1991) Cannibalism on eggs by walleye pollock, *Theragra chalcogramma*, in Shelikof Strait, Gulf of Alaska. *Mar. Ecol. Progr. Ser.* **71**: 207–218.

Brodeur, R.D. and Ware, D.M. (1992) Long term variability in zooplankton biomass in the subarctic Pacific Ocean. *Fish. Oceanogr.* **1**: 32–38.

Broecker, W.S. (1974) *Chemical oceanography*. Harcourt Brace Jovanovitch Inc., New York.

Broecker, W.S. and Peng, T-H. (1982) *Tracers in the sea.* Lamont Doherty Geological Observatory, Columbia University, Palisades, New York.

Broecker, W.S. and Peng, T-H. (1985) *Tracers in the sea.* Eldigio Press, Lamont, New York.

Bromley, P.J. (1988) Gastric evacuation in whiting, *Merlangius merlangus* (L.). *J. Fish. Biol.* **33**: 331–338.

Bromley, P.J. (1989) Evidence for density dependent growth in North Sea gadoids. *J. Fish. Biol.* **35** (Suppl. A): 117–123.

Bromley, P.J. and Last, J.M. (1990) Feeding in the trawl and the consequences for estimating food consumption in natural fish populations. ICES CM 1990/G35 (mimeo).

Brothers, E.B., Mathew, C.P. and Lasker, R. (1976) Daily growth increments in otoliths from larval and adult fishes. *Fish. Bull.* **74**: 1–8.

Brown, O.B., Cornillon, P.C., Emmerson, S.R. and Carle, H.M. (1986) Gulf Stream warm rings: a statistical study of their behavior. *Deep Sea Res.* **33**: 1459–1474.

Buchanan-Wollaston, H.J. (1926) Plaice egg production in 1920–21, treated as a statistical problem with comparison between the data from 1911, 1914 and 1921. *Fish. Invest. London,* **9**(2): 36 pp.

Bückmann, A. (1942) Die Untersuchungen der biologischen Anstalt über die Okologie der Heringsbrut in der südlichen Nordsee. *Helgo. wiss. Meeresunters* **3**: 1–17.

Bulmer, M.G. (1975) The statistical analysis of density dependence. *Biometrics* **31**: 901–911.

Burd, A.C. (1984) Density dependent growth in North Sea herring. I.C.E.S. 1984/H4 (mimeo).

Burd, A.C. (1985) Recent changes in the central and Southern North Sea herring stocks. *Can. J. Aquat. Sci.* **42** (Suppl.): 92–106.

Burd, A.C. and Parnell, W.G. (1971) The relationship between larval abundance and year class strength. *Rapp. Procès-Verb. Réun. Cons. Explor. Mer,* **164**: 30–36.

Butler, E.I., Corner, E.D.S. and Marshall, S.M. (1969) On the nutrition and metabolism of zooplankton. VI Feeding efficiency of *Calanus* in terms of nitrogen and phosphorus. *J. Mar. Biol. Assn UK,* **49**: 977–1002.

Butler, E.I., Knox, S. and Liddicoat, M.I. (1979) The relationship between inorganic and organic nutrients in the sea. *J. Mar. Biol. Assn UK,* **59**: 239–250.

Butler, J.L. (1991) Mortality and recruitment of Pacific sardine, *Sardinops sagax,* larvae in the Californian current. *Can. J. Fish. Aquat. Sci.* **48**: 1713–1723.

Butler, J.L. and de Mendiola, B.R. (1985) Growth of larval sardines off Peru. *CalCOFI Rept.* **26**: 113–118.

Cain, A.J. (1983) Concluding remarks. In *Protein polymorphisms: adaptive and taxonomic significance* (ed. Oxford, G.S. and Rollinson, D), pp. 391–396. Systematics Association Special Vol. 24, Dept Geol. Sci., Durham University.

Campana, S.E., Gagné, J.A. and Munro, J. (1987) Otolith microstructure of larval herring (*Clupea harengus*): image or reality. *Can. J. Fish. Aquat. Sci.* **44**: 1422–1429.

Campana, S.E. and Hurley, P.C.F. (1989) An age and temperature mediated growth model for cod (*Gadus morhua*) and haddock (*Melanogrammus aeglefinus*) larvae in the Gulf of Maine. *Can. J. Fish. Aquat. Sci.* **46**: 603–613.

Campana, S.E. and Nielsen, J.D. (1985) Microstructure of fish otoliths. *Can. J. Fish. Aquat. Sci.* **42**: 1014–1032.

Cavalli-Sforza, L.L. and Edwards, A.W.F. (1967) Phylogenetic analysis: models and estimation procedures. *Evolution,* **21**: 550–571.

Chambers, R.C. and Leggett, W. (1987) Size and age at metamorphosis in marine fishes: an analysis of laboratory reared winter flounder

(*Pseudopleuronectes americanus*) with a review of variation in other species. *Can. J. Fish. Aquat. Sci.* **44**: 1936–1947.

Chambers, R.C. and Leggett, W.C. (1992) Possible causes and consequences of variation in age and size at metamorphosis in flatfishes (Pleuronectiformes): an analysis at the individual, population and species levels. *Neth. J. Sea Res.* **29**: 7–24.

Chambers, R.C., Leggett, W.C. and Brown, J.A. (1988) Variation in and among early life history traits of laboratory reared winter flounder, *Pseudopleuronectes americanus. Mar. Ecol. Progr. Ser.* **47**: 1–15.

Charnock, H. and Philander, S.G.H. (1989) The dynamics of the coupled atmosphere and ocean. *Phil. Trans. Roy. Soc.* **329**(1604).

Checkley, D.M. (1984) Relation of growth to ingestion for larvae of Atlantic herring, *Clupea harengus* and other fish. *Mar. Ecol. Progr. Ser.* **18**: 215–224.

Colebrook, J.M. (1965) On the analysis of variation in the plankton, the environment and the fisheries. *Spec. Publ. Int. Commn, North West Atl. Fish.* **6**: 291–302.

Colebrook, J.M. (1978) Continuous plankton records: zooplankton and environment, North East Atlantic and North Sea, 1948–75. *Oceanologica Acta*, **1**: 9–23.

Colebrook, J.M. (1982) Continuous plankton records; seasonal variations in the distribution and abundance of plankton in the North Atlantic Ocean and the North Sea. *J. Plankt. Res.* **4**: 435–462.

Colebrook, J.M., Robinson, G.A., Hunt, H.G., Roskell, J., John, A.W.G., Bottrell, H.H., Lindley, J.A., Collins, N.R. and Halliday, N.C. (1984) Continuous plankton records: a possible reversal in the downward trend in the abundance of the plankton of the North Sea and North east Atlantic. *J. Cons. Int. Explor. Mer*, **41**: 304–306.

Cook, R.M. and Armstrong, D.W. (1986) Stock related effects in the recruitment of North Sea haddock and whiting. *J. Cons. Int. Explor. Mer*, **42**: 272–281.

Cooper, L.H.N. (1938) Phosphate in the English Channel, 1933–8 with a comparison with earlier years, 1916 and 1923–32. *J. Mar. Biol. Assn UK*, **23**: 171–178.

Costello, J.H., Strickler, J.R., Marrasé, C., Trager, G., Zeller, R. and Freise, H.J. (1990) Grazing in a turbulent environment: behavioral response of a calanoid copepod, *Centropages hamatus. Proc. Nat. Acad. Sci.* **87**: 1648–1652.

Cowles, T. and Strickler, J.R. (1983) Characterization of feeding activity patterns in the planktonic copepod *Centropages typicus* Krøyer under various food conditions. *Limnol. Oceanogr.* **28**: 106–115.

Crawford, R.J.M. and Shannon, L.V. (1988) Long term changes in the distribution of fish catches in the Benguela. In *Long term changes in marine fish populations* (ed. Wyatt, T. and Larraneta, M.G.), pp. 449–480. CSIC, Vigo.

Creutzberg, F. (1958) Use of tidal streams by migrating elvers (*Anguilla vulgaris.* Turt.). *Nature*, **181**: 857–858.

Creutzberg, F., Eltink, A.T.G.W. and Noort, G.J. van (1978) The migration of plaice larvae (*Pleuronectes platessa*) into the Western Wadden Sea. In *Proc. 12th Europ. Symp. Mar. Biol.* (ed. McLusky, D.S. and Borry, A.J.), pp. 243–251. Pergamon Press, Oxford and New York.

Cross, T.F. and Payne, R.H. (1978) Geographic variation in Atlantic cod

(*Gadus morhua*) off Eastern North America: a biochemical systematics approach. *J. Fish. Res. Bd Can.* **35**: 117–123.

Cullen, J. (1991) Hypotheses to explain high nutrient conditions in the open sea. *Limnol. Oceanogr.* **36**: 1578–1599.

Cullen, J., Yang, K. and MacIntyre, H.L. (1992) Nutrient limitation of marine photosynthesis. In *Primary productivity and biogeochemical cycles in the sea* (ed. Falkowski, P. and Woodhead, A.D.), pp. 69–80. Plenum Press, New York and London.

Cury, P. and Roy, C. (1989) Optimal environmental window and pelagic fish recruitment success in upwelling areas. *Can. J. Aquat. Sci.* **46**: 670–680.

Cushing, D.H. (1955) Production and a pelagic fishery. *Fish. Invest. London* **18**(7): 1–104.

Cushing, D.H. (1958) The effect of grazing in reducing the primary production: a review. *Rapp. Procès-Verb. Réun. Cons. Int. Explor. Mer*, **144**: 149–154.

Cushing, D.H. (1961) On the failure of the Plymouth herring fishery. *J. Mar. Biol. Assn UK*, **41**: 799–816.

Cushing, D.H. (1967) The grouping of herring populations. *J. Mar. Biol. Assn UK*, **47**: 193–208.

Cushing, D.H. (1968) *Fisheries biology: a study in population dynamics*. Univ. Wisconsin Press, 200 pp.

Cushing, D.H. (1969) The regularity of the spawning season of some fishes. *J. Cons. Int. Explor. Mer*, **33**: 81–92.

Cushing, D.H. (1971) Upwelling and the production of fish. *Adv. Mar. Biol.* **9**: 255–334.

Cushing, D.H. (1972) The production cycle and the numbers of marine fish. In *Symp. Zool. Soc. Lond. No. 29, Conservation and productivity of natural waters* (ed. R.W. Edwards and D.J. Garrod), pp. 213–232. Academic Press, London.

Cushing, D.H. (1974) The natural regulation of fish populations. In *Sea fisheries research* (ed. F.R. Harden Jones), pp. 399–412. Elek Science, London.

Cushing, D.H. (1975a) *Marine ecology and fisheries*. Cambridge University Press, Cambridge.

Cushing, D.H. (1975b) In praise of Petersen. *J. Cons. Int. Explor. Mer*, **36**: 277–291.

Cushing, D.H. (1979) The monitoring of biological effects: the separation of natural changes from those induced by pollution. *Phil. Trans. Roy. Soc.* **286**: 597–609.

Cushing, D.H. (1980) The decline of the herring stocks and the gadoid outburst. *J. Cons. Int. Explor. Mer*, **39**: 70–81.

Cushing, D.H. (1981) The effect of El Niño on the Peruvian anchoveta stock. In *Coastal Upwelling* (ed. F.A. Richards), pp. 449–457. Amer. Geophys. Union, Washington.

Cushing, D.A. (1982) *Climate and fisheries*. Academic Press, London and New York.

Cushing, D.H. (1983) Are fish larvae too dilute to affect the density of their food organisms? *J. Plankt. Res.* **5**: 847–854.

Cushing, D.H. (1984) The gadoid outburst. *J. Cons. Int. Explor. Mer*, **41**: 159–166.

Cushing, D.H. (1988a) *The provident sea*. Cambridge University Press, Cambridge.

Cushing, D.H. (1988b) The Northerly wind. In *Towards a theory of Biological-*

Physical interactions in the world ocean (ed. B. Rothschild), pp. 235–244. Kluwer, Dordrecht.

Cushing, D.H. (1989) A difference in structure between ecosystems in strongly stratified waters and in those that are only weakly stratified. *J. Plankt. Res.* **11**: 1–13.

Cushing, D.H. (1990*a*) Hydrographic containment of a spawning group of plaice in the Southern Bight of the North Sea. *Mar. Ecol. Progr. Ser.* **58**: 287–297.

Cushing, D.H. (1990*b*) Plankton production and year class strength in fish populations: an update of the match/mismatch hypothesis. *Adv. Mar. Biol.* **26**: 249–293.

Cushing, D.H. (1992) A short history of the Downs stock of herring. *ICES J. Mar. Sci.* **49**: 437–443.

Cushing, D.H. and Bridger, J.P. (1966) The stock of herring in the North Sea and changes due to fishing. *Fish. Invest. London* **25**(1): 1–123.

Cushing, D.H. and Dickson, R.R. (1976) Biological and hydrographic changes in British waters during the last thirty years. *Biol. Rev.* **41**(2): 221–258.

Cushing, D.H. and Harris, J.G.K. (1973) Stock and recruitment and the problem of density dependence. *Rapp. Procès-Verb. Réun. Cons. Int. Explor. Mer,* **164**: 142–55.

Cushing, D.H. and Horwood, J.W. (1977) Development of model of stock and recruitment. In *Fisheries mathematics* (ed. Steele, J.H.), pp. 21–35. Academic Press, London and New York.

Cushing, D.H. and Horwood, J.W. (1994) The growth and death of fish larvae. *J. Plankton Res.* **16**: 291–300.

Cushing, D.H. and Vucetic, T. (1963) Studies on a Calanus patch. III The quantities of food eaten by *Calanus finmarchicus. J. Mar. Biol. Assn UK,* **43**: 349–371.

Daan, N. (1978) Changes in cod stocks and cod fisheries in the North Sea. *Rapp. Procès-Verb. Réun. Cons. Int. Explor. Mer,* **172**: 39–57.

Daan, N., Rijnsdorp, A.D. and Overbeeke, G.R. van (1984) Predation by North Sea herring, *Clupea harengus* on the eggs of plaice, *Pleuronectes platessa* and cod, *Gadus morhua. Trans. Amer. Fish. Soc.* **114**: 499–506.

Daan, N., Bromley, P.J., Hislop, J.R.G. and Nielsen, N.A. (1990) Ecology of North Sea fish. *Neth. J. Sea Res.* **26**: 343–386.

Dagg, M.J. and Turner, J.T. (1982) The impact of copepod grazing on the phytoplankton of Georges' Bank and the New York Bight. *J. Fish. Res. Bd Can.* **39**: 979–990.

Dagg, M.J. and Walser, W.E. (1987) Ingestion, gut passage and egestion by the copepod *Neocalanus pulchrus. Limnol. Oceanogr.* **32**: 178–188.

Daley, R.J. and Hobbie, J.E. (1975) Direct counts of aquatic bacteria by a modified epifluorescence technique. *Limnol. Oceanogr.* **20**: 875–882.

Davies, J.M. and Payne, R. (1984) Supply of organic matter to the sediment in the northern North Sea during a spring phytoplankton bloom. *Mar. Biol.* **78**: 315–324.

Davis, C.O., Hollibaugh, J.T., Seibert, D.C.R., Thomas, W.H. and Harrison, P.J. (1980) Formation of resting spores by *Leptocylindrus danicus* (Bacillariophycae) in a controlled ecosystem. *J. Phycol.* **16**: 296–302.

Day, F. (1880–1894) *The fishes of Great Britain and Ireland.* Williams and Norgate, London.

de Baar, H.J.W., Buma, A.G.J., Nolting, R.F., Cadée, G.C., Jacques, G. and Tréguer, P.S. (1991) On iron limitation in the Southern Ocean:

experimental observations in the Weddell and Scotia seas. *Mar. Ecol. Progr. Ser.* **65**: 105–122.

de Groot, S.J. (1971) On the interrelationships between morphology of the alimentary tract, food and feeding behaviour in flatfishes (Pisces, Pleuronectiformes). *Neth. J. Sea Res.* **5**: 121–196.

de La Fontaine, Y. and Leggett, W.C. (1988) Predation by jellyfish on larval fish; an experimental evaluation employing in situ enclosures. *Can. J. Fish. Aquat. Sci.* **45**: 1173–1190.

de Ligny, W. (1969) Serological and biochemical studies on fish populations. *Oceanogr. Mar. Biol.* **7**: 411–513.

de Veen, J.F. (1962) On the subpopulations of plaice in the southern North Sea. *Intl Council Explor. Sea CM 1962*, **94**: 1–6.

de Veen, J.F. (1976) On changes in some biological parameters in the North Sea sole (*Solea solea*). *J. Cons. Int. Explor. Mer*, **37**: 60–90.

de Veen, J.F. (1978) On selective tidal transport in the migration of North Sea plaice (*Pleuronectes platessa* L.) and other flatfish species. *Neth. J. Sea Res.* **12**: 115–147.

Dempster, J.P. (1983) The natural control of populations of butterflies. *Biol. Rev.* **58**: 461–482.

Deriso, R.B. (1980) Harvesting strategies and parameter estimation for an age structured model. *Can. J. Fish. Aquat. Sci.* **17**: 268–282.

Deuser, W.G. and Ross, E.H. (1980) Seasonal changes in the flux of organic carbon in the deep Sargasso Sea. *Nature*, **232**: 364–365.

Deuser, W.G., Ross, E.H. and Anderson, R.F. (1981) Seasonality in the supply of sediment to the deep Sargasso Sea and implications for the rapid transfer of matter to the deep ocean. *Deep Sea Res.* 28A: 495–505.

Devold, F. (1963) The life history of the Atlanto-Scandian herring. *Rapp. Procès-Verb. Cons. Int. Explor. Mer*, **154**: 98–108.

Dickson, R.R. (1992) Hydrobiological variability in the I.C.E.S. area, 1980–1989. *Mar. Sci. Symp.* **195**: 1–10.

Dickson, R.R. and Brander, K.M. (1993) Effects of a changing windfield on cod stocks of the North Atlantic. *Fish Oceanogr.* **2**: 124–153.

Dickson, R.R., Kelly, P.M., Colebrook, J.M., Wooster, W.S. and Cushing, D.H. (1988*a*) North Winds and production in the eastern North Atlantic. *J. Plankt. Res.* **10**: 151–169.

Dickson, R.R. and Lamb, H.H. (1972) A review of hydrometeorological events in the North Atlantic. *Spec. Publ. 8. Int. Commn North West Atl. Fish.* 35–62.

Dickson, R.R., Lamb, H.H., Malmberg, S-A. and Colebrook, J.M. (1975) Climatic reversal in northern North Atlantic. *Nature* **256**(5517): 479–482.

Dickson, R.R., Malmberg, S-A, Jones, S.R. and Lee, A.J. (1985) An investigation of the earlier Great Salinity Anomaly of 1910–14 in waters west of the British Isles. Contributions to Council Meetings ICES CM 1985 Gen 4 (mimeo).

Dickson, R.R., Meincke, J., Malmberg, S-A. and Lee, A.J. (1988*b*) The Great Salinity Anomaly in the Northern North Atlantic, 1968–1982. *Progr. Oceanogr.* **20**: 103–151.

Dickson, R.R. and Namias, J. (1976) North American influences on the circulation and climate of the North Atlantic sector. *Monthly Weather Rev.* **104**: 1255–1265.

Dickson, R.R., Pope, J.G. and Holden, M.J. (1973) Environmental influences on the survival of North Sea cod. In *The early life history of fish* (ed.

Blaxter, J.H.S.), pp. 69–80. Springer, Berlin, Heidelberg and New York.

Dickson, R.R. and Reid, P.C. (1983) Local effects of wind speed and direction on the phytoplankton of the Southern Bight. *J. Plankt. Res.* **5**: 441–455.

Dietrich, G. (1953) Verteilung, Ausbreitung und Vermischung der Wasserkörper in der südwestlichen Nordsee auf Grund der Ergebnisse der 'Gauss' Fahrt im Februar–März 1952. *Ber. dt. wiss. Komm. Meeresforsch. NS*, **13**(2): 104–129.

Dortch, Q. (1990) The interaction between ammonium and nitrate uptake in phytoplankton. *Mar. Ecol. Prog. Ser.* **61**: 183–201.

Dortch, Q. and Conway, H.L. (1984) Interactions between nitrate and ammonium uptake: variation with growth rate, nitrogen source and species. *Mar. Biol.* **79**: 151–164.

Dortch, Q. and Postel, J.R. (1989) Biochemical indicators of N utilization by phytoplankton during upwelling off the Washington coast. *Limn. Oceanogr.* **34**: 758–773.

Droop, M.R. (1968) Vitamin B_{12} and marine ecology. IV The kinetics of uptake, growth and inhibition in *Monochrysis lutheri. J. Mar. Biol. Assn UK*, **48**: 689–733.

Droop, M.R. (1973) Steady state growth and ammonium uptake of a fast growing marine diatom. *Limnol. Oceanogr.* **23**: 695–730.

Droop, M.R. (1974) The nutrient status of algal cells in continuous culture. *J. Mar. Biol. Assn UK*, **54**: 825–833.

Dugdale, R.C. (1967) Nutrient limitation in the sea; dynamics, identification and significance. *Limnol. Oceanogr.* **12**: 685–695.

Dugdale, R.C. and Goering, J.J. (1967) Uptake of new and regenerated forms of nitrogen in primary productivity. *Limnol. Oceanogr.* **12**: 196–206.

Dunbar, R.B. and Wefer, G. (1984) Stable isotope fractionation in benthic foraminifera from the Peruvian continental margin. *Mar. Geol.* **59**: 215–225.

Dunn, J.R. and Matarese, A.C. (1987) A review of the early life history of the Northeast Pacific gadoid fishes. *Fish. Res.* **5**: 163–184.

Duran, M., Saiz, F., Lopez-Benito, M. and Margalef, R. (1956) El fitoplancton de la ria de Vigo de abril de 1954 a junio de 1955. *Invest. Pesq.* **4**: 67–96.

Dwyer, A.D., Bailey, K.M. and Livingston P.A. (1989) Feeding habits and daily ration of walleye pollock (*Theragra chalcogramma*) in the Eastern Bering Sea with special reference to cannibalism. *Can. J. Aquat. Sci.* **44**: 1972–1984.

Egger, J., Meyers, G. and Wright, P.B. (1981) Pressure, wind and cloudiness in the tropical Pacific related to the Southern Oscillation. *Monthly Weather Rev.* **109**: 1139–1149.

El Hag, A.G.D. (1986) Physiological studies on a coccoid marine blue-green alga (cyanobacterium). *Br. J. Phycol.* **21**: 315–319.

Ellertsen, E., Fossum, P., Solemdal, P., Sundby, S. and Tilseth, S. (1987) The effect of biological and physical factors on the survival of Arcto-Norwegian cod and the influence on recruitment variation. In *The effect of oceanographic conditions on distribution and population dynamics of commercial fish stocks in the Barents Sea*, (ed. Leong, H.), pp. 101–126. Institute of Marine Research, Bergen.

Ellertsen, B., Fossum, P., Solemdal, P. and Sundby, S. (1989) Relation between temperature and survival of eggs and first feeding larvae of North east Arctic cod (*Gadus morhua*). *Rapp. Procès-Verb. Réun. Cons. Int. Explor. Mer*, **191**: 209–219.

Elliott, A.J. and Clarke, T. (1991) Seasonal stratification in the northwest European Shelf Seas. *Cont. Shelf. Res.* **11**: 467–492.

Elliott, J.M. (1985) Population dynamics of migratory trout, *Salmo trutta*, in a Lake District stream, 1966–1983, and their implications for fishery management. *J. Fish. Biol.* **27**, Suppl. A: 35–43.

Enfield, D.B. (1989) El Niño, past and present. *Rev. Geophys.* **27**: 159–187.

Eppley, R.W. (1972) Temperature and phytoplankton growth in the sea. *Fish. Bull.* **70**: 1063–85.

Eppley, R.W. (1982) The PRPOOS Program: a study of plankton rate processes in oligotrophic oceans. *EOS*, **63**: 522.

Eppley, R.W. and Coatsworth, J.L. (1967) Uptake of nitrate and nitrite by *Ditylum brightwellii* – kinetics and mechanisms. *J. Phycol.* **4**: 151–156.

Eppley, R.W. and Peterson, B.J. (1979) Particulate organic matter flux and planktonic new production in the deep ocean. *Nature*, **282**: 677–680.

Eppley, R.W., Rogers, J.N. and McCarthy, J.J. (1969) Half saturation constants for uptake of nitrate and ammonium by marine phytoplankton. *Limnol. Oceanogr.* **14**: 912–920.

Evans, G.T. and Parslow, J.S. (1985) A model of annual plankton cycles. *Biol. Oceanogr.* **3**: 327–347.

Evans, R.H., Baker, K.H., Brown, O.B. and Smith, R.C. (1986) Chronology of Warm Core Ring 82B. *J. Geophys. Res.* **90**: 8803–8811.

Fairbairn, D.J. and Roff, D.A. (1980) Testing genetic models of isozyme variability without breeding data: can we depend on the chi^2? *Can. J. Fish. Aquat. Sci.* **37**: 1149–1159.

Fasham, M.J.R. (1985) Flow analysis of materials in the marine euphotic zone. *Can. Bull. Fish. Aquat. Sci.* **213**: 139–162.

Fasham, M.J.R. (1992) Modelling the ocean biota In *The global carbon cycle* (ed. Heimann, M.), pp. 457–504. Springer Verlag, Berlin.

Fasham, M.J.R., Ducklow, H.W. and McKelvie, S.M. (1990) A nitrogen-based model of plankton dynamics in the oceanic mixed layer. *J. Mar. Res.* **48**: 591–639.

Fasham, M.J.R., Sarmiento, J.L., Slater, R.D., Ducklow, H.W. and Williams R. (1993) Ecosystem behavior at Bermuda station 'S' and Ocean Weather Station 'India': a General Circulation Model and observational analysis. *Global Biogeochemical Cycles*, **7**: 379–415.

Fearnhead, P.G. (1975) On the formation of fronts by tidal mixing around the British Isles. *Deep Sea Res.* **22**: 311–321.

Fenchel, T. (1982a) Ecology of heterotrophic microflagellates. I. Some important forms and their functional morphology. *Mar. Ecol. Progr. Ser.* **8**: 211–223.

Fenchel, T. (1982b) Ecology of heterotrophic microflagellates. II Bioenergetics and growth. *Mar. Ecol. Progr. Ser.* **8**: 225–231.

Fenchel, T. (1982c) Ecology of heterotrophic microflagellates. III Adaptations to heterogeneous environments. *Mar. Ecol. Progr. Ser.* **9**: 25–33.

Fenchel, T. (1982d) Ecology of heterotrophic microflagellates. IV Quantitative occurrence and importance as bacterial consumers. *Mar. Ecol. Progr. Ser.* **9**: 35–42.

Fenchel, T. (1986) Protozoan filter feeding. *Progr. Protistology* **1**: 65–113.

Fitzwater, S.E., Knauer, G.A. and Martin, J.H. (1982) Metal contamination and its effect on primary production measurements. *Limnol. Oceanogr.* **27**: 544–551.

Flynn, K.J. (1989a) Nutrient limitation of marine microbial production: Fact or artefact. *Chemistry and Ecology*, **4**: 1–13.

Flynn, K.J. (1989*b*) Interaction between nutrient and predator limitation of production in the marine euphotic zone. *Chemistry and Ecology*, **4**: 21–36.

Flynn, K.J. (1990) The determination of nitrogen status in microalgae. *Mar. Ecol. Progr. Ser.* **61**: 297–317.

Flynn, K.J. and Butler, I. (1986) Nitrogen sources for the growth of marine microalgae: role of dissolved free amino acids. *Mar. Ecol. Prog. Ser.* **34**: 281–304.

Flynn, K.J., Dickson, D.M.J. and Al-Amoudi, O.A. (1989) The ratio of glutamine:glutamate in microalgae: a biomarker for N-status suitable for use at natural cell densities. *J. Plankt. Res.* **11**: 165–170.

Flynn, K.J. and Fielder, J. (1989) Changes in intracellular and extracellular amino acids during the predation of the chlorophyte *Dunaliella primolecta* by the heterotrophic dinoflagellate *Oxyrrhis marina* and the use of the glutamine/glutamate ratio as an indicator of nutrient status in mixed populations. *Mar. Ecol. Prog. Ser.* **53**: 117–127.

Flynn, K.J. and Syrett, P.J. (1986) Characteristics of the uptake system for L-lysine and L-arginine in *Phaeodactylum tricornutum*. *Mar. Biol.* **90**: 151–158.

Flynn, K.J. and Wright, C.R.N. (1986) The simultaneous assimilation of ammonia and L-arginine by the marine diatom, *Phaeodactylum tricornutum* Bohlin. *J. Exp. Mar. Biol. Ecol.* **95**: 257–269.

Foerster, R.E. (1937) The return from the sea of sockeye salmon (*Oncorhynchus nerka*) with special reference to percentage survival, sex proportion and progress of migration. *J. Biol. Bd. Can.* **3**: 26–42.

Foerster, R.E. (1968) The Sockeye salmon, *Oncorhynchus nerka*. *Bull. 162, Fish. Res. Bd. Can.*

Fogg, G.E. (1983) The ecological significance of extracellular products of phytoplankton photosynthesis. *Botanica Marina*, **26**: 3–14.

Fogg, G.E., Egan, B., Floodgate, G.D., Hoy, S., Jones, D.A., Kassab, J.Y., Lochte, K., Rees, E.I.S., Scrope-Howe, S., Turley, C.M. and Whitaker, C.J. (1985) Biological studies in the vicinity of a shallow sea tidal mixing front. *Phil. Trans. Roy. Soc.* **310**: 407–571.

Fonds, M. (1979) A seasonal fluctuation in growth rate of young plaice (*Pleuronectes platessa*) and sole (*Solea solea*) in the laboratory at constant temperature at a natural daylight cycle. In *Proc. 13th Europ. Mar. Biol. Symp.* (ed. Naylor, E. and Hartnoll, R.G.), pp. 151–156. Pergamon, Oxford.

Forsskahl, M., Laakonen, A., Leppanen, J-M, Niemi, A., Sundberg, A. and Tamelander, G. (1982) Seasonal cycle of production and sedimentation of organic matter at the entrance of the Gulf of Finland. *Neth. J. Sea Res.* **16**: 290–299.

Fortier, L. and Gagné, J. (1990) Larval herring dispersion, growth and survival in the St Lawrence estuary: match/mismatch or membership/vagrancy. *Can. J. Fish. Aquat. Sci.* **47**: 1898–1912.

Fortier, L. and Leggett, W.C. (1985) A drift study of larval fish survival. *Mar. Ecol. Progr. Ser.* **25**: 245–257.

Fortier, L., Leggett, W.C. and Gosselin, S. (1987) Pattern of larval emergence and their potential impact on stock differentiation in beach spawning capelin. *Can. J. Fish. Aquat. Sci.* **44**: 1326–1330.

Francisco, D.E., Mah, R.A. and Rabin, A.C. (1973) Acridine-orange epifluorescence technique for counting bacteria in natural waters. *Trans. Amer. Microsc. Soc.* **92**: 416–421.

Frank, K.T. and Leggett, W.C. (1981*a*) Wind regulation of emergence and early larval survival in capelin (*Mallotus villosus*). *Can. J. Fish. Aquat. Sci.* **38**: 215–223.

Frank, K.T. and Leggett, W.C. (1981*b*) Prediction of egg development and mortality rates in capelin (*Mallotus villosus*) from meteorological, hydrographic and biological factors. *Can. J. Fish. Aquat. Sci.* **38**: 1327–1338.

Frank, K.T. and Leggett, W.C. (1982) Coastal water mass replacement: its effect on zooplankton dynamics and predator prey complex associated with larval capelin (*Mallotus villosus*). *Can. J. Fish. Aquat. Sci.* **39**: 991–1003.

Frank, K.T. and Leggett, W.C. (1983) Multispecies larval fish associations; accident or adaptation. *Can. J. Fish. Aquat. Sci.* **40**: 754–763.

Frank, K.T. and Leggett, W.C. (1984) Selective exploitation of capelin (*Mallotus villosus*) eggs by winter flounder (*Pseudopleuronectes americanus*): capelin egg mortality rates and contribution to the annual growth of flounder. *Can. J. Fish. Aquat. Sci.* **41**: 1294–1302.

Frank, K.T. and Leggett, W.C. (1985) Reciprocal oscillations in densities of larval fish and potential predators: a reflection of present or past predation. *Can. J. Fish. Aquat. Sci.* **42**: 1841–1849.

Fransz, H.G. and Gieskes, W.W.C. (1984) The unbalance of phytoplankton and copepods in the North Sea. *Rapp. Procès-Verb. Réun. Cons. Int. Explor. Mer*, **183**: 218–225.

Friedman, M.M. and Strickler, J.R. (1975) Chemoreceptors and feeding in calanoid copepods (Arthropoda: Crustacea). *Proc. Nat. Acad. Sci.* **72**: 4185–4188.

Frost, B.W. (1972) Effects of size and concentration of food particles on the feeding behavior of the marine planktonic copepod *Calanus pacificus*. *Limnol. Oceanogr.* **17**: 805–815.

Frost, B.W. (1987) Grazing control of phytoplankton stock in the open subarctic Pacific Ocean: a model assessing the role of mesozooplankton particularly the large calanoid copepods, *Neocalanus* spp. *Mar. Ecol. Progr. Ser.* **39**: 49–68.

Frost, B.W. (1991) The role of grazing in the nutrient rich areas of the open sea. *Limnol. Oceanogr.* **36**: 1616–1630.

Frost, B.W., Landry, M.R. and Hassett, R.P. (1983) Feeding behavior of large calanoid copepods *Neocalanus cristatus* and *N. plumchrus* from the subarctic Pacific Ocean. *Deep Sea Res.* **30**: 1–13.

Furnas, M.J. (1982) An evaluation of two diffusion culture techniques for estimating phytoplankton growth rates *in situ*. *Mar. Biol.* **70**: 63–72.

Furnas, M.J. (1990) *In situ* growth rates of marine phytoplankton: approaches to measurement, community and species growth rates. *J. Plankt. Res.* **12**: 1117–1151.

Furnas, M.J. (1991) Net *in situ* growth rates of phytoplankton in an oligotrophic tropical shelf ecosystem. *Limnol. Oceanogr.* **36**: 13–29.

Garrod, D.J. (1983) On the variability of year class strength. *J. Cons. Int. Explor. Mer*, **41**: 63–66.

Garrod, D.J. (1988) North Atlantic cod: fisheries and management to 1986. In *Fish population dynamics*, Second Edition (ed. Gulland, J.A.), pp. 185–218. John Wiley, Chichester.

Garstang, W. (1900–3) The impoverishment of the sea. *J. Mar. Biol. Assn UK*, **6**: 1–70.

Gaston, K.J. and Lawton, J.H. (1987) A test of statistical techniques for detecting density dependence. *Oecologia* **74**: 404–410.

Gauldie, R.W. (1984) Allelic variation and fisheries management. *Fish. Res. Bull. NZ Min. Agric. Fish.* **26**: 1–35.

Gerritsen, J. and Strickler, J.R. (1977) Encounter probabilities and community structure in zooplankton: a mathematical model. *J. Fish. Res. Bd. Can.* **34**: 73–82.

Getz, W.M. and Swartzman, G.L. (1981) A probability transition matrix model for yield estimation in fisheries with highly variable recruitments. *Can. J. Fish. Aquat. Sci.* **38**: 847–855.

Gieskes, W.W.C. and Kraay, G.W. (1977) Continuous plankton records: changes in the plankton of the North Sea and its eutrophic Southern Bight from 1948 to 1975. *Neth. J. Sea Res.* **11**: 334–364.

Gieskes, W.W.C. and Kraay, G.W. (1980) Primary production and phytoplankton pigment measurement in the northern North Sea during FLEX 76. *Meteor Forsch. Ergebn.* **22**: 105–112.

Gieskes, W.W.C., Kraay, G.W. and Beers, M.A. (1979) Current ^{14}C methods for measuring primary production: gross underestimates in oceanic waters. *Neth. J. Sea Res.* **13**: 58–78.

Glover, R.S. (1979) Natural fluctuations of populations. *Ecotoxicology and Environmental Safety*, **3**: 190–203.

Godo, O.R. (1984) Migration, mingling and homing of Northeast Arctic cod from two separate spawning grounds. *Proceedings Soviet-Norwegian symposium reproduction and recruitment of the Arctic cod. Inst. Mar. Res. Bergen*, pp. 289–302.

Goldman, J.C. (1980) Physiological processes, nutrient availability, and the concept of relative growth rate in marine phytoplankton ecology. In *Primary production in the sea* (ed. Falkowski, P.G.), pp. 179–194. Plenum Press, New York and London.

Goldman, J.C. and Caron, D.A. (1985) Experimental studies on an omnivorous microflagellate; implications for grazing and nutrient regeneration in the marine microbial food chain. *Deep Sea Res.* **32**: 899–915.

Goldman, J.C. and McCarthy, J.J. (1978) Steady state growth and ammonium uptake of a fast growing marine diatom. *Limnol. Oceanogr.* **23**: 695–703.

Goldman, J.C., McCarthy, J.J. and Peavey, D.G. (1979) Growth rate influence on the chemical composition of phytoplankton in oceanic waters. *Nature*, **279**: 210–215.

Goode, G.B. (1884–1895) *The fisheries and fishing industries of the United States*, Vols I–V. Govt Printing Office, Washington.

Graham, M. (1924) The annual cycle in the life of the mature cod in the North Sea. *Fish. Invest.* **vi**(6): 1–77.

Graham, N.E. and White, W.B. (1988) The El Niño Cycle: A natural oscillator of the Pacific Ocean-atmosphere system. *Science*, **240**: 1293–1302.

Grall, J.R. (1972) Recherches quantitatives sur la production primaire du phytoplancton dans les parages de Roscoff. Thèse de doctorat ès Sciences Naturelles, Université de Paris.

Grande, K.D., Williams, P.J., Le, B., Marra, J., Purdie, D.A., Heinemann, K., Eppley, R.W. and Bender, M.L. (1989) Primary production in the North Pacific gyre: a comparison of the rates determined by the ^{14}C, O_2 concentration and ^{18}O methods. *Deep Sea Res.* **36**: 1621–1634.

Grant, W.S. and Utter, F.M. (1980) Biochemical genetic variation in Walleye pollock, *Theragra chalcogramma*: population structure in the southeastern Bering Sea and the Gulf of Alaska. *Can. J. Fish. Aquat. Sci.* **37**: 1093–1100.

Grant, W.S. and Utter, F.M. (1984) Biochemical population genetics of Pacific herring (*Clupea pallasi*). *Can. J. Fish. Aquat. Sci.* **41**: 856–864.

Gray, J. (1929) The kinetics of growth. *J. Exp. Biol.* **6**: 248–262.

Greer-Walker, M., Harden Jones, F.R. and Arnold, G.P. (1978) The movements of plaice (*Pleuronectes platessa* L.) tracked in the open sea. *J. Cons. Int. Explor. Mer*, **38**: 58–86.

Greer-Walker, M., Mitson, R.B. and Storeton-West, T. (1971) Trials with a transponding acoustic fish tag tracked with an electronic sector scanning sonar. *Nature*, **229**: 196–198.

Guillen, O. (1976) El sistema de la corriente peruana. Parte 1. Aspectos fisicos. *F.A.O. Fish. Inform.* **185**: 243–284.

Gulland, J.A. (1965) Estimation of mortality rate. Appendix to *Rep. Arctic Fish. Working Group Int. Council Explor. Sea*, pp. 231–241.

Gulland, J.A. and Williamson, G.R. (1962) Transatlantic journey of a tagged cod. *Nature*, **195**: 921.

Hagen, P.T. and Quinn, T.J. (1991) Long term growth dynamics of young Pacific halibut: evidence of temperature induced variation. *Fish. Res.* **11**: 283–306.

Haldane, J.B.S. (1953) Animal populations and their regulation. *New Biology*, **15**: 9–24.

Haldorson, L., Paul, A.J., Sterritt, D. and Watts, J. (1989) Annual and seasonal variation in growth of larval walleye pollock and flathead sole in a south eastern Alaskan Bay. *Rapp. Procès-Verb. Réun. Cons. Int. Explor. Mer*, **191**: 220–223.

Hansen, J., Johnson, D.A., Lacis, A., Lebedeff, S., Lee, P., Rind, D. and Russell, G. (1983) Observed temperature trends. *Science*, **220**: 874–875.

Hansen, P.M. (1968) Report on cod eggs and larvae. *ICNAF Spec.* No 7: Part 1 127–138.

Hansen, P.M., Jensen, A.F. and Tåning, A.V. (1935) Cod marking experiments in the waters of Greenland, 1924–33. *Meddel. fra Kommn. for Danmarks Fisk. og Havundersog. Serie Fiskeri*, **10**(1), 1–119.

Harding, D. and Talbot, J.W. (1973) Recent studies on the eggs and larvae of the plaice (*Pleuronectes platessa* L.) in the Southern Bight. *Rapp. Procès Verb. Cons. Réun. Int. Explor. Mer*, **164**: 261–269.

Harding, D., Nichols, J.H. and Tungate, D.S. (1978) The spawning of plaice (*Pleuronectes platessa* L.) in the Southern North Sea and English Channel. *Rapp. Procès Verb. Réun. Cons. Int. Explor. Mer*, **172**: 102–113.

Hardy, A.C. (1924) The herring in relation to its animate environment. Part 1 The food and feeding habits of the herring with special reference to the East Coast of England. *Fish. Invest. London* **II**: 7.3.1–53.

Hargraves, P.E. and French, F.W. (1983) Diatom resting spores: significance and strategies. In *Seasonal strategies of the algae* (ed. Fryxell, G.A.), pp. 49–68. Cambridge University Press, Cambridge.

Harris, G.P. (1978) Photosynthesis, productivity and growth; the physiological ecology of phytoplankton. *Arch. Hydrobiol. Ergeb. Limnol.* **10**: 1–171.

Harris, G.P. (1986) *Phytoplankton ecology; structure, function and fluctuation.* Chapman & Hall, London and New York.

Harris, G.P., Davis, P., Nunez, M. and Meyers, G. (1988) Interannual variability in climate and fisheries. *Nature*, **333**: 754–7.

Harris, J.G.K. (1975) The effect of density dependent mortality on the shape of the stock recruitment curve. *J. Cons. Int. Explor. Mer*, **36**: 144–149.

Harris, R.P. (1982a) Trophic interactions and production processes in natural zooplankton communities in enclosed water columns. In *Marine Mesocosms* (ed. Grice, G.D. and Reeve, M.R.), pp. 353–387. Springer, Berlin.

Harris, R.P. (1982*b*) Comparison of the feeding behaviour of *Calanus* and *Pseudocalanus* in two experimentally manipulated enclosed ecosystems. *J. Mar. Biol. Assoc. UK*, **62**: 71–91.

Hart, T.J. and Currie, R.I. (1960) The Benguela Current. *Disc. Rep.* **31**: 123–298.

Hartley, S.E. and Horne, M.T. (1984) Chromosome relationships in the genus *Salmo*. *Chromosoma*, **90**: 229–37.

Hasler, A.D. (1966) *Underwater guideposts; homing of salmon*. Univ. Wisconsin Press, Madison.

Hasler, A.D. and Scholz, A.T. (1980) Artificial imprinting: a procedure for conserving salmon stocks. In *Fish behaviour and its use in their capture and culture of fishes* (ed. Bardach, J.E., Magnusson, J.J., May, R.C. and Reinhart, J.M.), pp. 179–199. ICLARM, Manila.

Hassell, M.P. (1985) Insect natural enemies as regulating factors. *J. Anim. Ecol.* **54**: 323–334.

Hassell, M.P. (1986) Detecting density dependence. *Trends Ecol. Evol.* **1**: 90–93.

Hassell, M.P. (1987) Detecting regulation in patchily distributed animal populations. *J. Anim. Ecol.* **56**: 705–713.

Hassell, M.P., Latto, J. and May, R.M. (1989) Seeing the wood for the trees: detecting density dependence from existing life table studies. *J. Anim. Ecol.* **58**: 883–892.

Haury, L.R., Yamazaki, H. and Fey, C.L. (1992) Simultaneous measurements of small scale physical dynamics and zooplankton distributions. *J. Plankton Res.* **14**: 513–530.

Haury, L. and Weihs, D. (1976) Energetically efficient swimming behavior of negatively buoyant zooplankton. *Limnol. Oceanogr.* **21**: 797–803.

Head, E.J.H. and Horne, E.P.W. (1993) Pigment transformation and vertical flux in an area of convergence in the North Atlantic. *Deep Sea Res.* **40**: 329–346.

Heessen, H.J.L. and Rijnsdorp, A.D. (1989) Investigations on egg production and mortality of cod (*Gadus morhua* L.) and plaice (*Pleuronectes platessa* L.) in the southern and eastern North Sea in 1987 and 1988. *Rapp. Procès-Verb. Réun. Cons. Int. Explor. Mer*, **191**: 15–20.

Heincke, F. (1898) Naturgeschichte des Herings. *Abh. dt. Seefischerei* **2**: 136: 1–351.

Henderson, G.T.D. (1953) Continuous plankton records: the young fish and fish eggs, 1932–39 and 1946–49. *Bull. Mar. Ecol.* **3**: 215–252.

Henderson, G.T.D. (1961) Contribution towards a plankton atlas of the North eastern North Sea. V Young fish. *Bull. Mar. Ecol.* **5**(42): 105–111.

Hennemuth, R.C., Palmer, J.E. and Brown, B.E. (1980) A statistical description in eighteen selected fish stocks. *J. North Atl. Fish. Org.* **1**: 101–111.

Hentschel, E. and Wattenberg, H. (1931) Plankton und Phosphat in der Oberflächenschicht des Südatlantischen Ozeans. *Ann. Hydrog.* **58**: 273–277.

Hermann, F. (1967) Temperature variation in the West Greenland area since 1950. *Red Book Int. North West Atl. Fish.* **IV**: 70–85.

Heron, A.C. (1968) Reviews on zooplankton sampling methods: Plankton gauzes. In *Monographs on oceanographic methodology 2*, pp. 19–25. UNESCO, Paris.

Herrington, W.C. (1948) Limiting factors for fish populations: some theories and an example. *Bull. Bingh. Oceanogr. Coll.* **11**: 220–283.

Hickling, C.F. (1939) The distribution of phosphates in the south western area in April, 1936. *J. Mar. Biol. Assn UK*, **23**: 197–200.

Hilborn, R. and Walters, C.J. (1992) *Quantitative fisheries assessment: choice, dynamics and uncertainty*. Chapman and Hall, New York and London.

Hjort, J. (1914) Fluctuations in the great fisheries viewed in the light of biological research. *Rapp. Procès-Verb. Cons. Int. Explor. Mer*, **20**: 1–228.

Høglund, H. (1972) On the Bohuslan herring during the great herring period in the eighteenth century. *Rep. Inst. Mar. Res.* **20**: 1–86.

Holligan, P.M. and Harbour, D.S. (1977) The vertical distribution and succession of phytoplankton in the western English Channel in 1975 and 1976. *J. Mar. Biol. Assn UK*, **57**: 1075–1093.

Holligan, P.M., Harris, R.P., Newell, R.C., Harbour, D.S., Head, R.N., Linley, E.A.S., Lucas, M.I., Tranter, P.R.G. and Weekley, C.M. (1984*a*) Vertical distribution and partitioning of organic carbon in mixed frontal and stratified water of the English Channel. *Mar. Ecol. Prog. Ser.* **14**: 111–127.

Holligan, P.M., Williams, P.J., Le, B., Purdie, D. and Harris, R.P. (1984*b*) Photosynthesis, respiration and nitrogen supply of plankton populations in stratified, frontal and tidally mixed shelf waters. *Mar. Ecol. Prog. Ser.* **17**: 201–213.

Hollowed, A.B. and Wooster, W.S. (1992) Variability of winter ocean conditions and strong year classes of North east Pacific groundfish. *ICES Mar. Sci. Symp.* **195**: 433–444.

Hollowed, A.B., Bailey, K.M. and Wooster, W.S. (1987) Patterns of recruitment of marine fishes in the Northeast Pacific. *Biol. Oceanogr.* **5**: 99–131.

Horel, J.D. and Wallace J.M. (1981) Planetary scale atmospheric phenomena associated with the Southern Oscillation. *Monthly Weather Rev.* **109**: 813–829.

Horne, E.P.W., Loder, J.W., Harrison, W.G., Mohn, R., Lewis, M.R. and Platt, T. (1990) Nitrate supply and demand at the George's Bank tidal front. *Scient. Mar.* **53**: 145–158.

Horwood, J.W., Bannister, R.C.A. and Howlett, G.J. (1986) Comparative fecundity of North Sea Plaice (*Pleuronectes platessa* L.) *Proc. Roy. Soc. Lond. B* **228**: 401–431.

Houghton, R. and Flatman, S. (1984) The exploitation pattern, density dependent catchability and growth of cod (*Gadus morhua*) in the west central North Sea. *J. Cons. Int. Explor. Mer*, **39**: 271–287.

Houghton, R.G. and Harding, D. (1976) The plaice of the English Channel: spawning and migration. *J. Cons. Int. Explor. Mer*, **36**(3): 229–239.

Hovenkamp, F. (1989) Within season variation in growth of larval plaice (*Pleuronectes platessa*). *Rapp. Procès-Verb. Réun. Cons. Int. Explor. Mer*, **191**: 248–257.

Hovenkamp, F. (1990) Growth differences in larval plaice, *Pleuronectes platessa*, in the Southern Bight of the North Sea as indicated by otolith increments and RNA/DNA ratios. *Mar. Ecol. Progr. Ser.* **58**: 205–215.

Hovgård, H. and Buch, E. (1986) Fluctuations in the cod biomass of the West Greenland ecosystem in relation to climate. In *Variability and management of large marine ecosystems* (ed. Sherman, K. and Alexander,

L.M.), pp. 36–43. Westrion Press, Boulder, Colorado.

Hovgård, H. and Buch, E. (1990) Fluctuations in the cod biomass of the West Greenland Sea ecosystem in relation to climate. In *Large marine ecosystems: patterns processes and yields* (ed. Sherman, K., Alexander, L.M. and God, B.D.), pp. 36–43. American Association for the Advancement of Science, Washington DC.

Howard, L.O. and Fiske, W.F. (1911) The importation into the United States of the parasites of the Gipsy moth and the Browntail moth. *US Dept. Agr. Bur. Ent. Bull.* **91**: 1–312.

Hubold, G. (1978) Variations in growth rate and maturity of herring in the northern North Sea in the years, 1955–1973. *Rapp. Procès-Verb. Réun. Cons. Int. Explor. Mer*, **172**: 154–163.

Hughes, D.G. (1975) The development of seasonal stratification around the British Isles. M.Sc. Thesis, Univ. Coll. North Wales, Bangor, pp. 1–39.

Hughes, D.G. (1976) A simple method for predicting the occurrence of seasonal stratification and fronts in the North Sea and around the British Isles. ICES CM 1976/C: 1–8.

Hunter, J.R. (1972*a*) Behavior and survival of northern anchovy, *Engraulis mordax*, larvae. *CalCOFI Rept.* 19: 138–146.

Hunter, J.R. (1972*b*) Swimming and feeding behaviour of larval anchovy, *Engraulis mordax. Fish. Bull.* **70**: 821–838.

Hunter, J.R. (1976) Culture and growth of northern anchovy, *Engraulis mordax*, larvae. *Fish. Bull.* **74**: 81–88.

Hunter, J.R. and Kimbrell, C.A. (1980) Egg cannibalism in the northern anchovy *Engraulis mordax. Fish. Bull.* **78**: 811–816.

Hunter, J.R. and Thomas, G.L. (1974) Effect of prey distribution and density on the searching and feeding behavior of larval anchovy, *Engraulis mordax*. In *Early life history of fish* (ed. Blaxter, J.H.S.), pp. 559–574. Springer-Verlag, Berlin.

Huntley, M. and Boyd, C. (1984) Food limited growth of marine zooplankton. *Amer. Nat.* **124**: 455–478.

Husby, D.M. and Nelson, C.S. (1982) Turbulence and vertical stability in the California Current. *CalCOFI Rep.* **23**: 113–129.

Hutchings, J.A. and Myers, R.A. (1993) The effect of age on the seasonality of maturation and spawning of Atlantic cod, *Gadus morhua. Can. J. Fish. Aquat. Sci.* **50**: 2468–2474.

Ihssen, P.E., Booke, H.E., Casselman, J.M., Payne, N.R. and Utter, F. (1981) Stock identification: materials and methods. *Can. J. Fish. Aquat. Sci.* **38**: 1838–1855.

Iles, T.C. and Beverton, R.J.H. (1991) Mortality rates of O-group plaice (*Pleuronectes platessa* L.), dab (*Limanda limanda* L.) and turbot (*Scophthalmus maximus* L.) in European waters. 1 Statistical analysis of the data and estimation of parameters. *Neth. J. Sea Res.* **27**: 217–235.

Iles, T.D. and Sinclair, M. (1982) Atlantic Herring. Stock Discreteness and Abundance. *Science*, **215**: 627–633.

Incze, L. and Schumacher, J.D. (1986) Variability of the environment and selected fisheries resources in the Eastern Bering Sea ecosystem. In *Variability and management of large marine ecosystems* (ed. Sherman, K. and Alexander, L.M.), pp. 109–143. Westview Press, Boulder, Colorado.

Jakobsson, J. (1980) Exploitation of the Icelandic spring and summer spawning herring in relation to fisheries management, 1947–77. *Rapp. Procès-Verb. Réun. Cons. Int. Explor. Mer*, **177**: 23–42.

Jakobsson, J. (1992) Recent variability in the fisheries of the North Atlantic. *ICES Mar. Sci. Symp.* **195**: 291–315.

James, I.D. (1977) A model of the annual cycle of temperature in a frontal region of the Celtic Sea. *Est. Coast. Mar. Sci.* **5**: 339–353.

James, I.D. (1983) A three dimensional model of shallow sea fronts. In *North Sea dynamics* (ed. Sundermann, J. and Lenz, W.), pp. 173–184. Springer Verlag, Berlin, Heidelberg and New York.

Jamieson, A. (1966) The distribution of transferrin genes in cattle. *Heredity* **21**(2): 191–218

Jamieson, A. and Birley, A.J. (1989) The distribution of transferrin alleles in haddock stocks. *J. Cons. Int. Explor. Mer*, **45**: 248–262.

Jamieson, A. and Turner, R.J. (1978) The extended series of Tf alleles in Atlantic cod, *Gadus morhua* L. In *Marine organisms: genetics, ecology and evolution* (ed. Battaglia, B. and Beardmore, J.), pp. 699–729. Plenum Press, New York.

Jassby, A.D. and Platt, T. (1976) Mathematical formulation of the relationship between photosynthesis and light for phytoplankton. *Limnol. Oceanogr.* **21**: 540–547.

Jaworski, A. and Rijnsdorp, A.D. (1989) *Size selective mortality in plaice and cod eggs: a new approach to the study of mortality.* ICES CM 1989 L29. 10 pp.

Jenkins, G.P., Young, J.W. and Davis, T.L.O. (1991) Density dependence of larval growth of a marine fish, the southern bluefin tuna. *Can. J. Fish. Aquat. Sci.* **48**: 1358–1363.

Jenkins, W.J. (1982) Oxygen utilization rates in the North Atlantic subtropical gyre and primary production in oligotrophic systems. *Nature*, **300**: 246–248.

Jenkins, W.J. and Goldman, J.C. (1985) Seasonal oxygen cycling and primary production in the Sargasso Sea. *J. Mar. Res.* **43**: 465–491.

Johansen, A.C. (1926) On the remarkable quantities of haddock in the Baltic Sea during the winter 1925–6 and the causes leading to the same. *J. Cons. Int. Explor. Mer*, **1**: 140–156.

Johnson, P.W. and Sieburth, J. McN. (1979) Chroococcoid cyanobacteria in the sea: a ubiquitous and diverse phototropic biomass. *Limnol. Oceanogr.* **24**: 928–935.

Jones, C. (1985) Within season differences in growth of larval Atlantic herring, *Clupea harengus harengus*. *Fish. Bull.* **83**: 289–298.

Jones, F.R. Harden (1968) *Fish migration*. Edward Arnold, (Publishers) Ltd., London.

Jones, F.R. Harden (1980*a*) The nekton: production and migration patterns. In *Fundamentals of aquatic ecosystems* (ed. Barnes, R.S.K. and Mann, K.H.), pp. 119–142. Blackwell Scientific Publications, Oxford.

Jones, F.R. Harden (1980*b*) The migration of plaice (*Pleuronectes platessa*) in relation to the environment. In *Fish behavior and its use in the capture and culture of fishes* (ed. Bardach, J.E., Magnusson, J.J., May, R.C. and Reinhart, J.M.), pp. 383–399. ICLARM, Manila.

Jones, F.R. Harden (1981) Fish migration: strategy and tactics. In *Animal migration* (ed. Aidley, D.), pp. 139–165. Cambridge University Press, Cambridge.

Jones, F.R. Harden (1984) Acoustics and the fisheries: recent work with sectorscanning sonar at the Lowestoft Laboratory. In *Advanced concepts in ocean measurements for marine biology* (ed. Diemer, F.P., Vernberg, F.J. and Mirkes, D.Z.), pp. 409–421. Univ. South Carolina Press, Columbia.

Jones, F.R. Harden and Arnold, G.P. (1982) Acoustic telemetry and the marine fisheries. *Symp. Zool. Soc. Lond.* **49**: 75–93.

Jones, F.R. Harden, Arnold, G.P., Greer Walker, M. and Scholes, P. (1979) Selective tidal stream transport and the migration of plaice (*Pleuronectes platessa*) in the southern North Sea. *J. Cons. Int. Explor. Mer,* **38**: 331–337.

Jones, F.R. Harden, Greer Walker, M. and Arnold, G.P. (1978) Tactics of fish movement in relation to migration strategy and water circulation. In *Advances in oceanography* (eds. Charnock, H. and Deacon, G.E.R.), pp. 185–207. Plenum Press, New York and London.

Jones, F.R. Harden, Greer Walker, M. and Arnold, G.P. (1984) Tactics of fish movement in relation to migration strategy and water circulation. In *Advances in oceanography* (eds. Charnock, H. and Deacon, Sir George), pp. 185–207. Plenum Press, New York.

Jones, R. (1973) Density dependent regulation of the numbers of cod and haddock. *Rapp. Procès-Verb. Réun. Cons. Int. Explor. Mer,* **164**: 156–173.

Jones, R. and Hislop, J.R.G. (1978) Changes in North Sea haddock and whiting. *Rapp. Procès-Verb. Réun. Cons. Int. Explor. Mer,* **172**: 58–71.

Jørgensen, T. (1990) Long term changes in age at sexual maturity of North east Arctic cod. *J. Cons. Int. Explor. Mer,* **46**: 235–248.

Jørstad, K.E., King, D.P.F. and Naevdal, G. (1991) Population structure of Atlantic herring, *Clupea harengus* L. *J. Fish. Biol.* **39** (Suppl. A): 43–52.

Karahiri, M., Berghahn, R. and Westernhagen, H. (1989) Growth differences in O group plaice, *Pleuronectes platessa*, as revealed by otolith microstructure. *Mar. Ecol. Progr. Ser.* **55**: 15–22.

Kawasaki, T. (1983) Why do some pelagic fishes have wide fluctuations in their numbers? *FAO Fish Rept.* **291**(3): 1065–1080.

Kawasaki, T. and Omori, M. (1988) Fluctuations in the three major sardine stocks in the Pacific and the global trend in temperature. In *International symposium on long term changes in marine fish populations* (ed. Wyatt, T. and Larrañeta, M.G.), pp. 37–53, Instituto de Investigaciones Marinas, Vigo.

Kendall, A.W., Clarke, M.E., Yoklavitch, M.M. and Boehlert, G.W. (1987) Distribution, feeding and growth of larval walleye pollock (*Theragra chalcogramma*) in the Bering Sea with reference to spawning stock structure. *Fish. Bull.* **85**: 499–527.

Ketchum, B.H. (1939*a*) The development and restoration of deficiency in the phosphorus and nitrogen composition of unicellular plants. *J. Cell. Comp. Physiol.* **13**: 373–81.

Ketchum, B.H. (1939*b*) The absorption of phosphate and nitrate by illuminated cultures of *Nitzschia closterium*. *Amer. J. Bot.* **96**: 399–407.

Kimura, M. (1968) Evolutionary rate at the molecular level. *Nature* **217**: 624, 626.

Kimura, M. (1970) Formation of a false annulus on scales of Pacific sardines of known age. *CalCOFI Rept.* **14**: 73–75.

Kimura, M. (1977) The neutral theory of molecular evolution and polymorphism. *Scientia,* **112**: 687–707.

King, D.P.F., Ferguson, A. and Moffett, I.J.J. (1987) Aspects of the population genetics of the herring around the British Isles and in the Baltic Sea. *Fish. Res.* **6**: 38–52.

Kleerekorper, H. (1969) *Olfaction in fishes.* Indiana University Press, Bloomington and London.

Koehl, M.A.R. and Strickler, J.R. (1981) Copepod feeding currents: food capture at low Reynolds number. *Limnol. Oceanogr.* **26**: 1062–1073.

Kondo, K. (1988) Relationships between long term fluctuations in the Japanese
    sardine *Sardinops melanostictus* (Temminck and Schlegel) and
    oceanographic conditions. *International symposium on long term changes
    in marine fish populations* (ed. Wyatt, T. and Larrañeta, M.G.), pp.
    365–392, Instituto de Investigaciones Marinas, Vigo.
Kornfeld, I., Sidall, B.D. and Gagnon, P.S. (1982) Stock definition in Atlantic
    herring (*Clupea harengus harengus*): genetic evidence for discrete fall and
    spring spawning populations. *Can. J. Fish. Aquat. Sci.* **39**: 1610–1621.
Koslow, J.A., Thompson, K.R. and Silvert, W. (1987) Recruitment to the
    Northwest Atlantic cod (*Gadus morhua*) and haddock (*Melanogrammus
    aeglefinus*) stocks: Influence of stock size and climate. *Can. J. Fish. Aquat.
    Sci.* **44**: 26–39.
Kosoka, S. (1986) Relation of the migration of the Pacific sauries to oceanic
    fronts in the Northwest Pacific ocean. *Bull. Int. N. Pac. Fish. Commn*, **47**:
    229–246.
Kramer, D. and Zweifel, J.R. (1970) Growth of anchovy larvae (*Engraulis
    mordax* Girard) in the laboratory as influenced by temperature. *CalCOFI
    Rept.* **14**: 84–87.
Kremer, J.N. and Nixon, S.W. (1978) *A coastal marine ecosystem: simulation
    and analysis.* Springer Verlag, Berlin, Heidelberg and New York.
Lack, D. (1954) *The natural regulation of animal numbers.* Clarendon Press,
    Oxford.
Lampitt, R.S. (1985) Evidence for the seasonal deposition of detritus to the
    deep sea floor and its subsequent resuspension. *Deep Sea Res.* **32**: 885–897.
Lampitt, R.S. and Burnham, M.P. (1983) A free fall time lapse camera and
    current meter system, 'Bathysnap', with notes on the foraging behaviour
    of a bathyal decapod shrimp. *Deep Sea Res.* **30**: 1009–1017.
Landry, M.R. and Hassett, R.P. (1987) Time scales in behavioural, biochemical
    and energetic adaptations to food limiting conditions by a marine
    copepod. In *Food limitation and the structure of zooplankton communities*
    (ed. Lampert. W.), pp. 209–221. Ergebnisse der Limnologie, Plön.
Larrance, J.D. (1971) Primary productivity and related oceanographic data,
    Subarctic Pacific region 1966–1971. *Data Rep. NOAA*, 50.
Lasker, R. (1975) Field criteria for survival of anchovy larvae: the relation
    between inshore chlorophyll maximum layers and successful first feeding.
    Fish. Bull. **73**: 453–462.
Lasker, R. (1981) Factors contributing to variable recruitment of the northern
    anchovy (*Engraulis mordax*) in the California Current: contrasting years,
    1975 through 1978. *Rapp. Procès-Verb. Réun. Cons. Int. Explor. Mer*, **178**:
    375–388.
Lasker, R., Feder, H.M., Theilacker, G.H. and May, R.C. (1970) Feeding,
    growth and survival of *Engraulis mordax* larvae reared in the laboratory.
    Mar. Biol. **5**: 345–353.
Laurence, G.C. (1982) Nutrition and trophodynamics of larval fish, review,
    concepts, strategy, recommendations and options. *Fish Ecology III*,
    pp. 1–22. Coop. Inst. Mar. Atmos. Sci., University of Miami.
Laws, E.A., Bienfang, P.K., Ziemann, D.A. and Conquest, L.D. (1988)
    Phytoplankton population dynamics and the fate of production during the
    spring bloom in Auke Bay, Alaska. *Limnol. Oceanogr.* **33**: 57–65.
Laws, E.A., diTullio, G.R., Carder, K.L., Betzer, P.R. and Hawes, S. (1990)
    Primary production in the deep blue sea. *Deep Sea Res.* **37**: 715–730.
Laws, E.A., Redalje, D.G., Haas, L.W., Bienfang, P.K., Eppley, R.W.,
    Harrison, W.G., Karl, D.M. and Marra, J. (1984) High phytoplankton

growth and production rates in oligotrophic Hawaiian coastal waters. *Limnol. Oceanogr.* **29**: 1161–1169.

Lebed, N.I., Ponamarenko, I.Y. and Yaragina, N.A. (1983) *Some results of cod tagging in the Barents Sea in 1966–1982.* ICES CM 1983 G21 12 (mimeo).

Lee, A.J. and Folkard, A.R. (1969) Factors affecting turbidity in the Southern North Sea. *J. Cons. Int. Explor. Mer*, **32**: 291–302.

Le Fèvre, J. (1986) Aspects of the biology of frontal systems. *Adv. Mar. Biol.* **23**: 163–299.

Le Fèvre, J. and Grall, J.R. (1970) On the relationships of *Noctiluca* swarming off the west coast of Brittany with hydrological features and plankton characteristics of the environment. *J. Exp. Mar. Biol. Ecol.* **4**: 287–306.

Legeckis, R. (1978) A survey of worldwide sea surface temperature fronts detected by environmental satellites. *J. Geophys. Res.* **83**: 4501–4522.

Leggett, W.C., Frank, K.T. and Carscadden, J.T. (1984) Meteorological and hydrographic regulation of year class strength in capelin (*Mallotus villosus*). *Can. J. Fish. Aquat. Sci.* **41**: 1193–1201.

Leven, H. (1947) On a matching problem arising in genetics. *Ann. Math. Stat.* **20**: 91–94.

Li, W.K.W. (1983) Consideration of errors in estimating kinetic parameters based on Michaelis–Menten formalism in microbial ecology. *Limnol. Oceanogr.* **22**: 47–123.

Lighthill, J. (1976) Flagellar hydrodynamics. *SIAM Rev* **18**: 161–230.

Ljungman, A.V. (1880) The Great Bohuslan herring Fisheries. *Bull. US Bureau Fish.* 1880. Washington.

Lluch-Belda, D., Crawford, R.J.M., Kawasaki, T., MacCall, A.D., Parrish, R.H., Schwartzlose, R.A. and Smith, P.E. (1989) World-wide fluctuations of sardine and anchovy stocks: the regime problem. *South African J. Mar. Sci.* **8**: 195–205.

Lochte, K. and Turley, C.M. (1985) Heterotrophic activity and carbon flow via bacteria in waters associated with a tidal mixing front. In *Proceedings 19th European Marine Biology Symposium* (ed. Gibbs, P.E.), pp. 75–85. Cambridge University Press, Cambridge.

Lockwood, S.J. (1972) Density dependent mortality in O group plaice (*Pleuronectes platessa* L.) populations. *J. Cons. Int. Explor. Mer*, **39**: 148–153.

Lockwood, S.J. (1974) The settlement, distribution and movements of 0 group plaice, *Pleuronectes platessa* L., in Filey Bay, Yorkshire. *J. Fish. Biol.* **6**: 465–477.

Loder, J.W. and Greenberg, D.A. (1986) Predicted positions of tidal fronts in the Gulf of Maine region. *Cont. Shelf Res.* **6**: 397–414.

Loder, J.W. and Platt, T. (1985) Physical controls on phytoplankton production at tidal fronts (Keynote lecture). In *Proceedings 19th European Marine Biology* Symposium (ed. Gibbs, P.E.), pp. 3–21. Cambridge University Press, Cambridge.

Loeng, H., Nakken, O. and Raknes, A. (1983) Loddas utbredelse i Barentshavet i forhold til temperatur feltet i perioden 1974–1982. *Fisk. Hav.* **1983**(1): 1–17.

Lohmann, H. (1908) Untersuchungen zur Feststellung das vollständlich Gehaltes die Meeres an Plankton. *Wiss. Meeresunters Kiel* **9**: 1–122.

Longhurst, A.R. (1967) The pelagic phase of *Pleuroncodes planipes* Stimpson (Crustacea, Galatheidae) in the California Current *CalCOFI Rept.* **11**: 142–154.

Longhurst, A.R. and Harrison, W.G. (1989) The biological pump: profiles of plankton production and consumption in the upper ocean. *Progr. Oceanogr.* **22**: 47–123.

Lough, R.G. (1982) Age and growth of larval Atlantic herring, *Clupea harengus* L. in the Gulf of Maine–George's Bank region based on otolith growth increments. *Fish. Bull.* **80**: 187–199.

Lough, R.G. (1984) Larval fish trophodynamic studies on Georges' Bank: sampling strategy and initial results. *Flødevigen Rapportser* **1**: 393–434.

Lough, R.G. and Laurence, G.C. (1981) *Larval haddock and cod survival studies on Georges Bank*. ICES Larval Fish Ecology Working Group, Lowestoft July 1–3 (mimeo).

Lough, R.G., Pennington, M.R., Bolz, G.R. and Rosenberg, A.S. (1980) *A growth model of larval sea herring (*Clupea harengus*) in the Georges Bank Gulf of Maine area based on otolith growth measurements*. ICES CM 1980 H65.

Ludwig, D. and Walters, C.J. (1981) Measurement errors and uncertainty in parameter estimates for stock and recruitment. *J. Fish. Res. Bd. Can.* **38**: 711–720.

MacCall, A.D. (1979) Population estimates for the waning years of the Pacific sardine fishery. *CalCOFI Rep.* **20**: 72–82.

MacKendrick, A.G. (1926) Applications of mathematics to medical problems. *Proc. Edin. Math. Soc.* **40**: 98–130.

McCarthy, J.J. (1981) The kinetics of nutrient utilization. *Can. Bull. Fish. Aquat. Sci.* **210**: 211–233.

McCave, I.N. (1984) Size spectra and aggregation of suspended particles in the deep ocean. *Deep Sea Res.* **31**: 329–352.

Mace, P.M. (1994) Relationships between common biological reference points used as thresholds and targets of fisheries management. *Can. J. Fish. Aquat. Sci.* (in press).

McCracken, F. (1959) Cod tagging off northern New Brunswick in 1955 and 1956. *Progr. Rep. Atl. Cst. Sta. Fish. Res. Bd Can.*: 8–19.

McElligot, E.A. and Cross, T.F. (1991) Protein variation in wild Atlantic salmon, with particular reference to southern Ireland. *J. Fish. Biol.* **39** (Suppl A.): 35–42.

McKenzie, R.A. (1934) Cod movements on the Canadian Atlantic coast. *Contrib. Can. Biol.* **8**: 433–458.

McKenzie, R.A. (1956) Atlantic cod tagging off the Southern Canadian mainland. *Bull. Fish. Res. Bd Can.* **105**.

Mackereth, F.J. (1953) Phosphorus utilization by *Asterionella formosa* Hass. *J. Exp. Bot.* **4**: 296–313.

Maillet, G.L. and Checkley, D.M. (1991) Storm related variation in the growth rate of otoliths of larval Atlantic menhaden *Brevoortia tyrannus*: a time series analysis. *Mar. Ecol. Progr. Ser.* **79**: 1–16.

Malmberg, S-A., Hallgrimmsson, I. and Jakobsson, J. (1967) Report of the joint meeting on the distribution of herring in relation to hydrography and plankton. Seydisfjördur, 20–23 June, 1965. *Ann. Biol. Cons. Int. Explor. Mer*, **22**: 188–195.

Malmberg, S-A., Thoradottir, T.H. and Vilhjalmsson, H. (1968) Report on the joint meeting on Atlanto-Scandian herring distribution held at Akureyri 12–14 June, 1966. *Ann. Biol. Cons. Int. Explor. Mer*, **23**: 215–220.

Malmberg, S-A. and Vilhjalmsson, H. (1968) Report on the joint meeting of Icelandic, Norwegian and Soviet investigators on the Atlanto-Scandian herring distribution in relation to oceanographic conditions held at

Seydisfjördur, 5–6 June, 1968. *Ann. Biol. Cons. Int. Explor. Mer*, **25**: 260–265.

Malone, T.C., Hopkins, T.S., Falkowski, P.G. and Whitledge, T.E. (1983) Production and transport of phytoplankton biomass over the continental shelf of the New York Bight. *Cont. Shelf Res.* **1**: 305–337.

Marak, R.R. (1974) Food habits of larval cod, haddock and coalfish in the Gulf of Maine and Georges' Bank Area. *J. Cons. Int. Explor. Mer*, **25**: 147–157.

Marrasé, C., Costello, J.H., Granata, T. and Strickler, J.R. (1990) Grazing in a turbulent environment: energy dissipation, encounter rate and efficacy of feeding currents in *Centropages hamatus*. *Proc. Nat. Acad. Sci.* **87**: 1653–1657.

Margalef, R. (1958) Temporal succession and spatial heterogeneity in phytoplankton. In *Perspectives in marine biology* (ed. Buzzato-Traverso, A.A.), pp. 323–349. Univ. California Press, Berkeley, Los Angeles.

Margalef, R. (1967) Some concepts relative to the organization of plankton. *Oceanogr. Mar. Biol.* **5**: 257–290.

Margalef, R. (1968) *Perspectives in ecological theory*, pp. 1–111. University of Chicago Press, Chicago.

Margalef, R., Duran, M. and Saiz, F. (1955) El fitoplancton de la ria de Vigo de enero de 1953 a marzo de 1954. *Invest. Pesq.* **2**: 85–129.

Marr, J.C. (1960) The causes of major variations in the catch of Pacific sardine, *Sardinops caerulea* (Girard). In *Proc. world sci. meeting on the biology of sardines and related species.* Vol. 3, pp. 667–679.

Marra, J. and Heinemann, K.E. (1984) A comparison between noncontaminating and conventional incubation procedures in primary production measurements. *Limnol. Oceanogr.* **29**: 389–392.

Marra, J. and Heinemann, K.E. (1987) Primary production in the North Pacific Central Gyre: some new measurements based on ^{14}C. *Deep Sea Res.* **34**: 1821–1829.

Marshall, S.M. and Orr, A.P. (1930) A study of the spring diatom increase in Loch Striven. *J. Mar. Biol. Assn UK*, **16**: 853–878.

Martin, J.H., Fitzwater, S.E. and Gordon, R.M. (1990*a*) Iron deficiency limits phytoplankton growth in Antarctic waters. *Global Biogeochemical Cycles* **4**: 5–12.

Martin, J.H., Fitzwater, S.E. and Gordon, R.M. (1990*b*) Iron in Antarctic waters. *Nature* **345**: 156–158.

Martin, J.H. and Gordon, R.M. (1988) North east Pacific iron distribution in relation to phytoplankton productivity. *Deep Sea Res.* **35**: 177–196.

Martin, J.H., Gordon, R.M., Fitzwater, S.E. and Broenkow, W.W. (1989) Vertex: phytoplankton/iron studies in the Gulf of Alaska. *Deep Sea Res.* **36**: 649–680.

Martin, J.H., Knauer, G.A., Karl, D.M. and Broenkow, W.W. (1987) VERTEX: carbon cycling in the northeast Pacific. *Deep Sea Res.* **34**: 267–285.

Martin, W.R. and Jean, Y. (1964) Winter cod taggings off Cape Breton and on offshore Nova Scotia Banks 1959–62. *J. Fish. Res. Bd Can.* **21**: 215–238.

Maslov, N.A. (1972) Migrations of the Barents Sea cod. *Fish. Res. Bd Can. Transl.* **2129**: 1–43.

Matsumoto, W. (1966) Distribution and abundance of tuna larvae in the Pacific Ocean. In *Proc. Governor's Conf. on Central Pacific Research* (ed. Manier, T.A.), pp. 221–223. Dept. of the Interior.

May, R.M. (1977) Models of single populations. In *Theoretical ecology* (ed.

May, R.M.), pp. 4–25. Blackwell, Oxford.

May, R.M. (1986) Detecting density dependence in imaginary worlds. *Nature*, **338**: 16–18.

Medawar, P.B. (1945) Size, shape and age. In *Essays presented to D'Arcy Wentworth Thompson* (ed. Le Gros Clark, W. and Medawar, P.B.), pp. 157–187. Clarendon Press, Oxford.

Menzel, D.W. and Ryther, J.H. (1960) The annual cycle of primary production in the Sargasso Sea off Bermuda. *Deep Sea Res.* **6**: 351–367.

Mertz, G. and Myers, R.A. (1994) Match/mismatch prediction of spawning duration versus recruitment variability (in press).

Metcalfe, J.D., Holford, B.H. and Arnold, G.P. (1993) Orientation of plaice (*Pleuronectes platessa*) in the open sea: evidence for the use of external directional clues. *Mar. Biol.* **117**: 559–566.

Methot, R.D. (1981) Spatial covariation of daily growth rate of larval northern anchovy, *Engraulis mordax* and northern lampfish, *Stenotrachius leucopsarus*. *Rapp. Procès-Verb. Réun. Cons. Int. Explor. Mer*, **178**: 424–431.

Methot, R.D. and Kramer, D. (1979) Growth of Northern anchovy (*Engraulis mordax* Girard) larvae in the sea. *Fish. Bull.* **77**: 413–423.

Middtun, L., Nakken, O. and Raknes, A. (1981) Variasjoner i utbredelsen av Torsk i Barentshavet i perioden, 1977–81. *Fisken Hav.* **1981**(4): 1–16.

Miller, C.B., Cowles, T.J., Wiebe, P.H., Copley, N.J. and Grigg, H. (1991) Phenology in *Calanus finmarchicus:* hypotheses on control mechanisms. *Mar. Ecol. Progr. Ser.* **72**: 79–81.

Minas, H.J., Minas, M. and Packard, T.T. (1986) Productivity in upwelling areas deduced from hydrographic and chemical fields. *Limnol. Oceanogr.* **31**: 1182–1206.

Mitchell, J.M. (1864) *The herring, its natural history and national importance.* Edmonston and Douglas, Edinburgh.

Mitson, R.B. and Storeton-West, T.J. (1971) A transponding acoustic tag. *The Radio and Electronic Engineer* **41**: 483–489.

Mitson, R.B., Storeton-West, T.J. and Pearson, N.D. (1982) Trials of an acoustic transponding fish tag compass. *Biotelemetry* **9**: 69–79.

Moran, P.A.P. (1962) *The statistical processes of evolutionary theory.* Oxford University Press, Oxford.

Mork, J., Giskeodergard, R. and Sundnes, G. (1983) Haemoglobin polymorphism in *Gadus morhua*: genotypic differences in maturing age and within season gonad maturation. *Helgo. Meeres.* **36**: 313–322.

Mork, J., Giskeodegard, R. and Sundnes, G. (1984) Population genetic studies in cod (*Gadus morhua* L.) by means of the haemoglobin polymorphism: observations in a Norwegian coastal population. *Fisk. Dir. Skr. Hav.* **17**: 449–471.

Mork, J., Reuterwall, C., Ryman, N. and Stahl, G. (1982) Genetic variation in Atlantic cod (*Gadus morhua*): a quantitative estimate from a Norwegian coastal population. *Hereditas*, **96**: 55–61.

Mork, J., Ryman, N., Stahl, G., Utter, F. and Sundness, G. (1985) Genetic variation in Atlantic cod (*Gadus morhua*) throughout its range. *Can. J. Fish. Aquat. Sci.* **42**: 1580–1587.

Mork, J. and Sundnes, G. (1984) Haemoglobin polymorphism in *Gadus morhua*: genotypic differences in haematocrit. *Helgo. Meeres.* **38**: 201–206.

Mork, J. and Sundnes, G. (1985) Haemoglobin polymorphism in Atlantic cod (*Gadus morhua*): allele frequency variation between year classes in a Norwegian fjord stock. *Helgo. Meeres.* **39**: 55–62.

Morse, W., (1989) Catchability, growth and mortality of fish larvae. *Fish. Bull.* **87**: 417–446.

Mountford, M.D. (1988) Population regulation, density dependence and heterogeneity. *J. Anim. Ecol.* **57**: 845–858.

Muck, P. (1989) Major trends in the pelagic ecosystem off Peru and their implications for management. In *The Peruvian upwelling ecosystem: dynamics and interactions* (ed. Pauly, D., Muck, P., Mendo, J. and Tsukuyama, I.), pp. 386–403. ICLARM Conf. Proc. vol. 18, Manila.

Mullin, M.M. (1969) Production of zooplankton in the ocean: the present status and problems. *Oceanogr. Mar. Biol.* **7**: 293–314.

Mullin, M.M. and Brooks, E.R. (1970) Growth and metabolism of two planktonic marine copepods as influenced by temperature and type of food. In *Symposium on marine food chains* (ed. Steele, J.), pp. 74–95. Oliver and Boyd, Edinburgh.

Mullin, M.M. and Brooks, E.R. (1976) Some consequences of distributional heterogeneity of phytoplankton and zooplankton. *Limnol. Oceanogr.* **21**: 784–796.

Munawar, A. and Talling, J.F. (1986) Seasonality of freshwater phytoplankton. *Hydrobiologia*, **138**: 1–212.

Myers, R.A. (1991) Recruitment variability and range of three fish species. *NAFO Sci. Counc. Stud.* **16**: 21–24.

Myers, R.A., Barrowman, N.J. and Hutchings, J.A. (1994*b*) Depensatory recruitment and the collapse of fisheries. NAFO Sci. Council Rep. Doc. 1993.69.

Myers, R.A., Blanchard, W. and Thompson, K.R. (1990) Summary of North Atlantic fish recruitment, 1942–1987. *Can. Tech. Rep. Fish. Aquat. Sci.* 1743.

Myers, R.A., Bridson, J. and Barrowman, N.J. (1994*a*) Summary of world wide stock and recruitment data. *Can. Tech. Rep. Fish. Aquat. Sci.* (in press).

Myers, R.A. and Cadigan, N.G. (1993) Density dependent juvenile mortality in marine demersal fish. *Can. J. Fish. Aquat. Sci.* **50**: 1576–1590.

Myers, R.A. and Drinkwater, K. (1989) The influence of Gulf Stream warm core rings on recruitment of fish in the northwest Atlantic. *J. Mar. Res.* **47**: 635–656.

Myers, R.A., Mertz, G. and Bishop, C.A. (1993) Cod spawning in relation to physical and biological cycles of the northern north west Atlantic. *Fish. Oceanogr.* **2**: 154–165.

Mysak, L.A. (1986) El Niño, interannual variability and fisheries in the Northeast Pacific Ocean. *Can. J. Fish. Aquat. Sci.* **43**: 464–497.

Mysak, L.A., Hsieh, W.W. and Parsons, T.R. (1982) On the relationship between Interannual baroclinic waves and fish populations in the North East Pacific. *Biol. Oceanogr.* **2**: 63–103.

Nicholls, J. and Thompson, A.B. (1991) Selection of copepodites and nauplii. *J. Plankt. Res.* **13**: 661–672.

Nielsen, E.S. and Jensen, E.A. (1957) Primary oceanic production. The autotrophic production of organic matter in the oceans. *Galathea Rep.* **1**: 49–135.

Nielsen, E.S. and Hansen, K.V. (1959) Measurements with the carbon-14 technique of the respiration rate in natural populations of phytoplankton. *Deep Sea Res.* **5**: 222–233.

Nishikawa, Y., Honma, M., Ueyanagi, S. and Kikawa, S. (1985) Average distributions of larvae of oceanic species of scombroid fishes, 1956–81. *Publ. Far Seas Res. Lab.* **12**: 1–99.

Oceanographic Laboratory, Edinburgh (1973) Continuous plankton records: a plankton atlas of the North Atlantic and the North Sea. *Bull. Mar. Ecol.* **7**: 1–174.

Okubo, A. (1978) Advection-diffusion in the presence of surface convergence. In *Oceanic fronts in coastal processes.* Proceedings of a workshop held at the Marine Science Research Center, May 25–27, 1977 (ed. Bowman, M.J. and Esaias, W.E.), pp. 23–28. Springer Verlag, Berlin, Heidelberg and New York.

Omori, M. and Ikeda, T. (1984) *Methods in marine zooplankton ecology.* Chichester Wiley-Interscience, New York.

Ostvedt, O-J. (1955) Zooplankton investigations from Weather Ship M in the Norwegian Sea. *Hvalr. Skr.* **40**: 1–89.

Ottestad, P. (1960) Forecasting the annual yield in sea fisheries. *Nature,* **185**: 183.

Overholtz, W.J., Sissenwine, M.P. and Clark, S.M. (1984) Recruitment variability and its implications for managing and rebuilding the George's Bank haddock stock. ICES CM 1984, G77.

Overholtz, W.J., Sissenwine, M.P. and Clark, S.H. (1986) Recruitment variability and implication for managing and rebuilding the George's Bank haddock (*Melanogrammus aeglefinus*) stock. *Can. J. Fish. Aquat. Sci.* **43**: 748–753.

Paffenhöfer, G-A. (1988) Feeding rates and behavior of zooplankton. *Bull. Mar. Sci.* **43**: 430–455.

Paffenhöfer, G-A. and Lewis, K.D. (1992) Perceptive performance and feeding behaviour of calanoid copepods. *J. Plankt. Res.* **12**: 933–946.

Paffenhöfer, G-A., Strickler, J.R. and Alcaraz, M. (1982) Suspension feeding by herbivorous calanoid copepods: cinematographic studies. *Mar. Biol.* **67**: 193–199.

Page, F.H. and Frank, K.T. (1989) Spawning times and egg stage duration in North West Atlantic haddock (*Melanogrammus aeglefinus*) stocks with emphasis on George's and Brown's Bank. *Can. J. Fish. Aquat. Sci.* **46** (Suppl. 1): 68–81.

Pannella, G. (1971) Fish otoliths: daily growth layers and periodical patterns. *Science,* **173**: 1124–1127.

Parmanne, R. and Sjöblom, V. (1984) The abundance of spring spawning herring larvae around Finland in 1982 and 1983 and the correlation between zooplankton abundance and year class strength. ICES CM 1984/J18 (mimeo).

Parmanne, R. and Sjöblom. V. (1987) Possibility of using larval and zooplankton data in assessing the herring year class strength off the coast of Finland in 1974–86. ICES CM 1987/J19 (mimeo).

Parrish, R.H. and MacCall, A.D. (1978) Climatic variations and exploitation in the Pacific mackerel fishery. *Bull. Calif. Fish and Game* **167**: 1–109.

Parsons, T.R. and Anderson, G.C. (1970) Large scale studies of primary production in the North Pacific Ocean. *Deep Sea Res.* **17**: 765–776.

Parsons, T.R. and Lalli, C.M. (1988) Comparative Oceanic Ecology of the plankton communities of the subarctic Atlantic and Pacific Oceans. *Oceanogr. Mar. Biol. Ann. Rev.* **26**: 317–359.

Parsons, T.R. and Lebrasseur, R.J. (1968) The availability of food to different trophic levels in the marine food chain. In *Marine food chains* (ed. Steele, J.H.), pp. 325–343. Oliver and Boyd, Edinburgh.

Pauly, D. and Tsukayama, I. (1987) The Peruvian anchoveta and its upwelling system: three decades of change. *ICLARM Studies and Reviews,* **15**.

Pauly, D. and Tsukayama, I. (1989) *The Peruvian anchoveta and its upwelling system: dynamics and interactions.* (ed. Pauly, D., Muck, P., Mendo, J. and Tsukuyama, I.), Instituto del Mar del Peru, Deutsche Gesellschaft technische zusammenarbeit, International Center for Living Aquatic Resources Management.

Payne, R.H. (1980) The use of transferrin polymorphism to determine the stock composition of Atlantic salmon in the West Greenland fishery. *Rapp. Procès-Verb. Cons. Réun. Int. Explor. Mer,* **176**: 60–64.

Payne, R.H., Child, A.R. and Forrest, A. (1971) Geographical variation in the Atlantic salmon. *Nature,* **231**: 250–252.

Pearcy, W.G. (1983) Abiotic variations in regional environments. In *From year to year* (ed. Wooster, W.S.), pp. 143–165. Washington Sea Grant Program, Seattle.

Peinert, R., Saure, A., Stegmann, P., Stienen, C., Haardt, J. and Smetacek, V. (1982) Dynamics of primary production and sedimentation in a coastal ecosystem. *Neth. J. Sea Res.* **16**: 276–289.

Pella, J.J. and Milner, G.B. (1987) Use of genetic markers in stock composition analysis. In *Population genetics and Fishery Management* (ed. Ryman, N. and Utter, F.), pp. 247–276. Washington Sea Grant Program, Seattle and London.

Pella, J.J. and Tomlinson, P.K. (1969) A generalized stock production model. *Bull. Int. Amer. Trop. Tuna Commn,* **13**: 421–495.

Penney, R.W. and Evans, G.T. (1985) Growth histories of larval redfish (*Sebastes* spp.) on an offshore Atlantic fishing bank determined by otolith increment analysis. *Can. J. Fish. Aquat. Sci.* **42**: 1452–364.

Pepin, P. (1991) Effects of temperature and size on development, mortality and survival rates of the pelagic early life history stages of marine fish. *Can. J. Fish. Aquat. Sci.* **48**: 503–518.

Peterman, R. and Bradford, M.J. (1987) Wind speed and mortality rate of a marine fish, the northern anchovy, *Engraulis mordax. Science,* **235**: 354–356.

Peterman, R., Bradford, M.J., Lo, N.H.C. and Methot, R. (1988) Contribution of early life stages to interannual recruitment of northern anchovy (*Engraulis mordax*). *Can. J. Fish. Aquat. Sci.* **45**: 8–17.

Petersen, C.J.G. (1894) On the biology of our flatfishes and on the decrease of our flatfisheries. *Rep. Dansk. Biol. Sta.* **4**: 1–146.

Peterson, B.J. (1980) Aquatic primary productivity and the ^{14}C–CO_2 method. *Ann. Rev. Ecol. Syst.* **11**: 359–85.

Peterson, W.T. (1973) Upwelling indices and annual catches of Dungeness crab, *Cancer magister,* along the West Coast of the United States. *Fish. Bull.* **71**: 902–910.

Pingree, R.D. (1975) The advance and retreat of the thermocline on the continental shelf. *J. Mar. Biol. Assn UK,* **55**: 965–974.

Pingree, R.D. (1977) Mixing and stabilization of phytoplankton distributions on the North West European Continental Shelf. In *Spatial Pattern in Plankton Communities* (ed. Steele, J.H.), pp. 181–220. Plenum Press, New York and London.

Pingree, R.D. (1978) Cyclonic eddies and cross frontal mixing. *J. Mar. Biol. Assn UK,* **58**: 955–964.

Pingree, R.D. (1979) Baroclinic eddies bordering the Celtic Sea in late summer. *J. Mar. Biol. Assn UK,* **59**: 689–698.

Pingree, R.D., Forster, G.R. and Morrison, G.K. (1974) Turbulent convergent tidal fronts. *J. Mar. Biol. Assn UK,* **54**: 469–479.

Pingree, R.D. and Griffiths, D.K. (1978) Tidal fronts on the Shelf Seas around the British Isles. *J. Geophys. Res.* **83**: 4615–4622.

Pingree, R.D., Holligan, P.M., Mardell, G.T. and Head, R.N. (1976) The influence of physical stability on spring, summer and autumn phytoplankton blooms in the Celtic Sea. *J. Mar. Biol. Assn UK*, **56**: 845–873.

Pingree, R.D., Holligan, P.M. and Mardell, G.T. (1979) Phytoplankton growth and cyclonic eddies. *Nature*, **278**: 245–247.

Pingree, R.D., Maddock, L. and Butler, E.I. (1977) The influence of biological activity and physical stability in determining the chemical distributions of inorganic phosphate, silicate and nitrate. *J. Mar. Biol. Assn UK*, **57**: 1065–1073.

Pingree, R.D., Pugh, P.R., Holligan, P.M. and Forster, G.R. (1975) Summer phytoplankton blooms and red tides along tidal fronts in the approaches to the English Channel. *Nature*, **258**: 672–677.

Platt, T., Bird, D.F. and Sathyendranath, S. (1991) Critical depth and marine primary production. *Proc. Roy. Soc. B.* **264**: 205–217.

Platt, T., Gallegos, C.L. and Harrison, W.G. (1980) Photoinhibition of photosynthesis in natural assemblages of marine phytoplankton. *J. Mar. Res.* **38**: 687–701.

Platt, T. and Harrison, W.G. (1985) Biogenic fluxes of carbon and oxygen in the ocean. *Nature*, **318**: 55–58.

Platt, T., Harrison, W.G., Lewis, M.R., Li, W.K.W., Sathyendranath, S., Smith, R.E. and Vezina, A.F. (1989) Biological production of the oceans: the case for consensus. *Mar. Ecol. Progr. Ser.* **52**: 77–88.

Platt, T. and Jassby, A.D. (1976) The relationship between photosynthesis and light for natural assemblages of coastal marine phytoplankton. *J. Phycol.* **12**: 421–430.

Platt, T., Rao, D.V.S. and Irwin, B. (1983) Photosynthesis of picoplankton in the oligotrophic ocean. *Nature*, **301**: 702–704.

Platt, T. and Sathyendranath, S. (1988) Oceanic primary production: estimation by remote sensing at local and regional scales. *Science*, **241**: 1613–1620.

Platt, T., Sathyendranath, S. and Ravindran, P. (1990) Primary production by phytoplankton: analytic solutions for daily rates per unit area of water surface. *Proc. Roy. Soc. B*, **241**: 101–111.

Platt, T., Sathyendranath, S., Ulloa, O., Harrison, W.G., Hoepffner, N. and Goes, J. (1992) Nutrient control of phytoplankton photosynthesis in the Western North Atlantic. *Nature*, **356**: 229–231.

Policansky, D. (1982) Influence of age, size and temperature on metamorphosis of the starry flounder *Platichthys stellatus*. *J. Fish. Res. Bd Can.* **39**: 514–517.

Pollard, E. and Lakhani, K.H. (1987) The detection of density dependence from a series of annual censuses. *Ecology*, **68**: 2046–2055.

Pope, J.G. (1972) An investigation of the accuracy of virtual population analysis using cohort analysis. *Int. Commn North West Atl. Fish. Res. Bull.* **9**: 65–74.

Postma, H. and Rommets, J.W. (1979) Dissolved and particulate organic carbon in the North Equatorial Current of the Atlantic Ocean. *Neth. J. Sea Res.* **13**: 85–98.

Postolakii, A.I. (1966) Results of cod tagging in the Laborador and north Newfoundland bank regions, 1960–64. *Can. Tech. Ser.* **859**: 1–14.

Price, H.J., Paffenhöfer, G.-A. and Strickler, J.R. (1983) Modes of cell capture in calanoid copepods. *Limnol. Oceanogr.* **28**: 116–123.

Purcell, E.M. (1977) Life at low Reynolds numbers. *Amer. J. Phys.* **45**: 3–11.

Purdom, C.E. and Wyatt, T. (1969) Racial differences in Irish Sea and North Sea plaice (*Pleuronectes platessa*). *Nature*, **222**: 780–781.

Radovich, J. (1961) Relationships of some marine organisms of the Northeast Pacific to water temperatures particularly during 1957–59. *Fish. Bull. No 112*. State of California Dept. of Fish and Game.

Raitt, D.F.S. (1968) The population dynamics of the Norway pout in the North Sea. *Mar. Res.* **5**: 1–24.

Rauck, G. (1977) Two German plaice tagging experiments (1970) in the North Sea. *Arch. Fischwiss.* **28**: 57–64.

Rauck, G. and Zijlstra, J.J. (1978) On the nursery aspects of the Wadden Sea for some commercial fish species and possible long term changes. *Rapp. Procès-Verb. Réun. Cons. Int. Explor. Mer*, **172**: 266–275.

Raven, J.A. (1987) Physiological consequences of extremely small size for autotrophic organisms in the sea. *Can. Bull. Fish. Aquat. Sci.* **214**: 1–70.

Raymont, J.E.G. (1980) *Plankton and productivity in the oceans 2nd. Edn Vol. 1. Phytoplankton.* Pergamon Press, Oxford and New York.

Reddingius, J. (1971) Gambling for existence. *Acta Biotheor.* **20** (Suppl. 1): 1–208.

Redfield, A.C. (1934) On the proportions of organic derivatives in sea water and their relationship to the composition of plankton. In *James Johnstone memorial volume* (ed. Daniel, R.J.), pp. 176–192. University of Liverpool Press, Liverpool.

Reise, K. (1982) Long term changes in the macrobenthic invertebrate fauna of the Wadden Sea: are polychates about to take over? *Neth. J. Sea Res.* **16**: 29–36.

Reynolds, C.S. (1976a) Sinking movements of phytoplankton indicated by a single trapping method. I A *Fragilaria* population. *Brit. Phycol. J.* **11**: 279–91.

Reynolds, C.S. (1976b) Sinking movements of phytoplankton indicated by a single trapping method. II Vertical activity ranges in a stratified lake. *Brit. Phycol. J.* **11**: 293–303.

Reynolds, C.S. (1983) A physiological interpretation of the dynamic responses of populations of a planktonic diatom to physical variability of the environment. *New Phytol.* **95**: 41–53.

Reynolds, C.S. (1984) *The ecology of freshwater phytoplankton.* Cambridge University Press, Cambridge.

Reynolds, C.S., Morison, H.R. and Butterwick, C. (1982) The sedimentary flux of phytoplankton in the south basin of Windermere. *Limnol. Oceanogr.* **27**: 1162–1175.

Reynolds, C.S. and Wiseman, S.W. (1982) Sinking losses of phytoplankton in closed limnetic systems. *J. Plankt. Res.* **4**: 489–522.

Rice, A.L., Billett, D.S.M., Fry, F.J., John, A.W.G., Lampitt, R.S., Mantoura, R.F.C. and Morris, R.J. (1986) Seasonal deposition of phytodetritus to the deep sea floor. *Proc. Roy. Soc. Edin.* **88B**: 265–279.

Richardson, K., Lavin-Peregrina, M.F., Mitchelson, E.G. and Simpson, J.H. (1985) Seasonal distribution of chlorophyll in relation to physical structure in the western Irish Sea. *Oceanol. Acta*, **8**: 77–86.

Ricker, W.E. (1954) Stock and recruitment. *J. Fish. Res. Bd Can.* **11**: 559–623.

Ricker, W.E. (1958) Handbook of computations for biological statistics of fish populations. *Bull. Fish. Res. Bd Can.* **119**: 1–300.

Ricker, W.E. (1975) Computation and interpretation of biological statistics of fish populations. *Bull. Fish. Res. Bd Can.* **191**.

Ricker, W.E. (1979) Growth rates and models. In *Fish Physiology Vol. 8 Bioenergetics and Growth* (eds Hoar, W.S., Randall, D.J. and Brett, J.R.), pp. 677–743. Academic Press, New York.

Ricker, W.E. and Foerster, R.E. (1948) Computation of fish production. *Bull. Bingh. Oceanogr. Coll.* **11**(4): 173–211.

Richardson, K., Lagrin-Peragrina, M.F., Mitchelson, E.G. and Simpson, J.H. (1985) Seasonal distribution of chlorophyll a in relation to physical structure in the western Irish Sea. *Oceanol. Acta,* **8**: 77–86.

Riebesell, U. (1991*a*) Particle aggregation during a diatom bloom I. Physical aspects. *Mar. Ecol. Progr. Ser.* **69**: 273–280.

Riebesell, U. (1991*b*) Particle aggregation during a spring bloom II. Biological aspects. *Mar. Ecol. Progr. Ser.* **69**: 281–291.

Riemann, B. and Bosselmann, S. (1984) *Daphnia* grazing on natural populations of bacteria. *Verh. Internat. Verein. Limnol.* **22**: 795–799.

Riepma, H.W. (1980) Residual currents in the North Sea during the INOUT Phase of JONSDAP '76. First results. *Meteor Forsch. Ergebnisse A*, **22**(1): 19–32.

Rijnsdorp, A.D. (1990) The mechanism of energy allocation over reproduction and somatic growth in female North Sea plaice, *Pleuronectes platessa* L. *Neth. J. Sea Res.* **25**: 279–290.

Rijnsdorp, A.D. (1991) Changes in fecundity of female North Sea plaice (*Pleuronectes platessa* L.) between three periods since 1900. *ICES. J. Mar. Sci.* **48**: 253–280.

Rijnsdorp, A.D., Daan, N., van Beek, F.A. and Heessen, H.J.L. (1991) Reproductive variability in North Sea plaice, sole and cod. *J. Cons. Int. Explor. Mer*, **47**: 35–375.

Rijnsdorp, A.D., van Lent, F. and Groeneveld, K. (1983) Fecundity and the energetics of reproduction and growth of North Sea plaice (*Pleuronectes platessa* L.). ICES CM 1983 G31.

Rijnsdorp, A.D. and van Stralen, M. (1982) Selective tidal migration of plaice larvae (*Pleuronectes platessa* L.) in the eastern Scheldt and the western Wadden Sea. ICES CM 1982 G:31.

Rijnsdorp, A.D., van Stralen, M. and van der Veer, H.W. (1985) Selective tidal transport of North Sea plaice larvae *Pleuronectes platessa* in coastal nursery areas. *Trans. Amer. Fish. Soc.* **114**: 461–470.

Riley, G.A., Stommel, H. and Bumpus, D.F. (1949) Quantitative ecology of the plankton of the western North Atlantic. *Bull. Bingh. Oceanogr. Coll.* **12**: 1–169.

Robinson, G.A., Colebrook, J.M. and Cooper, G.A. (1975) The Continuous Plankton Recorder Survey: plankton in the ICNAF area, 1961–71 with special reference to 1971. *Res. Bull. ICNAF* **11**: 61–71.

Rogers, J.C. (1984) The association between the North Atlantic oscillation and the Southern Oscillation in the Northern hemisphere. *Monthly Weath. Rev.* **112**: 1999–2015.

Rogers, J.S. (1972) Measures of genetic similarity and genetic distance. In *Studies in genetics VII*, Univ. Texas Publ. 7213, pp. 145–53.

Rollefsen, G. (1956) The arctic cod. *Papers presented at international technical conference on conservation of living resources of the sea, Rome*, pp. 115–17. New York.

Rothschild, B.J. (1986) *Dynamics of marine fish populations*. Harvard University Press, Cambridge, Mass. and London.

Rothschild, B.J. and Mullen, A.J. (1985) The information content of stock and recruitment data and its non-parametric classification. *J. Cons. Int. Explor. Mer*, **42**: 116–24.

Rothschild, B.J. and Osborn, T.R. (1988) Small scale turbulence and plankton contact rates. *J. Plankt. Res.* **10**: 465–474.

Rowe, G.T., Smith, S., Falkowski, P.G., Whitledge, T.E., Theroux, R., Phoel, W. and Ducklow, H. (1986) Do continental shelves export organic matter? *Nature*, **324**: 559–561.

Royama, T. (1977) Population persistence and density dependence. *Ecol. Monogr.* **47**: 1–35.

Ruggles, C.P. and Ritter, J.A. (1980) Review of North American smolt tagging to assess the Atlantic salmon fishery off West Greenland. *Rapp. Procès-Verb. Réun. Cons. Int. Explor. Mer*, **176**: 82–92.

Russell, F.S. (1976) *The eggs and planktonic stages of British marine fishes.* Academic Press, Chichester, New York.

Ryman, N., Lagercrantz, U., Andersson, L. and Rosenberg, R. (1984) Lack of correspondence between genetic and morphologic variability patterns in Atlantic herring (*Clupea harengus*). *Heredity*, **53**: 687–704.

Ryther, J.H. and Dunstan, W.M. (1971) Nitrogen, phosphorus and eutrophication in the coastal marine environment. *Science*, **171**: 1008–1013.

Saino, T., Miyata, K. and Hattori, A. (1979) Primary productivity in the Bering and Chukchi Seas and in the northern north Pacific in 1978 summer. *Bull. Planktonol. Soc. Japan*, **26**: 96–103.

St Pierre, G. (1984) Spawning locations and seasons for Pacific halibut. *Bull. Int. Pac. Halibut Comm.* **70**: 1–46.

Salini, J. and Shaklee, J.B. (1988) Genetic structure of Barramundi (*Lates calcarifer*) stocks from Northern Australia. *Austr. J. Mar. Freshwater Res.* **39**: 317–329.

Santander, H. (1987) Relationship between anchoveta egg standing stock and parent biomass. *ICLARM Studies and Reviews* **15**: 179–207.

Savidge, G. (1976) A preliminary study of the distribution of chlorophyll a in the vicinity of fronts in the Celtic and western Irish Seas. *Est. & Coast. Mar. Sci.* **4**: 617–626.

Saville, A. (1959) Mesh selection in plankton nets. *J. Cons. Int. Explor. Mer*, **23**: 192–201.

Schaefer, M.B. (1954) Some aspects of the dynamics of populations important to the management of commercial fish stocks. *Bull. Inter-Amer. Trop. Tuna Commn*, **1**: 25–56.

Schaefer, M.B. (1957) A study of the dynamics of the fishery for yellowfin tuna. *Bull. Inter-Amer. Trop. Tuna Commn*, **2**: 245–285.

Schmidt, J. (1904) On the larval and post larval stages of the Atlantic species of Gadus. A monograph. *Medd. Kommn. Hav. Ser. Fisk. Bind*, **1**: 1–77.

Schnute, J. (1977) Improved estimates from the Schaefer model: theoretical considerations. *J. Fish. Res. Bd Can.* **34**: 583–603.

Schroeder, W.C. (1930) Migrations and other phases in the life history of the cod off southern New England. *Bull. US Bur. Fish.* **46**: 1–136.

Schweder, T. and Spjøtvoll, E. (1982) Plots of P-values to evaluate many tests simultaneously. *Biometrika* **69**: 493–502.

Scura, E.D. and Jerde C.W. (1977) Various species of phytoplankton as food for larval northern anchovy, *Engraulis mordax*, and relative nutritional value of the dinoflagellates *Gymnodinium splendens* and *Gonyaulax polyedra*. *Fish. Bull.* **75**: 577–583.

Seikai, T., Tanangonan, J.B. and Tanaka, M. (1986) Temperature influence on larval growth and metamorphosis of the Japanese flounder *Palalichthys olivaceus* in the laboratory. *Bull. Jap. Soc. Sci. Fish.* **52**: 977–982.

Serebryakov, V.P. (1991) Predicting year class strength under uncertainties related to survival in the early life history of some North Atlantic commercial species. *N.A.F.O. Sci. Council Studies*, **16**: 49–55.

Shackleton, L.Y. (1988) Fossil pilchard and anchovy scales – indicators of past fish populations off Namibia. In *Long term changes in marine fish populations* (ed. Wyatt, T. and Larrañeta, M.G.), pp. 55–68. CSIC, Vigo.

Shaklee, J.B., Klaybor, D.C., Young, S. and White, B.A. (1991) Genetic stock structure of odd year pink salmon (*Oncorhynchus gorbuscha* (Walbaum)) from Washington and British Columbia and potential mixed stock fisheries applications. *J. Fish. Biol.* **39** (Suppl. A): 21–34.

Shaklee, J.B. and Salini, J.B. (1985) Genetic variation and population subdivision in Australian barramundi, *Lates calcarifer* (Bloch). *Austr. J. Mar. Fresh. Sci.* **36**: 203–218.

Shelton, P.A. (1992) The shape of recruitment distributions. *Can. J. Fish. Aquat. Sci.* **49**: 1754–1761.

Shephard, M.P. and Withler, F.C. (1958) Spawning stock size and resultant production for Skeena sockeye. *J. Fish. Res. Bd Can.* **15**: 1007–1025.

Shepherd, J.G. (1982) A versatile new stock and recruitment relationship for fisheries and the construction of sustainable yield curves. *J. Cons. Int. Explor. Mer*, **40**: 189–199.

Shepherd, J.G. (1988) An exploratory method for the assessment of multispecies fisheries. *J. Cons. Int. Explor. Mer*, **44**: 189–199.

Shepherd, J.G. (1994) Second level models for larval growth and mortality, allowing for predator satiation and time and size-dependence of predation rates. I.C.E.S. CM 1991 Mini Symp. **10**: 1–9.

Shepherd, J.G. and Cushing, D.H. (1980) A mechanism for density dependent survival of larval fish as the basis of stock-recruitment relationship. *J. Cons. Int. Explor. Mer*, **39**: 160–167.

Shepherd, J.G. and Cushing, D.H. (1990) Regulation in fish populations: myth or mirage? *Proc. Roy. Soc. B*, **330**: 151–164.

Shepherd, J.G., Pope, J.G. and Cousens, R.D. (1984) Variations in fish stocks and hypotheses concerning their links with climate. *Rapp. Procès-Verb. Réun. Cons. Int. Explor. Mer*, **185**: 255–267.

Sherman, K.E. (1986) Measurement strategies for monitoring and forecasting variability in large marine ecosystems. In *Variability and management of large marine ecosystems* (ed. Sherman, K. and Alexander, L.M.), pp. 203–236. Westview Press, Boulder, Colorado.

Sherr, B.F. and Sherr, E.B. (1984) Role of heterotrophic protozoa in carbon and energy flow in aquatic ecosystems. In *Current problems in microbial ecology* (ed. Klug, M.J. and Reddy, C.A.), pp. 412–423. American Society of Microbiology, Washington DC.

Shuter, B.J. (1978) Size dependence of phosphorus and nitrogen subsistence quotas in unicellular microorganisms. *Limnol. Oceanogr.* **23**: 1248–1255.

Silver, M.W., Shanks, A.C. and Trent, J.D. (1978) Marine snow: microplankton habitat and source of small scale patchiness in pelagic populations. *Science*, **201**: 371–373.

Simkiss, K. (1974) Calcium metabolism in relation to ageing. In *Ageing of fish* (ed. Bagenal, T.B.), pp. 1–12. Unwin Brothers, Old Woking.

Simpson, A.C. (1959) The spawning of the plaice in the North Sea. *Fish. Invest. London* **22**(7).

Simpson, J.H. (1971) Density stratification and microstructure in the Irish Sea. *Deep Sea Res.* **18**: 309–319.

Simpson, J.H., Edelsten, D.J., Edwards, A., Morris, N.C.G. and Tett, P.B. (1979) The Islay front: physical structure and phytoplankton distribution. *Est. Coast. Mar. Sci.* **9**: 713–726.

Simpson, J.H., Hughes, D.G. and Morris, N.C.G. (1977) The relation of seasonal stratification to tidal mixing on the continental shelf. In *A Voyage of Discovery* (ed. Angel, M.), pp. 327–340. Pergamon Press, Oxford.

Simpson, J.H. and Hunter, J.R. (1974) Fronts in the Irish Sea. *Nature*, **250**: 404–406.

Sinclair, M. (1988) *Marine populations.* Washington Sea Grant Program, Seattle.

Sinclair, M., Tremblay, M.J. and Bernal, P. (1985) El Niño events, and variability in a Pacific mackerel (*Scomber japonicus*) survival index: Support for Hjort's second hypothesis. *Can. J. Fish. Aquat. Sci.* **42**: 602–608.

Sissenwine, M.P., Cohen, E.B. and Grosslein, M.D. (1984) Structure of the George's Bank ecosystem. *Rapp. Procès-Verb. Réun. Cons. Int. Explor. Mer*, **183**: 243–254.

Sissenwine, M.P. and Marchessault, G.D. (1985) New England groundfish management: a scientific perspective on theory and reality. In *Fisheries management issues and options* (ed. Frady, T.), pp. 255–278. University of Alaska Sea Grant Rep. 84–1.

Sissenwine, M.P. and Shepherd, J.G. (1987) An alternative perspective on recruitment overfishing and biological reference points. *Can. J. Fish. Aquat. Sci.* **44**: 913–918.

Skud, B. (1977) Drift, migration and intermingling of Pacific halibut stocks. *Int. Pac. Hal. Commn Sci. Rep.* **63**.

Smetacek, V.S. (1985) Role of sinking in diatom life history cycles: ecological, evolutionary and geological significance. *Mar. Biol.* **84**: 239–252.

Smetacek, V. and Passow, U. (1990) Spring bloom initiation and Sverdrup's critical depth model. *Limnol. Oceanogr.* **35**: 228–234.

Smetacek, V., Scharek, R. and Nottig, E.-M. (1990) Seasonal and regional variation in the Pelagial and its relationship to the life history of krill. In *Antarctic ecosystems. Ecological change and conservation* (ed. Kerry, K.R. and Hempel, G.), pp. 103–114. Springer Verlag, Berlin and Heidelberg.

Smetacek, V.S., von Bondungen, B., Knoppers, B., Peinert, R., Pollehne, F., Stegmann, P. and Zeitschel, B. (1984) Seasonal stages characterizing the annual cycle of an inshore pelagic system. *Rapp. Procès-Verb. Réun. Cons. Int. Explor. Mer*, **183**: 126–135.

Smetacek, V., von Bröckel, K., Zeitschel, B. and Zenk, W. (1978) Sedimentation of particulate matter during a phytoplankton spring bloom in relation to the hydrographical regime. *Mar. Biol.* **47**: 211–226.

Smith, A.D.M. and Walters, C.J. (1981) Adaptive management of stock recruitment systems. *Can. J. Aquat. Sci.* **38**: 690–703.

Smith, H.S. (1935) The role of biotic factors in the determination of population densities. *J. Econ. Entom.* **28**: 873–898.

Smith, P.E. (1978) Biological effects of ocean variability: Time and space scales of biological response. *Rapp. Procès-Verb. Réun. Cons. Int. Explor. Mer*, **173**: 117–127.

Smith, P.E. (1985) A case history of an anti El Niño to El Niño transition on plankton and nekton distribution and abundance. In *El Niño North* (ed.

Wooster, W.S. and Fluharty, D.C.), pp. 121–142. Washington Sea Grant Program, Seattle.

Smith, P.E. and Moser, H.G. (1983) Recurrent groups of larval fish species in the California Current Area. *CalCOFI Rep.* **24**: 152–164.

Smith, P.E. and Moser, H.G. (1988) CalCOFI Time series: an overview of fishes. *CalCOFI Rep.* **29**: 66–80.

Smith, P.J., Jamieson, A. and Birley, A.J. (1990) Electrophoretic studies and the stock concept in marine teleosts. *J. Cons. Int. Explor. Mer*, **47**: 231–245.

Smith, R.J.F. (1985) *The control of fish migration.* Springer Verlag, Berlin, Heidelberg and New York.

Smith, S.E. (1936) Environmental control of photosynthesis. *Proc. Nat. Acad. Sci.* **22**: 504–511.

Solomon, M.E. (1949) The natural control of animal populations. *J. Anim. Ecol.* **18**: 1–35.

Solow, A.R. and Steele, J.H. (1990) On sample size, statistical power and the detection of density dependence. *J. Anim. Ecol.* **59**: 1073–1076.

Somerton, D.A. and Kobayashi, D.R. (1989) A method of correcting catches of fish larvae for the size selection of plankton nets. *Fish. Bull.* **87**: 447–455.

Soutar, A. and Isaacs, J.D. (1974) Abundance of pelagic fish during the 19th and 20th centuries as recorded in anaerobic sediments off California. *Fish. Bull.* **72**: 257–75.

Southward, A.J. and Mattacola, A.D. (1980) Occurrence of Norway pout *Trisopterus Esmarki* (Nilsson) and blue whiting *Micromesistius poutassou* (Risso) in the western English Channel off Plymouth. *J. Mar. Biol. Assn UK* **60**: 39–44.

Southward, G.M. (1967) Growth of Pacific halibut *Rep. Int. Pac. Halibut Commn* **43**.

Stahl, G. (1987) Genetic population structure of Atlantic salmon. In *Population genetics and fisheries management* (ed. Ryman, N. and Utter, F.), pp. 121–140. Washington Sea Grant Program, Seattle and London.

Steele, J.H. (1974) *The structure of marine ecosystems.* Harvard University Press, Harvard.

Steele, J.H. and Edwards, R.R.C. (1970) The ecology of 0 group plaice and common dab in Loch Ewe. IV Dynamics of the plaice and dab populations. *J. Exp. Mar. Biol.* **4**: 174–187.

Steele, J.H. and Menzel, D.W. (1962) Conditions for maximum primary production in the mixed layer. *Deep Sea Res.* **9**: 39–49.

Steeman Nielsen, E. and Hansen, K.V. (1959) Measurements with the C14 technique of the respiration rates in two natural populations of phytoplankton. *Deep Sea Res.* **5**: 222–233.

Stephenson, R.L. and Kornfeld, I. (1990) Reappearance of spawning Atlantic herring (*Clupea harengus harengus*) on Georges Bank: population resurgence not recolonization. *Can. J. Fish. Aquat. Sci.* **47**: 1060–1064.

Strass, V. and Woods, J.D. (1988) Horizontal and seasonal variation of density and chlorophyl profiles between the Azores and Greenland. In *Toward a theory on biological-physical interactions in the world ocean* (ed. Rothschild, B.J.), pp. 113–156. Kluwer, Dordrecht.

Strickler, J.R. (1982) Calanoid copepods, feeding currents and the role of gravity. *Science*, **218**: 158–160.

Strickler, J.R. (1985) Feeding currents in calanoid copepods: two new hypotheses. In *Physiological adaptations of marine animals* (ed. Laverack, M.S.), pp. 459–485. Company of Biologists, Cambridge.

Strubberg, A.C. (1916) Marking experiments with cod at the Faroes. *Medd. Komm. fra Hav. Ser. Fisk.* **5**(2): 1–125.

Strubberg, A.C. (1922) Marking experiments with cod (*Gadus callarias* L.) in Danish waters, 1905–1913. *Medd. Komm. Hav. Fisk.* **8**: 1–59.

Struhsacker, P. and Uchiyama, J.H. (1976) Age and growth of the nehu, *Stolephorus purpureus* (Pisces: Engraulidae), from the Hawaiian Islands as indicated by daily growth increments of sagittae. *Fish. Bull.* **74**: 9–17.

Sverdrup, H.U. (1953) On conditions for the vernal blooming of phytoplankton. *J. Cons. Int. Explor. Mer*, **18**: 287–295.

Sundby, S., Bjørke, H., Soldal, A.V. and Olsen, S. (1989) Mortality rates during the early life stages and year class strength of North East Arctic cod (*Gadus morhua* L.). *Rapp. Procès-Verb. Réun. Cons. Int. Explor. Mer*, **191**: 15–38.

Sundby, S. and Fossum, P. (1990) Feeding conditions of North east Arctic cod larvae compared to the Rothschild–Osborn theory on small scale turbulence and plankton contact rates. *J. Plankt. Res.* **12**: 1153–1162.

Swain, A. (1980) Tagging of salmon smolts in European rivers with special reference to recaptures off West Greenland in 1972 and earlier years. *Rapp. Procès-Verb. Réun. Cons. Int. Explor. Mer*, **176**: 93–113.

Swartzman, G.L., Getz, W.M., Francis, R.C., Haar, R.T. and Rose, K. (1983) A management analysis of the Pacific whiting (*Merluccius productus*) fishery using an age structured stochastic recruitment model. *Can. J. Fish. Aquat. Sci.* **40**: 524–539

Taggart, C.T. and Leggett, W.C. (1987*a*) Short term mortality in post emerged larval capelin, *Mallotus villosus*. I Analysis of short term multiple estimates. *Mar. Ecol. Progr. Ser.* **41**: 205–217.

Taggart, C.T. and Leggett, W.C. (1987*b*) Short term mortality in post emerged larval capelin, *Mallotus villosus*. II Importance of food and predator density and density dependence. *Mar. Ecol. Progr. Ser.* **41**: 219–229.

Taggart, C.T. and Leggett, W.C. (1987*c*) Wind forced hydrodynamics and their interaction with larval fish and plankton abundance: a time series analysis of physical–biological data. *Mar. Ecol. Progr. Ser.* **44**: 438–451.

Tait, J.B. (1957) Hydrography of the Faroe–Shetland Channel, 1927–52. *Mar. Res. Scot.* **2**: 1–309.

Tait, J.B. and Martin, J.H.A. (1965) Inferential biological effects of long term hydrographical trends deduced from investigations in the Faroe–Shetland Channel. *ICNAF Spec. Publ.* **6**: 855–858.

Takahashi, K. (1986) Seasonal fluxes of pelagic diatoms in the subarctic Pacific, 1982–1983. *Deep Sea Res.* **33**: 1225–1251.

Takahashi, M. and Bienfang, B.K. (1983) Size structure of phytoplankton biomass and photosynthesis in subtropical Hawaiian waters. *Mar. Biol.* **76**: 203–211.

Takeshita, K. and Hayashi, K. (1991) Interaction between horse mackerel and common mackerel in the East China Sea, suggested by their catch trends and distribution areas. In *Long term variability of pelagic fish populations and their environment*, pp. 173–179. Instituto de Investigaciones Marinas, Vigo.

Tåning, A.V. (1937) Some features in the migration of Cod. *J. Cons. Int. Explor. Mer*, **12**: 3–35.

Templeman, W. (1948) The life history of the caplin (*Mallotus villosus* O.F. Muller) in Newfoundland waters. *Bull. Newfoundland Govt. Laboratory*, **17**: 1–151.

Templeman, W. (1974) Migrations and intermingling of Atlantic cod (*Gadus morhua*) stocks of the Newfoundland area. *J. Fish. Res. Bd Can.* **31**: 1073–1092.

Templeman, W. (1979) Migration and intermingling of stocks of Atlantic cod, *Gadus morhua*, of the Newfoundland and adjacent areas from tagging in 1962–1966. *Res. Bull. ICNAF*, **14**: 5–50.

Templeman, W. and Andrews, G.L. (1956) Jellied condition in the American plaice *Hippoglossoides platessoides* (Fabricius). *J. Fish. Res. Bd Can.* **13**(2): 147–182.

Theilacker, G. and Dorsey, K. (1980) Larval fish diversity, a summary of laboratory and field research. *UNESCO Intergovernmental Oceanographical Commission Workshop Rept.* **28**: 105–142.

Thompson, H. (1943) A biological and economic study of cod (*Gadus callarias* L.). *Res. Bull.* **14**. Dept. Nat. Resources St John's, Newfoundland.

Thompson, W.F. and Herrington, W.C. (1930) Life history of the Pacific halibut. 1. Marking experiments. *Rep. Intern. Fish. Commn*, **2**: 1–137.

Thompson, W.F. and van Cleve R. (1936) Life history of the Pacific halibut. 2. Distribution and early life history. *Rep. Int. Fish. Commn*, **9**.

Thoradottir, T. (1977) Primary production in North Icelandic waters in relation to recent climatic changes. In *Polar oceans* (ed. Dunbar, M.J.), pp. 655–665. Arctic Institute of North America, Calgary, Alberta.

Thorpe, J.E. and Mitchell, K.A. (1981) Stocks of Atlantic salmon (*Salmo salar*) in Britain and Ireland: discreteness and current management. *Can. J. Fish. Aquat. Sci.* **38**: 1576–1590.

Thresher, R.E., Bruce, B.E., Furlani, D.M. and Gunn, J.S. (1988) Distribution, advection and growth of the southern temperate gadoid *Macruronus novaezelandiae* (Teleostei, Merluccidae) in Australian coastal waters. *Fish. Bull.* **87**: 29–48.

Thresher, R.E., Harris, G.P., Gunn, J.S. and Clementson, L.A. (1989) Phytoplankton production pulses and episodic settlement of a temperate marine fish. *Nature*, **341**: 641–643.

Tijssen, S.B. (1979) Diurnal oxygen rhythm and primary production in the mixed layer of the Atlantic Ocean at 20 N. *Neth. J. Sea Res.* **13**: 79–84.

Townsend, D.W. and Graham, J.J. (1981) Growth and age structure of larval Atlantic herring, *Clupea harengus harengus*, in the Sheepscot estuary, Maine, as determined by daily growth increments in otoliths. *Fish. Bull.* **79**: 123–130.

Townsend, D.W., Radtke, R.L., Morrison, M.A. and Folsom, S.D. (1989) Recruitment implications of larval herring overwintering distributions in the Gulf of Maine inferred using a new otolith technique. *Mar. Ecol. Progr. Ser.* **55**: 1–13.

Trent, J.D., Shanks, A.L. and Silver, M.W. (1978) In situ and laboratory measurements on macroscopic aggregates in Monterey Bay, California. *Limn. Oceanogr.* **23**: 626–635.

Uda, M. (1937) Researches on 'Siome' or Current Rip in the seas and oceans. *Geophys. Mag.* **11**: 307–392.

Uda, M. (1959) Oceanographic seminars, 2. Water mass boundaries – 'Siome' frontal theory in oceanography. *Fish. Res. Bd Can. MS Rep. Ser. (Oceanogr. and Limnol.)* **51**: 10–20.

Utter, F.M., Compton, D., Grant, S., Milner, G., Seeb, G., and Wishard, L. (1980) Population structures of indigenous salmonid species of the Pacific North West. In *salmonid ecosystems of the North Pacific* (ed. McNeil, W.J., and Himsworth, D.C.), pp. 285–304. Washington Sea Grant, Seattle.

Utter, F., Milner, G., Stahl, G. and Teel, G. (1989) Genetic population structure of chinook salmon *Oncorhynchus tshawytscha* in the Pacific Northwest. *Fish. Bull.* **87**: 239–264.

Valdes, E.S., Shelton, P.A., Armstrong, M.J. and Field, J.G. (1987) Cannibalism in South African anchovy: egg mortality and egg consumption rates. *S. Afr. J. Mar. Sci.* **5**: 613–622.

van der Veer, H.W. (1986) Immigration, settlement and density dependent mortality of a larval and a post larval 0 group plaice (*Pleuronectes platessa*) population in the Wadden Sea. *Mar. Ecol. Progr. Ser.* **29**: 223–236.

van der Veer, H.W. and Bergman, M.J.N. (1987) Predation by crustaceans on a newly settled 0 group plaice (*Pleuronectes platessa* L.) population in the western Wadden Sea. *Mar. Ecol. Progr. Ser.* **35**: 203–215.

van der veer, H.W., Bergman, M.J.N., Dapper, R. and Witte, J.I.J. (1991) Population dynamics of an intertidal 0-group flounder *Platichthys flesus* population in the western Dutch Wadden Sea. *Mar. Ecol. Progr. Ser.* **73**: 141–148.

van Loon, H. and Madden, R.A. (1981) The Southern Oscillation. Part I Global associations with pressure and temperature in the northern winter. *Monthly Weath. Rev.* **109**: 1150–1162.

van Loon, H. and Rogers, J.C. (1978) The seesaw in winter temperatures between Greenland and Northern Europe Part I. General description. *Monthly Weath. Rev.* **106**: 296–310.

van Loon, H. and Rogers, J.C. (1981) The Southern Oscillation Part II: Associations with changes in the middle troposphere in the northern winter. *Monthly Weath. Rev.* **109**: 1163–1168.

Varley, G.C. and Gradwell, G.R. (1960) Key factors in population studies. *J. Anim. Ecol.* **29**: 399–401.

Varley, G.C. and Gradwell, G.R. (1968) Population models for the winter moth. In *Insect abundance* (ed. Southwood, T.R.E.), *Symp. Roy. Entomol. Soc.* **4**: 132–142.

Venrick, E.L., McGowan, J.A., Cayan, D.R. and Hayward, T.L. (1987) Climate and chlorophyll a: long term trends in the Central North Pacific Ocean. *Science*, **238**: 70–72.

Victor, B.C. (1986) Delayed metamorphosis with reduced larval growth in a coral reef fish (*Thalassoma bifasciatum*). *Can. J. Fish. Aquat. Sci.* **43**: 1208–1213.

Vithayasai, C. (1969) Exact critical values of the Hardy–Weinberg test statistic for two alleles. *Commun. Stat.* **1**: 229–242.

Vollenweider, R. (1965) Calculation models of photosynthesis-depth curves and some implications regarding day rate estimates in primary production measurements. *Memorie dell'Istituto Italiano di Idrobiologia*, **18**: 425–457.

Vollenweider, R.A. (1975) Input–output models with special reference to the phosphorus loading concept in limnology. *Schweiz. Z. Hydrol.* **37**: 53–84.

von Bodungen, B., von Brockel, K., Smetacek, V. and Zeitschel, B. (1981) Growth and sedimentation of the phytoplankton spring bloom in the Bornholm Sea (Baltic Sea). *Kiel. Meeresforsch. Sonderh.* **5**: 49–60.

von Foerster, H. (1959) Some remarks on changing populations. In *The Kinetics of cellular proliferation* (ed. Stohlman, F.), pp. 382–407. Grune and Stratton, New York.

Walker, G.T. (1923) World Weather I. *Mem. Ind. Meteorol. Dept.* **24**: 75–131.

Walker, G.T. and Bliss, E.W. (1932) *World Weather V Mem. Roy. Met. Soc.* **4**: 53–84.

Walline, P. (1983) Growth and ingestion rates of larval fish populations in the coastal waters of Israel. *J. Plankton Res.* **9**: 91–102.

Walsby, A.E. and Reynolds, C.S. (1980) Sinking and floating. In *The physiological ecology of phytoplankton* (ed. Morris, I.), pp. 371–412. Blackwell Scientific Publications, Oxford, London.

Walsh, J.J. (1975) A spatial simulation model of the Peru upwelling ecosystem. *Deep Sea Research*, **22**: 201–236.

Walsh, J.J. (1983) Death in the sea: Enigmatic phytoplankton losses. *Progr. Oceanogr.* **12**: 1–86.

Walsh, J.J., Biscaye, P.E. and Ganady, G.T. (1988a) The 1983–84 Shelf-Edge Exchange Processes (SEEP)-1 experiment: hypotheses and highlights. *Cont. Shelf Res.* **8**: 435–456.

Walsh, J.J., Dieterle, D.A. and Meyers, M.B. (1988b) A simulation analysis of the fate of phytoplankton within the Mid-Atlantic Bight. *Cont. Shelf Res.* **8**: 757–788.

Ware, D.M. and Thomson, R.E. (1991) Link between long term variability in upwelling and fish production in the northeast Pacific Ocean. *Can. J. Fish. Aquat. Sci.* **48**: 2296–2306.

Warlen, S.M. (1988) Age and growth of larval gulf menhaden, *Brevoortia patronus*, in the northern Gulf of Mexico. *Fish. Bull.* **83**: 475–481.

Warnes, S. (1989) Spawning migrations of North-east Arctic cod. ICES CM 1989/G51: 1–19.

Watanabe, Y. (1988) Bias corrections of the number of saury larvae and juveniles captured by surface ring net tows. *Bull. Tohuko Reg. Fish. Lab.* **50**: 49–58.

Watanabe, Y. and Lo, N.C.H. (1989) Larval production and mortality of Pacific saury, *Cololabis saira*, in the northwestern Pacific Ocean. *Fish. Bull.* **87**: 601–613.

Waterbury, J.B., Watson, S.W., Guillard, R.R.L. and Brand, L.E. (1979) Widespread occurrence of a unicellular marine planktonic cyanobacterium. *Nature*, **277**: 293–294.

Waterbury, J.B., Watson, S.W., Valois, F.W. and Franks, D.G. (1987) Biological and ecological characterization of the marine unicellular cyanobacterium *Synechococcus. Can. Bull. Fish. Aquat. Sci.* **214**: 71–120.

Weichart, G. (1980) Chemical changes and primary production in the Fladen Ground area (North Sea) during the first phase of a spring phytoplankton bloom. *Meteor. Forsch-Ergebnisse A*, **22**: 79–86.

Welschmeyer, N.A. and Lorenzen, C.J. (1984) Carbon-14 labelling of phytoplankton chlorophyll carbon: determination of specific growth rates. *Limn. Oceanogr.* **29**: 135–145.

Welschmeyer, N.A. and Lorenzen, C.J. (1985) Chlorophyll budgets: Zooplankton grazing and phytoplankton growth in a temperate fjord and the Central Pacific Gyres. *Limnol. Oceanogr.* **30**: 1–21.

Wheeler, P.A. and Kirchman, D.L. (1986) Utilization of inorganic and organic nitrogen by bacteria in marine systems. *Limnol. Oceanogr.* **31**: 998–1009.

Wiborg, K.F. (1948) Experiments with the Clarke-Bumpus plankton sampler and with a plankton pump in the Lofoten area in northern Norway. *Fisk. Dir. Skr.* **9**(2): 1–22.

Williams, P.J., Le B., Heinemann, K.E., Marra, J. and Purdie, D.A. (1983) Comparison of ^{14}C and O_2 measurements of phytoplankton production in oligotrophic waters. *Nature*, **305**: 49–50.

Williams, R. and Robinson, G.A. (1973) Primary production at Ocean Weather Station India (59.00 N, 19.00 W) in the North Atlantic. *Bull. Mar. Ecol.* **8**: 115–121.

Wisby, W.J. and Hasler, A.D. (1958) Effect of olfactory occlusion on migrating silver salmon (*O. kisutch*). *J. Fish. Res. Bd Can.* **11**: 472–78.

Wolf, K.D. and Woods, J.D. (1988) Lagrangian simulation of primary production in the physical environment – the deep chlorophyll maximum and nutricline. In *Toward a Theory on Biological–Physical Interactions in the World Ocean* (ed. Rothschild, B.J.), pp. 51–70. Kluwer, Dordrecht.

Wood, A.M. (1985) Adaptation of photosynthetic apparatus of marine ultraphytoplankton adapted to natural light fields. *Nature*, **316**: 253–255.

Wood, S.N. and Nisbet, R.M. (1991) Estimation of mortality rates in size structured populations. *Lecture notes in biomathematics* **90**: 1–101. Springer Verlag, Berlin.

Woods, E.A. (1991) A two layer model of chlorophyll and nutrients and its application to the North West European continental shelf seas. *UCES Rep.* **U91**(8): 1–111.

Woods, J.D. and Barkmann, W. (1993*a*) The plankton multiplier – positive feedback in the greenhouse. *J. Plankt. Res.* **15**: 1053–1074.

Woods, J.D. and Barkmann, W. (1993*b*) Diatom demography in winter – simulated by the Lagrangian Ensemble method. *Fish. Oceanogr.* **2**: 202–222.

Woods, J.D. and Barkmann, W. (1994) Diatom demography in winter – simulation by the Lagrangian ensemble method. *Fish. Oceanogr.* in press.

Woods, J.D. and Onken, R. (1982) Diurnal variation and primary production in the ocean–preliminary results of a Lagrangian ensemble model. *J. Plankt. Res.* **4**: 735–756.

Wootton, R. (1977) Effect of food limitation during the breeding season on the size, body components and egg production of female sticklebacks (*Gasterosteus aculeatus*). *J. Anim. Ecol.* **46**: 823–834.

Wyatt, T. and Perez-Gandoras, G. (1988) El trasporto de Ekman y los rendimientos de sardina en el occidente iberico. In *Long term changes in marine fish populations* (ed. Wyatt, T. and Larrañeta, M.G.), pp. 125–138. Instituto de Investigaciones Marinas, Vigo.

Yamanaka, I. (1978) Oceanography in tuna research. *Rapp. Procès-Verb. Réun Cons. Int. Explor. Mer*, **173**: 203–211.

Yoklavich, M.M. and Bailey, K.M. (1990) Hatching period, growth and survival of young walleye pollock, *Theragra chalcogramma* as determined from otolith analysis. *Mar. Ecol. Progr. Ser.* **64**: 12–23.

Zeitschel, B., Diekmann, P. and Uhlmann, L. (1978) A new multi-sample sediment trap. *Mar. Biol.* **45**: 285–288.

Zijlstra, J.J. (1972) On the importance of the Wadden Sea as a nursery area in relation to the conservation of the southern North Sea fishery resources. *Symp. Zool. Soc. London*, **29**: 233–258.

Zijlstra, J.J., Dapper, R. and Witte, J.Ij. (1982) Settlement, growth and mortality of post larval plaice (*Pleuronectes platessa* L.) in the western Wadden Sea. *Neth. J. Sea Res.* **15**: 260–272.

Zupanovitch, S. (1968) Causes of fluctuation in sardine catches along the eastern coast of the Adriatic Sea. *Anali. Jadranskog Instituta* **4**: 401–489.

Zweifel, J.R. and Lasker, R. (1976) Prehatch and post hatch growth of fishes – a general model. *Fish. Bull.* **74**: 609–621.

Index

Page numbers for tables and figures are in italics.